Historical Geography and Geosciences

This book series serves as a broad platform for contributions in the field of Historical Geography and related Geoscience areas. The series welcomes proposals on the history and dynamics of place and space and their influence on past, present and future geographies including historical GIS, cartography and mapping, climatology, climate history, meteorology and atmospheric sciences, environmental geography, hydrology, geology, oceanography, water management, instrumentation, geographical traditions, historical geography of urban areas, settlements and landscapes, historical regional studies, history of geography and historic geographers and geoscientists among other topically related areas and other interdisciplinary approaches. Contributions on past (extreme) weather events or natural disasters including regional and global reanalysis studies also fit into the series.
Publishing a broad portfolio of peer-reviewed scientific books Historical Geography and Geosciences contains research monographs, edited volumes, advanced and undergraduate level textbooks, as well as conference proceedings. This series appeals to scientists, practitioners and students in the fields of geography and history as well as related disciplines, with exceptional titles that are attractive to a popular science audience.
If you are interested in contributing to this book series, please contact the Publisher.

More information about this series at http://www.springer.com/series/15611

Charles Travis · Francis Ludlow ·
Ferenc Gyuris
Editors

Historical Geography, GIScience and Textual Analysis

Landscapes of Time and Place

Springer

Editors
Charles Travis
Department of History
University of Texas
Arlington, TX, USA

Trinity Centre for Environmental Humanities
Trinity College Dublin
Dublin, Ireland

Francis Ludlow
Department of History
Trinity College Dublin
Dublin, Ireland

Ferenc Gyuris
Department of Regional Science
Eötvös Loránd University
Budapest, Hungary

ISSN 2520-1379 ISSN 2520-1387 (electronic)
Historical Geography and Geosciences
ISBN 978-3-030-37568-3 ISBN 978-3-030-37569-0 (eBook)
https://doi.org/10.1007/978-3-030-37569-0

This Springer imprint is published by the registered company Springer Nature Switzerland AG
The registered company address is: Gewerbestrasse 11, 6330 Cham, Switzerland

Introduction

History is not the past, but a map of the past drawn from a particular point of view to be useful to the modern traveler.[1]
Henry Glassie, *Passing Time in Ballymenone*

In 1992, Stanley W. Trimble, conjuring the specter of Carl Ortwin Sauer, spoke of the "Great Retreat" of historical geography from engagement with physical phenomena.[2] Indeed, research in the field during the late twentieth-century "cultural turn" focused its gaze through the lenses of structuralist/post-structuralist, post-colonial, gender, ethnic, critical, and literary theory, trimming many of the sails in the Anglo-American historical geography fleet. This is not to say that we should not keep the waters navigated during the "cultural turn" charted in our logbooks. There was no need to throw the physical phenomena baby out with the bathtub water at the end of the last century, and there is no need now to abandon perspectives and insights acquired during the cultural turn. We need to embrace all theories, methods, and tools accrued since the field's pre-modern incarnation in order to transform historical geography's teaching and research practices in the twenty-first century. It can be argued that in the West, the practice the ancient Greeks designated γεωγραφία (*earth-writing*, trans. *geographia*) emerged as bridge between Strabo's (63 B–23AD) graphic and poetic impressions of place, and Herodotus' (484 BC–425 BC) chronicling of events and regional cultures. In addition, Eratosthenes, (276 BC–194 BC) keeper of the Great Library at Alexandria, calculated the spherical nature of the earth, and contributed with the efforts of Euclid (300 BC–200 BC), Ptolemy (100–170 AD), and Al-Idris (AD 1100–1165) to infusing *geographia* with the language of mathematics, the theorems of geometry and the calculations of terrestrial, astronomical and oceanic navigation.

[1]Henry Glassie, *Passing Time in Ballymenone: Culture and History of an Ulster Community*. Philadelphia: University of Pennsylvania Press, 1982., pg. 621.

[2]Stanley W. Trimble, 1992, Preface, in eds. Lary M. Dilsaver and Craig E. Colten, The American environment: Interpretations of past geographies. Rowman & Littlefield., pg. xvi.

Like the mutually reinforcing struts of architecture that allow a suspension bridge to seemingly float in the air above a body of water, practices in these classical fields coalesced over time to provide support and stability to *geographia* as a discipline. Though lost to the Dark Ages, this early 'transdisciplinary' knowledge system with roots in the Mediterranean basin, was rediscovered during the Florentine Renaissance. Practices of *geographia* infused the permeable borders of late medieval and early modern European literary, historical, scientific and cartographical discourse. Modern historical geography's genealogy is protean, and as the study of place and period, it has traditionally focused on the *geographia* of the past, and how these "foreign countries" (as David Lowenthal would phrase it) are parsed and plotted into the geographical present. In the seventeenth century, Philipp Clüver published *Germaniae antiquae libri tres* (1616) based on the writings of the Roman historian Tacitus, and this text stands as one of the founding works of early modern historical geography. However, for the most part until the nineteenth century, practices in historical geography drew upon Biblical hermeneutics and antiquarian geographies of classical Greco-Roman literature and history.[3]

Prior to introducing this volume's sixteen chapters, the following sections provide a *précis* of sorts, on interchanges between geography and history over the past 100 years.

Environmental Determinism and the *Fin De Siècle*

At the close of the nineteenth century, "End of Geography" debates dominated historical and geographical thought. With the earth seemingly mapped in its entirety, *environmental determinism* emerged as a theory promoting that human culture and agency were strongly dictated by the physical environment. It has been argued that its origins lie in a pseudo-scientific reading of Charles Darwin's study of human evolution in *On the Origins of Species* (1859). Providing a rather crude and racist analysis of people and place, *environmental determinism* (also called *geographical determinism*) focused its myopic eye on cultural groups of color, inhabiting regions located between the Tropics of Cancer and Capricorn. It was also deployed in ethnographic studies on indigenous and diasporic peoples in the Americas, Asia, Oceania, and in colonized European regions such as Ireland, Scotland, and Scandinavia. German zoologist and biologist Friedrich Ratzel's work *Anthropogeographie* (1882–1892) and *Der Lebensraum* (1904) drew upon the theory to define culture as an organism seeking a "living-space". *Lebensraum* was partially based upon observations made during his travels in the Western United States, where Ratzel perceived an "annihilating struggle, the prize for which was the land, the space," between Anglo-European settlers and indigenous high plain tribes.[4] In addition, British geographer Halford Mackinder's "World Island" and "Heartland"

[3]W. Gordon East. 1965. *The Geography Behind History*. New York: Norton & Company, Inc., pg. 2.

[4]Sven Lindqvist, 1996. *Exterminate all the Brutes*. New Press.

theories in *The Geographical Pivot of History* (1904) have been critiqued as having a taint of *fin de siècle* deterministic thought.[5] However, it was Ellen Churchill Semple (a student of Ratzel's) and her book *Influences of Geographic Environment* (1911) that introduced the theory of environmental determinism to American academics. Ratzel's influence can be seen in historian Frederick J. Turner's *The Significance of the Frontier in American History* (1894), a lecture that still casts a long shadow over spatial historiography in the United States. Drawn to period geographical thought, Turner played with the idea that regional environments shaped the stages of human cultural progress, an idea challenged by the doyen of the Berkeley School of Cultural Geography geographer Carl Ortwin Sauer in *Historical Geography and the Western Frontier* (1969).

The Roles of Human and Historical Agency

Though Turner scrutinized physiographic maps, demographic and census data to formulate his grand thesis on American historical processes, he did very little actual fieldwork in the 'frontier' regions he was writing about. Indeed, many historians conflate the map as a result of various processes, but a map is merely a subjective snapshot in time. Drawing on British historical geographer H. C. Darby's analogy, a map can be seen as a 'still' taken from a frame in a spool of film. However, during the *fin de siècle*, more nuanced works in historical and cultural geography belied the influences of *environmental* and *geographical determinism.* August Meitzen published a statistically granulated regional study, *Settlement and agriculture of the west and east Germans, the Celts, Romans, Finns and Slavs* in 1895. Paul Vidal de la Blache's, *Atlas General Histoire et Geographie* (1912) and Lucien Febvre's *La Terre et l'évolution humaine, introduction géographique à l'histoire* (1922) repudiated deterministic readings with the theory of *possibilism.* In their works, human and environmental phenomena are posited as interactive forces, strongly influenced by a region's cultural dispositions and unique physiography. As doyens of the French *Annales* school of history, de la Blache and Febvre's studies of regional *payes* were temporally dynamic, and deployed cartography as an effective tool for analysis and scholarly communication.

In turn, Sauer, who began his college studies in geology, was particularly influenced by the Germanic *Landschaff* tradition and the *Annales* school. In a series of field-based studies in northern Mexico, Sauer illustrated the roles played by human agency and time in shaping environments. In *The Morphology of Landscape* (1923), Sauer described geography as "a critical system which embraces the phenomenology of landscape, in order to grasp in all of its meaning and color the varied terrestrial scene."[6] He further stated

[5]Gerry Kearns, 2009. *Geopolitics and empire: The legacy of Halford Mackinder*. Oxford University Press on Demand.

[6]Carl O. Sauer, 1969. The Morphology of Landscape. *Land and life*. Univ. of California Press., pg. 321.

that "we cannot form an idea of landscape except in terms of its time relations as well as it space relations. It is in continuous process of development, or of dissolution and replacement."[7] Attempting to strike a balance on the interplay between human agency and environmental processes, Sauer invoked de la Blache's warning of perceiving "the earth as 'the scene on which the activity of man unfolds itself,' without reflecting that this scene itself is living."[8]

In *Forward to Historical Geography* (1941), Sauer moving on from positions discussed in the *Morphology of Landscape* described the "reconstruction of past culture areas" as "a slow task of detective work," and argued that historical geographers needed to acquire the "ability to see the land with the eyes of [. . .] former occupants" and place themselves as a "member of the cultural group and time being studied."[9] To do this, historical geographers needed to combine knowledge of how a culture functioned, with a command of existing historical data and an intimate familiarity with physical terrain.[10] In *Agricultural Origins and Dispersals* (1952), Sauer situated the map as a "visual hypothesis" in order to consider questions on human agency, cultural processes, and physical landscapes.

In contrast to Sauer's panoramic perspective, H. C. Darby, doyen of historical geography in Britain, drilled down through the historical and cultural sediments of the English landscape and created a method we can coin as "longitudinal geohistoric-stratigraphy". Darby's major works, the *Draining of the Fens* (1940) and *The Domesday Geography of Midland England* illustrate his richly empirical, if not theoretically descriptive regional focus.[11] In addition, William George Hoskins argued in *The Making of an English Landscape* (1955) that "a landscape historian needs to use botany, physical geography and natural history as well as historical knowledge to interpret any given scene fully."[12]

In 1949, the French historian Fernand Braudel published *La Méditerranée et le Monde Méditerranéen à l'Epoque de Philippe II* (1949) introducing a new lexicon and temporal dynamic to describe multiple cycles and overlapping scales of time: the *longue durée* (spanning centuries, and almost imperceptible), *histoire conjucturelle* (the slow, but perceptible rhythm of social history), and *histoire événémentielle* (the time of daily life, and the historical event).[13]

---

[7]Ibid., 333.

[8]Ibid., 321.

[9]Carl O. Sauer, 1941. Foreword to historical geography. *Annals of the Association of American Geographers*, 31(1), pp. 1–24, pgs. 10; 23.

[10]Ibid., pg. 4.

[11]Darby, H. C. (1953). On the relations of geography and history. *Transactions and Papers* (Institute of British Geographers), (19), 1–11.

[12]W. G. Hoskins, 1955. *The making of the English landscape.*

[13]F. Braudel, 1949. *La M'editerran'ee et le monde M'editerran'een 'a l'epoque de Philippe II.* Colins.

Computing, Cultural, and Spatial Turns

Published on the cusp of geography's Quantitative Revolution (1950s and 1960s), Braudel's ideas did not readily diffuse into historical geography. While many historical geographers huddled with humanistic geographers to form a response to the Quantitative Revolution's skeletal landscapes of spatial statistics, a few did engage in its methods by applying FORTRAN computer language to perform punch card analyses of historical data.[14] These efforts foreshadowed early developments in the digital humanities, and it is important to distinguish the use of quantitative methods from attempts to reframe geography as a nomothetic, rather than idiographic discipline. One response to the revolution was Clarence Glacken's *Traces on the Rhodesian Shore* (1967) illustrating how changes in environmental perception from classical Greco-Roman culture, to the Age of Enlightenment shaped Western history and culture.[15] In history, scholars like Emmanuel Le Roy Ladurie (a doctoral student of Braudel's) published a regional, quantitative study *The Peasants of Languedoc* (1966). Ladurie extolled the virtue of computing methods in *The Territory of the Historian* (1973). However, one of the most influential twentieth-century geographers to consider questions on computation, time, and environment was Torsten Hagerstrand. His works *The Computer and the Geographer* (1967) *Space, time and human conditions* (1975) and *Survival and Arena: on the life-history of individuals in relation to their geographical environment* (1978) cartographically operationalized Braudel's cycles of history. Illustrating a prescience that currently informs GIScience and the digital humanities, Hägerstrand introduced geocoding to human and critical geography by plotting people's daily *lifepaths*, in 3D space–time cube/prism diagrams.

In 1971, H. C. Prince argued that the focus of historical geographers should be at the center of three interpenetrating worlds: the *real*—as represented in documents, the landscape, and maps; the *abstract*—models of past spatial order; and the *perceived*, which he claimed were "past worlds, seen through the eyes of contemporaries, perceived according to their culturally acquired preferences and prejudices, shaped in the images of their assumed worlds."[16] In 1975, the *Journal of Historical Geography*, founded by H. C. Darby in Britain, provided an outlet for work on empire, commerce, landscape, theory, identity, race, class, and gender. In 1993, *Historical*

[14]R. Johnston, and J. D. Sidaway, 2015. *Geography and geographers: Anglo-American human geography since 1945*. Routledge.

[15]C. J. Glacken,1967. *Traces on the Rhodian Shore: Nature and Culture in Western Thought from Ancient Times to the End of the Eighteenth Century*. Berkeley: University of California Press.

[16]H. C. Prince, 1971. Real, imagined and abstract worlds of the past. *Progress in Geography* 3, 1–86.

Geography, published in the United States, followed suit. The cultural and spatial turns of the 1980s and 1990s, promoted debates on subjects such as decolonization and postmodernism, the crisis of representation and identity politics, and produced critiques of 'traditional' historical and cultural geography practices. Engagements with critical, Marxist, feminist and post-structural theory, rather than solely empirical studies of cultural regions, landscapes, and physiographic surveys were called for.

In particular, Allan Pred's *Lost words and lost worlds: Modernity and the language of everyday life in late nineteenth century Stockholm* (1990) deployed time-geographic montage, historical and linguistic methods to map the daily life path of a Swedish docker named Sörmlands-Nisse. In conveying the "phantasmagoric" world of Stockholm through the prisms of language and literature, Pred argued that "the use of poetic forms is not an end in itself, but an attempt to exploit the physicality of the text, to exploit the landscape of the page."[17]

Other major works that integrated historical and geographical methods, with arguably wider impacts than disciplinary journals, include Paul Carter's *The road to botany bay: An essay in spatial history* (1987) detailing the colonization of Australia with a clear epistemological distinction between the "geographer's space" and the practice of "spatial history." Additionally, David Lowenthal's *The Past is a Foreign Country* (1985), provided an exegesis on the late-Renaissance segregation of historical and fictional narratives, and its influence on questions of historical validity and the veracity of source materials. Lowenthal asserted that historians who claim a unique fidelity to the past and fiction writers who claim exemption from such fidelity, are both deluding themselves and their readers.[18] In *The Landscape of History* (2002), John Lewis Gaddis asked, "What if we were to think of history as a kind of mapping,"[19] by linking the ancient practice of mapmaking with the archetypal three-part conception of time (past, present, and future). Both try to identify patterns and manage infinitely complex subjects by imposing abstract grids—longitude and latitude to frame space and hours and days to mark time.[20]

Historical Geography and Geographical History

While historical and cultural geography practices seemed to cleave away from engagements with the physical environment toward the end of the twentieth century, there remained a persistent strand of work integrating environmental, and ecological methods, fieldwork, and various theoretical lenses. Donald Meinig's *The Shaping of America: A Geographical*

[17]A. Pred, 1990. *Lost words and lost worlds: Modernity and the language of everyday life in late nineteenth-century Stockholm*. Cambridge University Press.

[18]D. Lowenthal, 1990. *The Past is a Foreign Country*. 1985. Cambridge: Cambridge UP.

[19]J. L. Gaddis, *The Landscape of History: How Historians Map the Past* (Oxford: Oxford University Press, 2002), 32.

[20]Ibid.

Perspective on 500 Years of History (1986, 1993, 1998, 2004) stands as a monumental work, harking back to the tradition of a Braudelian *grand project*. Meinig's panoramic, four-volume study mapped out the regional stages of settlement, national development, and post-colonial creation of vernacular landscapes in the United States from 1492 to 2000. In Ireland, E. Estyn Evan's *The Personality of Ireland: Habitat, Heritage and History* (1973) inspired by the work of H. J. Fleur and Sauer, integrated geography, history, anthropology, geology, and ecology thereby signalling a shift in modern Irish historical–geographical practices. Commensurate work can be found in John Andrews' *The Paper Landscape* (1975) a topographical exploration of the 1836 Ordnance Survey of Ireland. William Nolan's *Fassadinin: Land, Settlement and Society in Southeast Ireland c.1600–1850* (1979) and Annegret Simms' *Continuity and Change: Settlement and Society in Medieval Ireland c. 500–1500* (1986) blend tropes in history, geography, biography, and landscape studies, and Patrick Duffy *Exploring the history and heritage of Irish landscapes* (2007) provides a historical look at the island's cultural landscapes by mining local source data on physical and built environments.

E. Cotton Mather's *The American Great Plains* (1972) parsed this "inarticulated" region of the "West" physiographically and culturally, while down in the cotton and bible belt of the U.S.A., Sam Bowers Hilliard's *Hogmeat and Hoecake: Food Supply in the Old South, 1840–1860* (1972) blended demographic techniques and research on environmental conditions to study the production and diffusion of Antebellum food staples. Carville V. Earle's *The Evolution of a Tidewater Settlement System: All Hallow's Parish, Maryland, 1670–1783* (1975) revisited assumptions on how the colonial tobacco economy impacted the soils and affected the physical landscape of the Atlantic seaboard parish. In *Geographical inquiry and American historical problems* (1992), Earle teased out distinctions between historical geography and geographical history. The former "focuses upon those relationships which have shaped the evolution of place and landscape,"[21] while the latter focuses upon "those relationships which have shaped human affairs in the past."[22]

Jared Diamond's Pulitzer-Prize winning *Guns, Germs and Steel: The Fates of Human Societies* (1997) an international best seller, drew on tropes in Anglo-American historical geography. However, his study was met with a furious backlash from social scientists, historical, human, and critical geographers who argued that *Guns, Germs, and Steel* reintroduced *environmental determinism* into the public imagination. Diamond responded by making forceful and accurate distinctions between late nineteenth-century racialized discourses, and his own consideration of the interplay between geographical factors, world regions and human agency.

Mike Davis also focused on historical, geographical and environmental factors in *Late Victorian Holocausts: El Niño Famines and the Making of the*

---

[21]Alan R. H. Baker (2007) Classifying Geographical History, *The Professional Geographer*, 59:3, 344–356, pg. 347.

[22]Ibid.

Third World (2000), providing a convincing critical-Marxist reading of the El Niño-Southern Oscillation, and its relations to colonial practices and indigenous famine in the late nineteenth century.

Integrations: Historical Geography, GIScience, and Textual Analysis

Geocomputing labs at the University of Edinburgh, and Harvard University in the 1960s, '70s, and '80s facilitated the entrance of mainstream GIS platforms into academic, public, and private worlds in the 1990s. Initially, GIS was the provenance of the earth sciences, environmental studies, and urban planning, but in the 1990s and 2000s it was employed in historical geography research projects such as the *Great Britain Historical GIS* at Portsmouth University, and the *China Historical GIS* at Harvard University. Although many historians, geographers, and geoscientists regard geographical information science (GIS) as a mapping practice, its platforms have evolved into new types of visual database technology and interactive media. As a database technology, GIS spatially parses and itemizes attribute data (as a row of statistics, a string of text, an image, a movie) linking coordinates to representations of the locations to which the data refers.[23]

In this light, historical geography and textual analysis methods can help address what Donald G. Janelle has described as "the underlying complexities in the human organization of space that present methodological problems for GIS in linking empirical research questions with alternative theoretical frameworks."[24]

Key texts that have grappled with such GIS problems include Anne Kelly Knowles' *Past Time, Past Place: GIS for history* (2002), Ian Gregory and Paul Ell's *Historical GIS: Techniques, methodologies and scholarship* (2003), Meghan Cope and Sara Elwood's *Qualitative GIS: A Mixed Methods Approach* (2009), and Alexander von Lunen and Charles Travis' *History and GIS: Epistemologies, Considerations, Reflections* (2012). In particular, a volume edited by David Bodenhamer, John Corrigan, and Trevor M. Harris *The Spatial Humanities: GIS and the future of humanities scholarship* provides methodological insights into addressing Jannelle's concerns.

In the fields of literary, textual, rhetorical analysis, Franco Moretti's *Maps, Graphs and Trees* (2005) and *Distant Reading* (2013) provide Marxist and quantitative mapping and systemic analysis of the modern novel, and complement Barbara Piatti's *Literary Atlas of Europe*, and Ian Gregory and David Cooper's *Mapping the Lake Districts* online literary GIS projects. William Least Heat-Moon's *PrairyErth: A Deep Map* (1991) deployed a stratigraphic literary heuristic to explore scales of time and senses of place in

[23]Ian N. Gregory, and R. G. Healey, "Historical GIS: structuring, mapping and analysing geographies of the past." *Progress in Human Geography*, 31(5), 2007, pp. 638–653.

[24]D. G. Janelle, "Time-space. In Geography" in: N. J. Smelser and P. B. Baltes, eds. *International Encyclopedia of the Social and Behavioral Sciences.* Amsterdam: Pergamon-Elsevier Science, 2001, pp. 15746–15749.

Chase County, Kansas. Lastly, Robert J Mayhew (2011) coined the term *geohistoriography*, as a means to conceive of history and geography as rhetorical disciplines, with the latter providing a spatial discursive framework to shape the narrative scales of the former.[25] Mayhew's neologism curiously echoes Philip J. Ethington's (2007) observation that: "The past cannot exist in time: only in space. Histories representing the past represent the places (*topoi*) of human action. Knowledge of the past, therefore, is literally cartographic: a mapping of the places of history indexed to the coordinates of spacetime."[26]

In conclusion, triangulating methods in historical geography, GIScience, and textual analysis may help us conceive of new ontologies and epsitemologies to understand the dynamics of human and environmental phenomena in the past, the present, and our possible and speculative futures.

Chapter Contents

The chapters comprising Part I *Landscape, Time, Text* illustrate the hermeneutic potentials afforded by blending methods in historical geography, GIScience, and textual analysis. Chapter 1*Ghost Cathedral of the Blackland Prairie* by Charles Travis and Javier Reyes draws on data from the IMPUS Historical GIS database and other sources, contextualized by the "text maps" of the film, *Places in the Heart* (1984), novel *Einstein's Bridge* (1997), and the National Academy of Sciences and National Academy of Engineering site location report for a Federal government Superconducting Super Collider (SSC) tunnel ring project. The socio-cultural and historical contextualizations provided by these cinematic, imaginary, and engineering "text maps" guide a GIS extirpation of the local and deep history of Waxahachie, Texas by juxtaposing a cartographical *filmscape* of the 1930s Texas Blackland Prairie with a *sideshadow* model of the SSC project, abandoned by the Federal government in the 1990s.

Emily Lethbridge's Chap. 2, *Digital Mapping and the Narrative Stratigraphy of Iceland*, outlines the ongoing Icelandic Saga Map (ISM) project linking Iceland's rich medieval textual corpus with the country's eight geographical regions. The project facilitates a better understanding of the various functions the island's historical and geological landscape fulfills in situating the medieval sagas and other works, as well as encouraging reflection on the role that Icelandic landscapes have played in the transmission and reception of works such as the *Njáls Saga* over a milenial time period.

Chapter 3, *Dead Men Tell Tales: History and Science at Duffy's Cut* by William E. Watson, J. Francis Watson, and Earl H. Schandelmeir, discusses

[25]R. J. Mayhew, 2011. "Historical geography, 2009–2010: Geohistoriography, the forgotten Braudel and the place of nominalism." *Progress in Human Geography*, 35(3), 2011, pp. 409–421.

[26]P. J. Ethington, 2007. "Placing the past: Groundwork for a spatial theory of history."*Rethinking History*, 11(4), 2011, pp. 465–493.

the GIS mapping of an 1832 mass grave of Irish immigrant railroaders and their by homocide at Mile 59 of the Philadelphia and Columbia Railroad track in Chester County, Pennsylvania. Stories of the worker's death due to a Cholera outbreak were originally disseminated by a few contemporary newspapers, in addition, stories abounded in railroad and local folklore, which told of Irish ghosts dancing on graves close to the railroad fill which laborers were building at the time of their deaths.

Chapter 4*"Please Mention the Green Book:" The Negro Motorist Green Book as Critical GIS*, by Ethan Bottone maps out the Jim Crow era travel guide created for and by African-Americans, which has received much recent popular and academic scrutiny. Consisting of almost 30 editions published between 1936 and 1966, the *Green Book* features thousands of addresses for businesses that catered to African-Americans during a period of institutionalized discrimination and segregation. By comparing the spatial data of the *Green Book* to historical census data, trends in urban neighborhood composition explain how and why African-American travel patterns shifted within New Orleans' neighborhood.

Part II, titled *Cultures, Networks and Mobilities* carries on from the previous section to explore GIScience mappings of historical spatial practices that led to discrimination, marginalization, migration and strategic mobility. Chapter 5*Queer Cartographies: Urban Redevelopment and the Changing Sexual Geography of Postwar San Francisco* by Damon Scott reflects on the methodological and historiographic insights gained using GIS to reconstruct the impact of urban renewal on "hangouts for homosexuals" on the San Francisco waterfront during the 1950s and early 1960s. These stigmatized places—which entered the popular-spatial imaginary through contemporaneous local press coverage of bar raids—were initially documented as historical sites by LGBTQ community archivists who culled basic information about the physical location, approximate years of operation, and business type from oral history transcripts, newspaper clippings, early gay bar directories, and nightlife columns in the city's first gay periodicals dating to the early 1960s.

In Chap. 6, Don Lafreniere and Jason Gilliland's *Revisiting the Walking City: A Geospatial Examination of the Journey to Work* collates city directories and decennial censuses through the use of probabilistic record linkage techniques. Their method uncovers the relationship between work and home for over 5,000 workers in London, Ontario in 1881. A GIS network-derived journey to work model recreates journeys that consider the many natural and built environment barriers that influenced the paths and distances workers traveled on a daily basis.

In Chap. 7, *Corruption and Development of Atlanta Streetcar Lines in the Nineteenth Century: A Historical GIS Perspective,* S. Wright Kennedy employs historical geographic systems (HGIS) to trace the formative years of public transportation in Atlanta. This study finds evidence that Atlanta city leaders engaged in corrupt bargains to benefit themselves with little regard for the lasting effects on the city's transportation network and spatial development. Streetcar networks in postbellum Atlanta created demographic patterns that persist until today. By illustrating how HGIS can enable historians to distil new value from sources underutilized in urban history, this chapter shows how such sources can be collated by a GIS into an extensive archive of historical urban processes.

This section's concluding Chap. 8, *"A Brother Orangeman the World Over": Migration and the Geography of the Orange Order in the United States*, is a case study by Cory Wells and Charles Travis focused on the Orange Order, a Protestant organization established in County Armagh in 1795 and its diasporic resettlement patterns from the Province of Ulster, in Ireland to North America. Settlement patterns were typically anchored by Orange Lodges, established as nodes on trans-national migration networks that reached from Northern Ireland to the United Kingdom, North America, Australia, New Zealand, and West Africa. The chapter's case study sources data from the *The International Bureau of 'Orange' Information*, a body established in 1903 to gather information on the Order's diasporic communities.

The volume's third section, Part III *Climate, Weather, Environment*, provides unique examples of topics generally explored from a geoscience perspective, but which tease out socio-cultural, political, historical, and economic dimensions of human-environmental phenomena. Chapter 9 by Jase Bernhardt titled *Mining Weather and Climate Data from the Diary of a Forty-Niner* provides an especially compelling case study of how such documentation can be employed to determine spatiotemporal patterns in historical meteorology. Gideon Nichols, a farmer in Long Island, New York, elected to venture across the United States in 1849 to partake in the California Gold Rush.

Chapter 10, *Unmappable Variables: GIS and the Complicated Historical Geography of Water in the Rio Grande Project*, by Daniel R. Beene and K. Maria D. Lane provides a GIScience/discourse study of on the U.S. Bureau of Reclamation's Rio Grande Project which radically changed irrigation practices and the legal structure of water deliveries along the entire reach of the watershed. Using a sequential exploratory research design, the chapter explores how water governance has precipitated ecological change and the loss of cultural knowledge, political change, the exacerbation of inequalities, and legal conflict leading to political battles between New Mexico and Texas, and in communities downstream of the Elephant Butte Dam.

Chris Hewitt's *Supplying the Conquest: A Geospatial Visualization and Interpretation of Available Environmental Resources at the Battle of Hastings in 1066* in Chap. 11 presents a never before considered GIScience interpretation of environmental resources available to English and Norman armies at the Battle of Hastings in 1044. The mapping and discussion of the battle's historical and geographical contexts explores the types of natural and man-made resources were available to the two combating armies. Findings contribute to existing literature on the battle in terms of the volume and location of abundant resources including food, water, iron, salt, and timber available in the Hastings area.

Robert Legg, Francis Ludlow, and Charles Travis' Chap. 12, *Mapping the Irish* Rath *(Ringfort): Landscape and Settlement Patterns in the Early Medieval Period*, discusses a GIScience and geostatistical study of ringfort locations in the Irish Midlands sourced from the Irish Government's Department of the Environment, Heritage and Local Government archeological site spatial database. This study also draws on Irish medieval archeological and historical database records and fieldwork analysis to consider the relations between different environmental contexts and types of ringfort settlement patterns.

The volume's final section Part IV *Place, Philology, History* provides case studies and theoretical considerations on the mapping of historical landscapes from a variety of perspectives. Chapter 13, Ryan Horne's *Mapping Power: Using HGIS and Linked Open Data to Study Ancient Greek Garrison Communities*, discusses in the context of the *Pleiades* project how the institution of the *phrourarchia* was a critical component of Greek civic and military identity. The chapter's mapping illustrates how this institution was central to the control of cities within the Athenian Empire in fifth century BCE and to maintaining isolated outposts on the border of the Parthian Empire in the second century CE.

Gordon Cromley and Chris Post's Chap. 14, *The Preservation of Paradox: Bismarck Towers as National Metaphor and Local Reality*, explores the network of memorials to Otto von Bismarck, first minister-president of the Kingdom of Prussia and later the first Chancellor of the German Empire. Representing tensions between place and wider geographic discourses, approximately 240 Bismarck Tower memorials were built between 1867 and 1935. While each tower is grounded within a specific local narrative, they were meant to provide a link to a larger discourse on national identity within the community.

Chapter 15, *Mapping the Historical Transformation of Beijing's Regional Naming System*, by Yong Yu focuses on the introduction of humanist and Marxist philosophies, and a shifting trend from empirical examination to theoretical discussions in historical geography using GIS visualization. Through spatial analysis, this chapter illustrates how the naming of regions in Beijing has been transformed from the humanistic and practical ways during the Ming Dynasty (1368–1644), to the social constructionist ways during the Qing Dynasty (1644–1912), and then to the scientific ways during the Republic of China (1912–1949).

Chapter 16, *Geographical Enrichment of Historical Landscapes: Spatial Integration, Geo-Narrative, Spatial Narrative, and Deep Mapping*, by May Yuan reviews cartographic and phenomenological views of landscape, and discusses four geospatial approaches to enrich historical landscapes geographically. Montello's categorization of *figurative*, *vista*, *environmental*, and *geographic* spaces helps to transition landscape concepts between phenomenological and cartographic perspectives. In addition, GIST methods of spatial integration, geo-narrative, spatial narrative, and deep mapping incorporate geographic features, feelings, events, and senses to enrich the representation, analysis, and communication of historical landscapes. Furthermore, recent developments in unmanned aviation vehicle (UAV) surveys, virtual reality, and augmented reality hold great potential for further geographic engagements with and enrichments of historical landscapes through the holistic synthesis of cartographic and phenomenological perspectives.

Conclusion

Chapters in this volume illustrate a variety of approaches blending methods in historical geography, GIScience, and textual analysis that explore a number of topics, subjects, and themes, and perhaps answer the spirit of a recent question posed by John Hudson in 2015:

> What would historical geographers of old been able to accomplish had the Excel spreadsheet been available to them? (One can imagine the delight Andrew Clark would have had using a spreadsheet to calculate sheep-swine ratios in Nova Scotia). Historical data are available to all of us, free of charge, and in downloadable form that can be copied, cut, and pasted straight into a GIS mapping program, but there is little hint that this matters in historical geography.[27]

Indeed, not only scholars in historical geography but GIScience and textual analysis need to move beyond their own methodological silos to address collective questions larger than their disciplinary foci can envision alone.

In conclusion, by triangulating the methods of these heretofore distinct modes of inquiry, new means to explore the broad realms of human-environmental phenomena emerge. Emphasizing that the *caesura* between empirical/descriptive and interpretive/aesthetic disciplines is but a fiction, the chapters featured in this edited volume suggest alternatives, remedies, and ways forward.

Acknowledgements

Special thanks are extended to Michael Leuchner for commissioning this volume, Alexander von Lunen for helping select and critique contributions to this volume, and Andy Milson for providing feedback, advice, and support. In addition, we are grateful for the participants of the Association of American Geographer's 2019 Conference Historical Geography and the Geosciences Panel who contributed invaluable insight that helped shape the themes of this volume: Dorothy Sack, Craig Colten, Poul Holm, and Robert Doe. Also, thanks are extended to the History Department at the University of Texas, Arlington and the Trinity Centre for Environmental Humanities, Trinity College, The University of Dublin.

Charles Travis
e-mail: charles.travis@uta.edu

[27]John C. Hudson. 2015 Colten & Buckley reviewed by Hudson, *Historical Geography* B4–B5, Vol. 43.

Contents

Part I

Landscape, Time, Text

Ghost Cathedral of the Blackland Prairie: Waxahachie, Texas, *Places in the Heart* and the Superconducting Super Collider

1

Charles Travis and Javier Reyes

Abstract

Methods blending approaches in historical geography, GIScience and text can aid in extirpating how various perceptions of landscape (cultural and physical), identity (individual and collective) and sense of place (historical and contemporary) coalesce around a specific locale, or place. Data drawn from the IMPUS Historical GIS database and other sources were contextualized in this chapter's exploration by the use of the following "text maps": the film, *Places in the Heart* (1984), the novel *Einstein's Bridge* (1997) and the National Academy of Sciences and National Academy of Engineering site location report for a Federal government superconducting super collider (SSC) tunnel ring project. Drawing on the socio cultural and historical contextualization provided by these cinematic, imaginary and engineering "text maps", a GIS deep mapping of the local and deep history of Waxahachie, Texas, juxtaposes a cartographical *filmscape* of the 1930s Texas Blackland Prairie with a *sideshadow* model of the SSC project, abandoned by the Federal government in the 1990s.

Keywords

Waxahachie · *Filmscape* · Local history · *Places in the Heart* · Superconducting super collider · GIScience *sideshadow*

1.1 Introduction

In *The Making of an English Landscape* (1955), William George Hoskins argued that "a landscape historian needs to use botany, physical geography and natural history as well as historical knowledge to interpret any given scene fully."[1] Drawing on Hoskins's perspective, and other tropes in traditional and "new" cultural geography, this chapter re-considers how GIScience abstract framings of space and time can be applied in conjunction with the subjective and non-Euclidean place-mapping techniques employed by historical, cultural and literary scholars. The choice of Waxahachie, in Ellis County Texas as the location of this chapter's deep local history, borrows from William Least

C. Travis (✉) · J. Reyes
Department of History, UT Arlington, Waxahachie, USA
e-mail: charles.travis@uta.edu

J. Reyes
e-mail: javier.reyes2@mavs.uta.edu

[1]Hoskins, W. G., 1955. *The making of the English landscape*.

C. Travis et al. (eds.), *Historical Geography, GIScience and Textual Analysis*, Historical Geography and Geosciences, https://doi.org/10.1007/978-3-030-37569-0_1

Heat Moon's survey of Chase County, Kansas in *PrairyErth: A Deep Map* (1991), described as:

> […] the work of an encyclopaedist without an alphabet. It is arranged geographically, quadrant by quadrant, around the county, "my arbitrary pattern … that of a Japanese reading a book, up to down, right to left" The grid that Heat-Moon imagines for his book is part actual, part imaginary -it corresponds to the 12 U.S. Geological Survey maps that cover the center of the county, and it resembles the grid that 'an archaeologist lays over ground he will excavate.'[2]

This mélange of techniques applied in an exegetical use of GIS to illuminate how various perceptions of landscape (cultural and physical), identity (individual and collective) and sense of place (historical and contemporary) coalesce around a specific locale, or place. Perceptions of Waxahachie, Texas framed by the texts of Robert Benton, native-born writer and filmmaker, John G. Cramer, a physicist and hard science fiction writer, and the text of a proposal drafted for the Federal government by the National Academy of Sciences and National Academy of Engineering (NASNAE) to construct a super-conducting super collider (SSC) will be explored in this chapter using GIS. *Places in the Heart* (1984) is a film written, directed and set in Waxahachie by Robert Benton and won the 1985 Academy Award for Best Original Screenplay. *Einstein's Bridge* (1997), a hard science fiction novel by the physicist John G. Cramer, is parsed in conjunction with the NASNAE's plans for the Federal SSC tunnel ring to be bored into the bedrock under Ellis County. Drawing on the historical and cultural contexts provided by these cinematic, imaginary and engineering texts, the chapter's discussion will juxtapose a GIS cartographical *filmscape* mapping of the 1930s Texas Blackland Prairie with a *sideshadow* GIS model case study of the SSC project, abandoned by the Federal government in the 1990s.

1.1.1 Waxahachie Local History and the Apocrypha of Place Naming

The seat of Ellis County Texas, Waxahachie is located where the tail ends of the Appalachian and Deep Southern cultural landscapes entwine and taper out on to the cotton and cornfield plains of the east and central Texas Blackland Prairie, named after the black and deep dark-gray alkaline clay of its soil. Before Anglo-European settlement this physical landscape was molded by lightning strike wildfires and migratory plains bison whose massive herds stamped down and fertilized the clay to stimulate growth of prairie tallgrasses. Unfolding an old Texaco road atlas, with oil smudges and coffee ring stains on its creased and faded pages, one can locate the town just about 30 miles southwest of Dallas. Waxahachie can be located just beneath where Interstate Highway 35 East crosses US Highway 287 on the road-map—just like the "X" of its third letter. However, trying to pin down the etymology of its name is a descent into apocrypha. Officially, the town's name is purported to signify "buffalo" and is taken from Waxahachie (or Buffalo) Creek which rises a mile north of neighboring Midlothian at 32°29′ N, 96°59′. The creek parallels a rail-road track and meanders south skirting the western edge of Waxahachie's historic district before parting ways with the rails. Following twists and turns southward to Navarro County, the creek returns to mother earth at 32°13′ N, 96°35′ W, three miles south of the Bardwell Reservoir. In contrast, the rail tracks forge on into the landscapes of antebellum memory, carrying ghost trains of cotton bales from the haunted cargo terminals of Dallas and Waxahachie to the port of Houston on the Gulf of Mexico.

Various sources seek the origin of Waxahachie's place name from the Muscogee Creek word *wakvhvce*, the Alabama-Coushatta word *waakasi hachi* and the Wichita word *waks'ahe:*

[2]Klinkenborg, V. 1991. Cameos of Kansas: PRAIRYERTH (a deep map) By William Least Heat-Moon. *Los Angeles Times*. 20 October 1991. http://articles.latimes.com/1991-10-20/books/bk-20_1_chase-county.

ts'i. Rough translations of these tongues parse Waxahachie to mean "cow or calf's tail."[3] An *Ethnolinguistic Map of Texas*, dated to 1776 (Fig. 1.1) by L. L. M. Skeels locates Waxahachie in Wichita tribal territory. However, Wichita linguists dispute that the town name refers to "buffalo" with other linguists noting that *hatchie* was an indigenous term for "creek" or "river" used in parts of Tennessee, the Carolinas, Mississippi and Alabama. In the early nineteenth century, the region was the nomadic home of the Tonkawa (*tichkan-watich*, "the most human of people.") Anglo settlement and the US Government's Native America policy forcibly relocated the tribe to the Washita River Reservation in Oklahoma. The first recorded white settler in the area was a Methodist preacher who squatted illegally along the Waxahachie Creek. The first "legal" settler was a farmer named William R. Howe, who arrived in 1843, followed by Emery Rogers, a settler from Alabama, who is recorded as the town's first official citizen. After being granted a homestead in 1847, Rogers built a cabin, and planted a field, and by

> […] early June, when the cornstalks were just beginning to put on their first true leaves, the Rogers family found their prairie land alive with buffalo. Everywhere they turned, not 100 yards from their cabin's front door, buffalo gathered. The large animals had migrated each year from the west for hundreds of years to feed, mate and rest along the waterways of Waxahachie Creek.[4]

Regardless of the town's place name apocrypha, Ellis County and Waxahachie as its seat became Cartesian facts on 23 December 1849. On that date, two bills introduced by Edward Hampton Tarrant having passed the third session of the Texas State Legislature were signed into law by Governor Peter Bell.[5] Three years previously, on 29 December 1845, Texas had entered the Union as a slave state. Initially sparking the Mexican-American War (1846–1848), the long run consequence of slave-statehood raised racial and political strife, and secession fever in the region. Figure 1.2 illustrates the 1860 US Census' topography of slave holding in the central and eastern parts of Texas prior to the US Civil War (1861–1865).

During a period of drought from July to September 1860, Waxahachie was seized by the hysteria of a panic that gripped north Texas. Fires causing hundreds of thousands of dollars in damages broke out in Dallas and a dozen other towns. While the origins of the fires remain a mystery, they were initially ascribed to defective phosphorus matches. Charles Pryor, editor of the *Dallas Herald* newspaper who lost his printing press to fire, fanned the flames of public hysteria by suggesting the fires were set by slaves and white abolitionist allies. On 31 July the *Houston Telegraph* newspaper reported:

> A letter from a merchant in Waxahachie … relates that a plot was discovered to carry out murders and destroy the town on Sunday July 22 similar to the Dallas affair. Two white men were hanged on Sat., July 21, and some twenty-odd negroes were to be hung also next week.[6]

Historian Donald E. Reynolds argues that the fires were accidental, but the conflagrations allowed secessionists to whip up anti-Union sentiment in the period, leading up to the Civil War. Lynching, violence and terror lasted into the autumn of 1860, and contributed to the political

[3]Geographic Identifiers: 2010 Census Summary File 1 (G001): Waxahachie city, *Texas American Factfinder*. U. S. Census Bureau; Felty, M. L. Waxahachie, *Handbook of Texas*. Texas State Historical Association. Retrieved 5 June 2014.

[4]Stott, K. M., 2002. *Waxahachie: Where Cotton Reigned King*. Arcadia Publishing.

[5]Hamilton, J., 2019. Our Waxahachie Heritage – How did Waxahachie get its name? *Waxahachie Sun*, 5 March 2019. https://www.waxahachiesun.com/news/local/our-waxahachie-heritage-how-did-waxahachie-get-itsname/article_a8a5fce0-3f9f-11e9-8bef-0b3d47578f1a.html.

[6]*State Gazette*, August ii, 1860, quoting the *Houston Telegraph*, July 31; Reynolds, D. E. 2007, *Texas Terror: The Slave Insurrection Panic of 1860 and the Secession of the Lower South*. Baton Rouge: Louisiana State University Press.

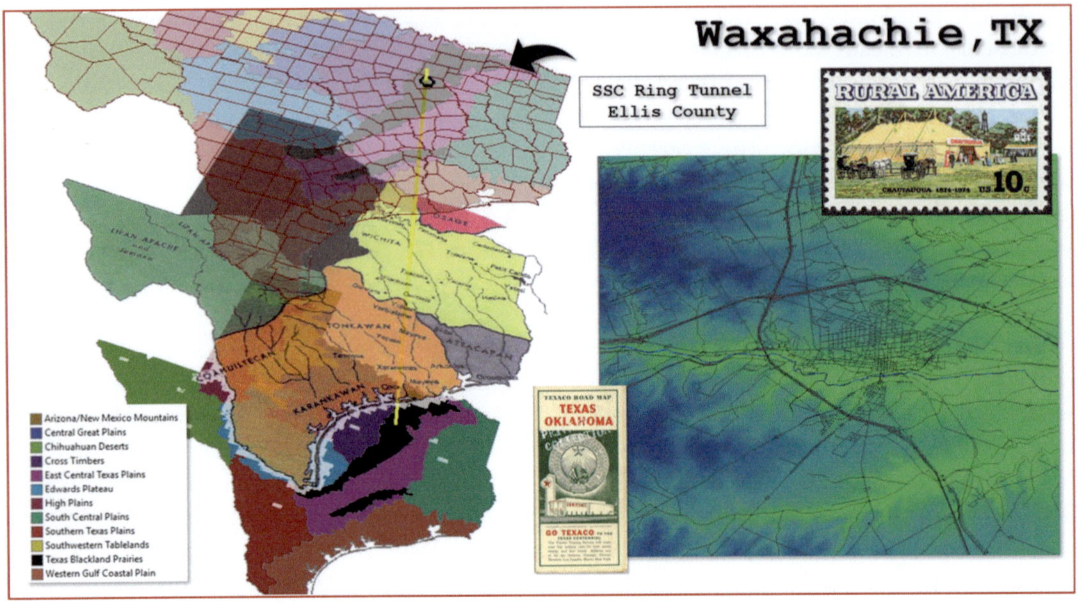

Fig. 1.1 **Local History Deep Map**: yellow line indicates location of Waxahachie (**Image Left**) Lower map, *Texas Natural Regions*, Middle map, *Indigenous Languages 1776*, Upper map, *Texas County Map with proposed location of SSC Ring Tunnel Project*. (**Image Right**) US Postage Stamp celebrating Rural America Chautauqua Movement. *Digital Elevation Topography of Waxahachie Creek Area* overlaid with *TXDOT Highways 35 and 287 and Waxahachie town street grid* (*Sources* Natural Regions, EPSG 4269 Texas Natural Resources Information System, Public Domain (Creative Commons CC0); Skeels, L.L.M. 1972. *An Ethnohistorical Survey of Texas Indians*, Texas Historical Survey Committee, Office of State Archeologist, Report No. 22, Austin, Texas; DEM Raster, grdn33w097_1, United States Geological Service; 35-276 HWY and Waxahachie Street Grid, GIS Appraisal Information, Ellis Appraisal District, TxDOT Roadways. TNRIS. SSC Ring Tunnel, Javier Reyes, 2018. US stamp honoring the 100th anniversary of the Chautauqua movement (Public Domain)) (GIS images created by Charles Travis)

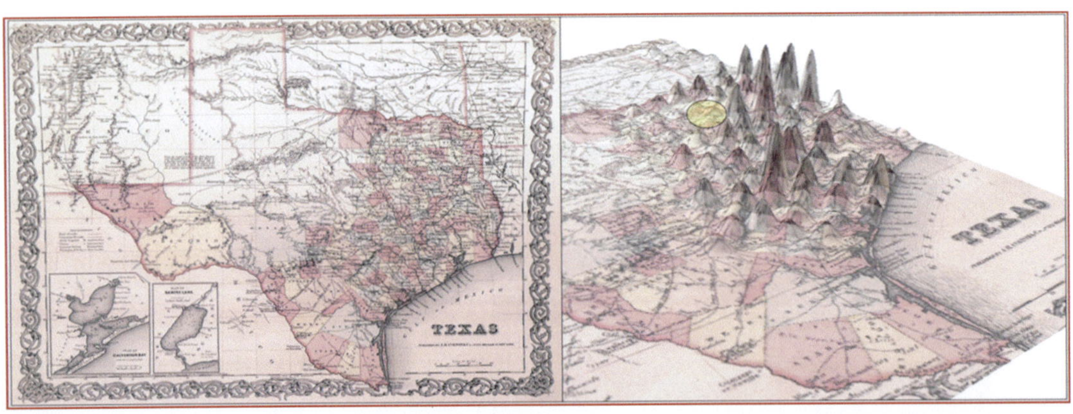

Fig. 1.2 **Map Left** J. H. Colton, *Texas*, 1854. **Image Right**: *The Topography of Slavery in Texas, 1860* (*Sources* J. H. Colton, *Texas*, color lithograph, 33 × 41 cm., 185[4], in *Colton's Atlas of the World* (New York and London: J. H. Colton and Company, 1855), plate 35. Source: *The University of Texas at Arlington Libraries Special Collections*; 3D kernel density graphing of slave holding per Texas county, 1860 on 'rubber sheeted' Colton 1854 *Texas*, map. Height extrusions illustrate locations with greatest densities of slaves according to the 1860s U.S. Census). Yellow sphere indicates approximate location of Ellis County (GIS image created by Charles Travis)

atmosphere that led to the February 1861 referendum vote on Texas' secession from the Union.[7] After the Civil War, reconstruction and the growth of rail-road networks brought a measure of prosperity to Waxahachie. During this time the southern and the border states of the broken Confederacy supplied a stream of migrants who by 1873 revived Ellis County's Blackland Prairie cotton culture.[8] The Texas State Legislature passed a bill in 1875 granting rail-road investors the right to build a track to Waxahachie, leading to a spike in the town's population. Within a year, over 5000 bales of cotton and 140 carloads of lumber were shipped along this new rail-road line, spurring the growth of the town economy and the construction of its ornate Victorian homes, numerous businesses and small factories.[9] By 1902, Waxahachie boasted a fully operating cotton textile mill, the home of Trinity University and a population of over 4000 citizens. In 1912, an electric inter-urban line running through town connected Dallas with Waco. Soon, the Oak Lawn Academy, a center of higher education for African-Americans, and the Chautauqua Summer Assembly, organized by the Texas Synod of the Cumberland Presbyterian Church, brought national speakers such as William Jennings Bryan and Will Rogers to Waxahachie. In the 1930s, despite the Wall Street Crash of 1929, the town's population was nearly 8000 and the home to a high school, four elementary schools, three banks, three cottonseed oil mills, five cotton gins, two daily and two weekly newspapers, and over 280 businesses and factories, including garment, broom, ice and ice cream manufacturers.[10] Industrialization during the Second World War, and the national economic boom that followed, created jobs in poultry processing, clothing, furniture and fiberglass manufacturing. During the 1970s, Waxahachie's population increased to more than 13,000 and by 2000 was nearly 20,000 (Fig. 1.3).

Waxahachie carries the appellation "Gingerbread City" and is listed on the National Register of Historic Places due to the period Victorian architecture (c.a. 1860–1900) that graces 300 of the town's most ornate homes and buildings.[11] In 1988, the area around Waxahachie was chosen as the site for the superconducting super collider (SSC) which was to be the world's largest sub-atomic particle accelerator, with a planned ring circumference of 54.1 miles (87.1 km). More than a dozen tunnels were bored into the Austin Chalk sediment underlying Ellis County before the project was canceled by Congress in the 1990s.[12]

1.2 GIS, Maps and Cinematography

Like art, historical geography is a social practice which ends in representation.[13]

1.2.1 Historical Geography and the Cartographic Illusion

Giuliana Bruno describes film making as a form of "modern cartography," which transforms images into different types of visual and spatial architecture.[14] In addition, Tom Conley argues that movies should be considered as maps, because they act as visual structures to shape the

[7]Ibid.; Phillips, M., 2009. Texas Terror: The Slave Insurrection Panic of 1860 and the Secession of the Lower South. *The Journal of Southern History*, 75(2), p. 448.

[8]Stott, Ibid.

[9]Felty, Ibid.; Crocket, G. L. 1932. *Two Centuries in East Texas*. Dallas: Southwest, facsimile reprod., 1962; Hawkins, E. D., et al., 1972. *History of Ellis County, Texas*. Waco: Texian; *Memorial and Biographical History of Ellis County*, 1892. Chicago: Lewis, rpt., as *Ellis County History*, Fort Worth, 1972. Historical Publishers; Richardson, T. C., 1940, *East Texas. Its History and Its Makers*. 4 vols., New York: Lewis Historical Publishing. Vertical Files, Dolph Briscoe Center for American History, University of Texas at Austin.

[10]Ibid.

[11]Ibid.

[12]Staff, Wire services Q & A: Texas supercollider project scrapped. *St. Petersburg Times*. 29 December 2009.

[13]J. B. Harley, 1989. Historical geography and the cartographic illusion. *Journal of Historical Geography*, 15(1), pp. 80–91, p. 80.

[14]Bruno, G. 2002. *The atlas of emotion: Journeys in art, architecture, and film*. New York: Verso, pp. 8–9.

Fig. 1.3 The 1900 Chaska House, one of dozens of "Gingerbread City" Victorian homes in Waxahachie, the seat of Ellis County, Texas (Wikimedia Commons)

imaginations of spectators. Accordingly, films can be used as tools for deciphering the worlds they depict. Films can be parsed through each frame and image. As they gradually unfold and change the scenery, they assemble subjective worlds for the eyes of the audience.[15] Les Roberts observes that "the idea that a film could function as a map, indeed could even be a map, is itself not new, as illustrated by Walter Benjamin's musings on the possibility of a cinematographic map of Paris."[16] Roberts notes that *cinematic cartography* focuses on "locational properties of the moving image, and the concomitant processes of spatial and temporal navigation—historiographical, ontological, geographical, archaeological, architectural—that these make possible."[17] From a technical perspective, the ability to apply *Python* and *Java* coding languages, and integrate artificial intelligence/deep learning, and R methods in GIS is creating a new, reflexive type of interactive visual media that can communicate "spatial/platial" data in a variety of formats. In this light, cinematography defined as "the process of taking ideas, words, actions, emotional subtext, tone and all other forms of non-verbal communication and *rendering them in visual terms*,"[18] and emerging GIS methods are becoming very similar.

1.2.2 Texas Cinematic Geographies

J. T. Caldwell observes that historical recollection, contextualization and rehabilitation are made possible through the medium of film.[19] Cinema captures the milieu of particular eras, while also preserving the historical events and cultural attitudes of past generations. Film can also serve as an archive, which can preserve in this age of global warming, a visual record of physical landscapes from the recent past, prior to drastic changes created by natural and anthropogenic causes (Fig 1.4). In the case of Texas, James Dean's sullen rancher-cum-oilman in *Giant* (1956) is framed by the iron orchards of oil wells dotting the bleak Chihuahuan Deserts in

[15]Baetens, J. 2008. Cartographic Cinema by Tom Conley Review. *Leonardo*, 41: 3, pp. 295–296, The MIT Press, p. 295.

[16]Roberts, L., 2012. Cinematic cartography: Projecting place through film. In *Mapping Cultures* (pp. 68–84). Palgrave Macmillan, London.

[17]Ibid.

[18]Brown, B. 2002. *Cinematography*. Oxford: Focal Press, ix.

[19]Caldwell, J. T. 2008. *Production culture: Industrial reflexivity and critical practice in film and television*. Chapel Hill, NC: Duke University Press.

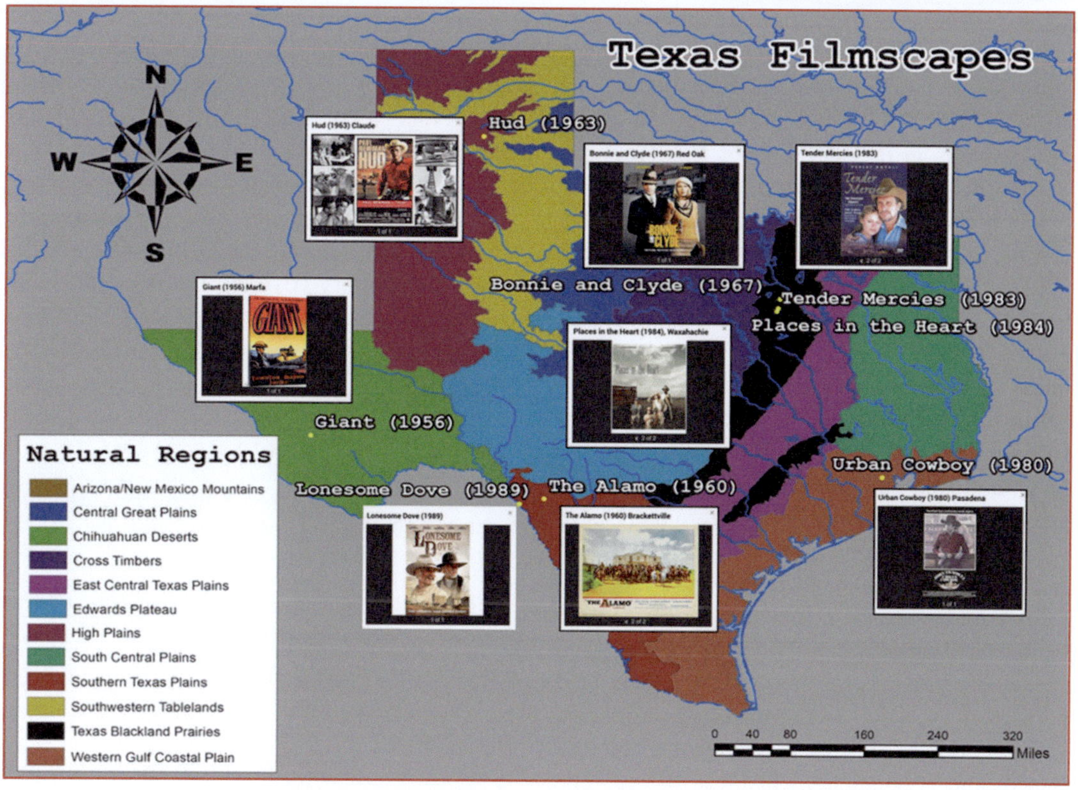

Fig. 1.4 Texas physical geography *filmscapes* and human geography representations (**clockwise from top**) *Hud* (1963) Southwestern Tablelands; *Bonnie and Clyde* (1967) and *Places in the Heart* (1984) Blackland Prairies; *Urban Cowboy* (1980) Western Gulf Coastal Plain; *The Alamo* (1960) and *Lonesome Dove* (1989) Southern Texas Plains; *Giant* (1956) Chihuahuan Deserts; *Tender Mercies* (1983) Blackland Prairie (GIS Map by Charles Travis)

the west of the state. *The Alamo* (1960) and the opening of *Lonesome Dove* (1989), both set in the nineteenth century, depict dry and dusty Southern Texas Plains, as they may have existed before Anglo and European settlement. *Urban Cowboy* (1980) set in the marshy Western Gulf Coastal Plain on which Houston sits, juxtaposes film-shots of oil refinery towers and fires with rodeo arenas, mechanical bulls and the working-class culture of Gilley's now defunct country and western music club. A trio of films, *Bonnie and Clyde* (1969), *Tender Mercies* (1983) and *Places in the Heart* (1984) depict the poor, small, agrarian communities of north Texas dotting the eastern plains of the Blackland Prairie region. The actor Robert Duvall, playing an alcoholic ex-country musician in *Tender Mercies*, stated that he was interested in depicting the lives of ordinary Texans "without parodying them, … [as] many Hollywood films tend to do."[20] The human and physical geographies of east Texas are framed in the film by Duvall's "aging face, a road map of dead ends and dry gulches."[21] The film's director Bruce Beresford drew parallels between the landscapes of Texas and those of his native Australia. Though *Tender Mercies* was shot around the Waxahachie area, the film's setting is nameless; however, for Beresford, Ellis County's terrain reminded him of the antipodean bush country, and the rural culture of Australian Outback communities.[22]

[20]Duvall, R. 2002. *Miracles & Mercies.*

[21]Corliss, Richard. 1983. Heart of Texas. *Time*, p. 63.

[22]Slawson, Judith. 1985. *Robert Duvall: Hollywood Maverick.* New York City: St. Martin's Press.

1.2.3 *Places in the Heart* (1984)

From Thomas Jefferson's agrarian democratic vision to Ronald Reagan's idyllic western range, there remains in the nostalgia of the American psyche a tenacious idea that national identity is somehow rooted in the terrains of the small town past, the farm and the family.[23] Such a sentiment has situated Waxahachie as a lure for film-makers: "Stately gingerbread homes that line tree-shaded streets and a restored town square make [the location] a perfect setting for turn-of-the-century-period movies."[24] In addition, at the edge of town "enormous churches and small cottages aside the capillaries of old farm-to-market roads that lead to rare patches of velour pasture among dry scrubland and Blackland Prairie, all under blue horizons and the skeletal tree lines of Texan winters."[25] Robert Benton, Waxahachie native, and writer and director of *Places in the Heart* (1984) fashioned the setting of his movie from such elements, leavened by an insider's knowledge of the nuances and contradictions of local history. The film's opening montage establishes period and place (Fig. 1.5).

The film's first image is a longshot of the historic Waxahachie court house clock tower standing sentinel-like over the east Texas prairie. The image intimates the insularity of rural life during the Great Depression of the 1930s. Subsequent shots provide glimpses of a Sunday congregation standing on the steps of a church, a woman living in a period automobile with a sign protesting her eviction leaning against its front grill, white and black families praying before a meal, skeletal structures of cotton ginning machinery with the sharp-edged fans of a windmill pump perched like a tin rooster, and an idyllic carpet of Texas bluebonnets flowering under a lone, leafless dendritic tree on a hill. The opening montage frames the film with an "idealized look of the work by some of the better, now-anonymous painters who, supported by Federal subsidies during the Depression, traveled around the country covering the walls of public buildings, in small towns and large."[26] Film geographer Christopher Lukinbeal notes that:

> It is montage that converts space into place, that allows for place-making to occur. Place is made through narration of and for the subject, not a whole coherent image, but a sutured, fragmented image unified into a cohesive form. Place-making therefore is constructed out of the gaze, the look, and point-of-view shots.[27]

The final image of *Places in the Heart's* title sequence gazes upon the rural home where heroine Edna Spaulding (played by Sally Field) is serving Sunday dinner to her husband and children (Fig. 1.6).

The "sense of place" built up by the previous images in the montage gains cohesion and cements the prime location from where the human and physical geographies of the story will unfold and be plotted, charted and navigated. Before the meal is even completed, Edna Spaulding becomes a widow when her husband, the town sheriff, is called away and accidentally shot trying to arrest a young drunk black man who has been firing a gun in the town railyard. Her husband is laid out on the dining room table, as the body of the lynched black youth is dragged by men in a truck on the dusty road in front of Edna's house. After 15 years of marriage, she finds herself with two small children, Frank Jr. and Possum, less than $200 in the bank and living on a farm threatened by foreclosure.

[23]Adams, W., 1985. American Gothic: country, the river, places in the heart. *The Antioch Review*, 43(2), pp. 217–224, p. 218.

[24]Evans, M. 2018. *Three Oscar-Winning Films Helped Make Waxahachie 'The Best Little Hollywood In Texas'*, 26 Feb 2018. https://artandseek.org/2018/02/26/three-oscar-winning-films-helped-make-Waxahachie-the-best-little-hollywood-in-texas/.

[25]Quirk, T. 2013. How Texas Lost the World's Largest Super Collider, *Texas Monthly*, 21 October 2013. https://www.texasmonthly.com/articles/how-texas-lost-the-worlds-largest-super-collider/.

[26]Specht, A. R. and Rutherford, T., 2015. The pastoral fantasy on the silver screen: The influence of film on American cultural memory of the agrarian landscape. *Journal of Applied Communications*, 99(1), p. 28.

[27]Lukinbeal, C., 2010. Mobilizing the cartographic paradox: tracing the aspect of cartography and prospect of cinema. *ETD-Educação Temática Digital*, 11(2), pp. 1–32, p. 19.

Fig. 1.5 Phenomenological impressions juxtaposed with Cartesian representations. Waxahachie, Texas, *Places in the Heart* (1984) opening montage *filmscape* overlain on contemporary TXDOT transportation grid, and USGS Dem (Benton, R., 2001. *Places in the Heart*. Columbia TriStar Home Entertainment) (Created by Charles Travis)

Fig. 1.6 *Places in the Heart* movie poster and film still of final title sequence montage image (Wikimedia Commons) (Benton, R., 2001. Places in the Heart)

Edna is soon bullied into a show of good faith against her mortgage by the bank's manager by taking his brother-in-law, Mr. Will, a blind, misanthrope broom-maker in as a boarder. Edna seeks work from her sister Margaret, a beautician who supports a child and a philandering husband, Wayne, but the dearth of customers makes this impossible.

The film's other main character is Moses-"Moze" (played by Danny Glover), a black transient in the Jim Crow south, who stops by Edna's farm one evening looking for work, is offered a meal, and steals her silver cutlery. Caught down at the railyard trying to hop a freight car, Moze is brought back to the farm by a deputy sheriff. In a ruse, Edna explains that Moze was taking the silver to her sister, and thought dubious about the story, the deputy departs. Edna then asks Moze to help her grow and sell cotton, so she can pay her mortgage in return for his bed and board. She also assures Moze that if he ever steals from her again, she will shoot him herself. The black field hand, the blind boarder, Edna's children, her sister Margaret and her husband Wayne, estranged after the discovery of his affair, take part in the planting and harvesting of the cotton crop. Their efforts soon become a race against time, migrant cotton pickers and local industrial farmers to win a $100 ginning prize and make the farm's mortgage payment. The collective effort transforms the farm into a home for an unlikely cast of marginalized characters. Landscape, identity and sense of place are accentuated in the film by period music and radio broadcasts of baseball scores and commodity prices. Cinematographically-framed set pieces of Saturday night country dances and sheltering in a root cellar from the destructive wrath of a tornado also set the tone. At the film's end, the Ku-Klux-Klan, led by the owner of the cotton gin, who, riled by his efforts to help Edna win the $100 prize, drive Moze beaten and bruised from Waxahachie. Peter Lev states that the film serves as:

> […] a critique and correction of the macho stereotypes of Texas presented by *Giant*, *Dallas* and hundreds of Texas / Western movies. *Places in the Heart* […] shows 1930s Texas to be a rural, agrarian, sexually restrictive and racially oppressive society […] it demonstrates that Texas is not all cowboys; *Places in the Heart* may also propose a new paradigm for Texas, a redefinition of the frontier [and] the struggle to survive, to raise a family, and even to reform society without male dominance.[28]

The last scene of the film serves as a requiem for long-lost period, and a reconciliation of the social conflicts arising from segregated spaces, which dilutes the "wholeness" of place in a small rural 1930s eastern Texas town: a Sunday church service is attended by all of the film's characters (including Edna's husband and the youth who killed him, and was in return lynched) and they pass and share receptacles of consecrated bread and wine with each other in a final communion. *Places in the Heart's* final scene telegraphs back to the opening montage visualized in GIS and illustrates Jane Stadler, Peta Mitchell and S. Carleton's argument that film geographies bring critical distinctions between concepts of land and landscape and can help illustrate theoretically informed understandings of the production of space and place as a confluence of spectacle, metaphor and cultural artifact.[29] The integration of GIS and film visualization and analysis methods create a "prism" through which "imagined places" depicted by the text of a film, and "actual spaces" plotted in the text of a map converge and can be connected to the wider geo-historical constellations of time and place.[30] While many films offer counter-factual, and alternative worlds for the viewer to enter, the next section will discuss the creation of a GIS model of a structure that, if it had been completed, would have formed an indelible aspect of the sense of place, in not only Waxahachie, but in the wider region of Ellis County.

1.3 Mapping the Ghost Cathedral of the Blackland Prairie

On 10 November 1988, the US Government's Department of Energy, after considering sites in 25 states, announced that the superconducting super collider (SSC), a ring tunnel project designed to find the "God Particle," would be located under the Blackland Prairie cotton and corn fields surrounding Waxahachie, Texas.[31] The Dallas-Fort Worth metroplex steeled itself for a massive influx of SSC project workers, scientists, and other personnel and their families.

The aim of the project was to identify the *Higgs boson* or "God" particle, a sub-atomic

[28]Lev, P., 1986. Regional cinema and the films of Texas. *Journal of Film and Video*, pp. 60–65, p. 64.

[29]Stadler, J., Mitchell, P., & Carleton, S. 2016. *Imagined Landscapes: Geovisualizing Australian Spatial Narratives*. Indiana University Press, p. 15.

[30]Roberts, *Cinematic cartography*.

[31]Westfall, C. L., 1989. The site contest for Fermilab. *Phys. Today*, 42, pp. 44–52.

Fig. 1.7 1995 photograph of the SSC's collider ring's 14-foot-diameter tunnel blasted about 250 feet deep through Texas rock (Public Domain—US Government)

object and purported by theories espoused in modern physics to holding the key to unlocking the secrets of the universe's origins. Super colliders were deemed necessary research tools to uncover the "God" particle through the observation of sub-atomic collisions. The SSC project in Waxahachie was to join European Organization for Nuclear Research (CERN) and Fermilab, outside of Chicago, Illinois as part of a global hadron collider constellation (Fig. 1.7).

However, the SSC project, much touted under President Ronald Reagan and George H.W. Bush's administrations, was shelved under President Bill Clinton in favor of the Spacelab and Human Genome projects. If it had been completed, the Texas SSC would have been the world's largest and most powerful sub-atomic particle accelerator, boasting a tunnel ring with a circumference of 54.1 miles (87.1 km). The dry language of the *Siting the Superconducting Super Collider* (1988) study commissioned by the National Academy of Sciences and National Academy of Engineering (NASNAE) reported that under the rolling prairie of Waxahachie, the "geology is excellent for construction."[32] Indeed, as Trevor Quirk notes:

> Ellis County happens to lie atop the Austin Chalk, a cretaceous geological formation that mirrors the curve of the Gulf, arching from Mississippi to Mexico. It is a tunneller's dream; easy to bore and shape without compromising stability. And considering that the SSC was one of the world's largest tunneling projects, one that would set myriad construction records, the geological makeup was important.[33]

More than a dozen tunnels were bored into the Austin Chalk sediment underlying Ellis County before the project was canceled by Congress in 1998.[34] Local stories and myths about the abandoned Cathedral of the Prairies lying deep in the bedrock of north-eastern Texas blame the SSC "ghost tunnels" for the unchanging climate that seems to drape the farmlands and dusty, deserted backroads around Waxahachie. In 2012, the Higgs boson was discovered at the CERN super collider in Geneva, Switzerland. This finding:

> […] required $9 billion, ten years of study, thousands of careers, and a seventeen-mile collider ring bored out of the earth on the Franco-Swiss border. At fourteen Teraelectron-volts (TeV), it is the most energetic super collider ever built, and also one of the largest, most complex scientific experiments in history. Many have called it a modern-day cathedral.[35]

According to rumors that still circulate among the denizens of the county, the cost to mothball the SSC project far exceeded the costs projected to complete it. Whether or not this is true remains up to debate and to the Congressional Budgeting Office. However, 14 miles of the SSC "ghost" tunnel, partially filled with rainwater, still lies dormant under the prairie landscape outside of

[32]National Academy of Sciences and National Academy of Engineering. 1988. *Siting the Superconducting Super Collider*. Washington, DC: The National Academies Press.

[33]Quirk, Ibid.

[34]Staff, Wire services, *Q & A*.

[35]Quirk, Ibid.

Waxahachie. In 1999, a film production company purchased rights at the SSC facility to produce the movie *Universal Soldier II*, a direct-to-video movie depicting the site as "the home of a supercomputer that puppeteers a small legion of hostile robots."[36] The shell of the site is currently owned by a chemical blending plant.

1.3.1 *Einstein's Bridge* (1997)

The SSC project spawned a literary treatment, *Einstein's Bridge* (1997) a counterfactual history/hard science-fiction novel set in Waxahachie by the physicist John G. Cramer. Set between 1987 and 2004, its story begins in a counter-factual past (where the SSC has been constructed and is up and running, but attracts a hostile alien species leading to the earth's destruction) and ends via the device of a time-travel wormhole in the space of contemporary, recorded history. *Einstein's Bridge* opens under the US presidential administration of the re-elected George H. W. Bush with a group of physicists running experiments at the SSC in Waxahachie (visualized by Reyes' model in Fig. 1.8). Preliminary trial runs of the SSC produce an unidentifiable object, which they call a *Snark* (after Lewis Carroll's *The Hunting of the Snark*) which emits a radio-active pulse pattern in a prime number sequence of 2-3-5-7-11-13-17-19-23-29-31-37. The pulse was sent from *The Concantation of Individuals*, a benevolent alien species confederation represented by the Tunnel Maker, whose "Einsteinian bridge" imparts knowledge to the physicists that is beyond their human conceptions. The bridge is soon discovered by the *Hive Mind*, a parasitic alien species, who arrive to harvest the earth's resources and destroy it in the process. Before this occurs, the physicists are enabled by the Tunnel Maker to travel through a wormhole, and land in an alternative, near past (reflecting events in recent US history), where by helping Arkansas Governor Bill Clinton Win the 1992 US Presidential election, facilitate the cancellation of the SSC project, thus avoiding the destruction of the earth by the *Hive Mind*.[37]

1.3.2 The Ghost Cathedral: The SSC Ring Tunnel GIS Model

A GIS model of the proposed SSC ring tunnel was created by drawing on materials and data from the NASNAE report, the Ellis Appraisal District website and the Texas Department of Transportation website (TXDOT, 2015).[38] Geo-data of boundaries, city locations and roadways data helped situate the model's location and provided a means to illustrate the true scale the SSC project size, allowing for speculation on potential impacts on Ellis County's environmental, economic and urban development.

The maps in Fig. 1.8 illustrate the proposed outline of the SSC. A similar image was converted into a raster file, and then geo-referenced to align with contemporary street, road and highway grids in Ellis County. Major streets and highway intersections were used as primary reference points. A 2-mile buffer zone was created to identify towns that were not technically within the central boundary of the SSC ring tunnel. The buffer zone was merged with the SSC polygon layer to generate a more accurate image representation of land usage. Data obtained on the Ellis Appraisal District resources website provided a point-file of all of Texas cities including those located in Ellis County. An intersect of this city point layer with the buffer zone and SSC polygon revealed the number and locations of towns directly impacted by the SSC's proposed construction. The 2D map representation of the SSC was transformed into a 3D model in order to better visualize how big the scale and scope of the project was meant to be. A 2D outline shapefile was imported as a Cad file. The scale 3D model of the SSC was created by importing this resulting Cad file into *Trimble SketchUp* and

[36]Ibid.

[37]Cramer, J., 1997. *Einstein's Bridge*. Harper Collins.

[38]Data Downloads—GIS and Appraisal Information. Ellis Appraisal District, TxDOT Roadways. TNRIS.

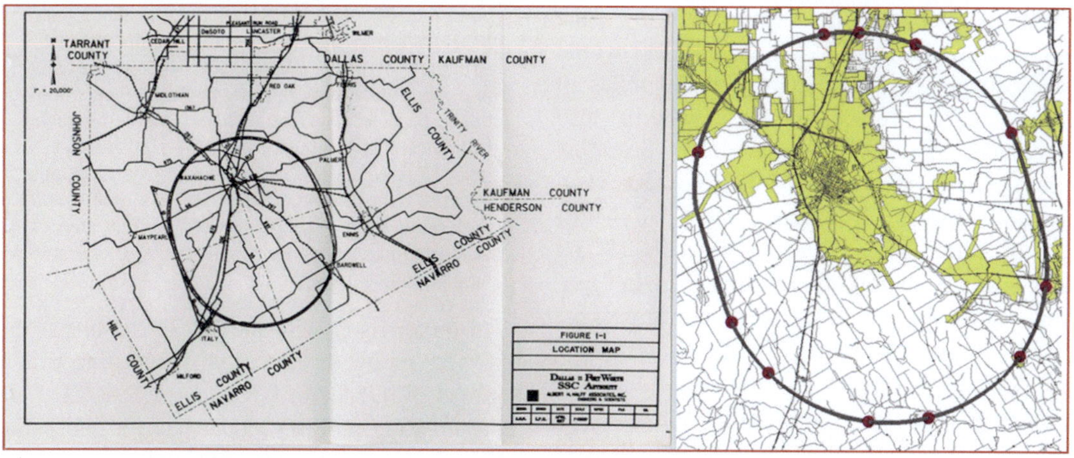

Fig. 1.8 Image Left: Location map, Dallas-Fort Worth Superconducting Super Collider Authority, SSP Collection, University of North Texas, Special Collections (Schütz, A. T. 2015. *A Look into Tomorrow for Waxahachie, Discovering the Southwest Metroplex*, UNT Libraries Library). **Image Right:** GIS model image of SSC Ring Tunnel with locations of access points encircling the town of Waxahachie (GIS model created by Javier Reyes)

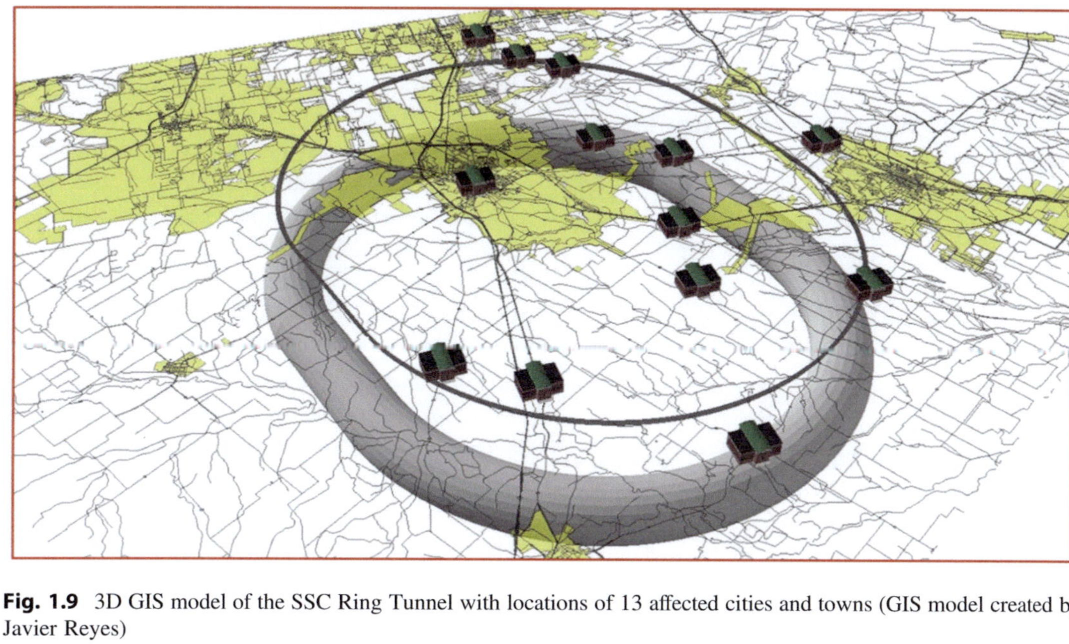

Fig. 1.9 3D GIS model of the SSC Ring Tunnel with locations of 13 affected cities and towns (GIS model created by Javier Reyes)

engaging the Extrusion and Follow Me tools. The SSC model *SketchUp* file was then imported into ArcScene, its coordinates corrected and altitude value adjusted to situate the model ring into its subterranean location. TXDOT 2015 GIS vector shape files of roadways and town boundaries were overlaid on SSC ring tunnel area, and the coordinates of the cities intersecting with the project space were plotted (Fig. 1.9).

The resulting GIS model of the SSC ring tunnel circumference measures 59.479 miles, 4 miles longer than the planned circumference of

54.121 miles/87.1 km, estimated by the NASNAE. Additionally, the SSC project would have occupied approximately 276.490 miles of Ellis County. Reyes' model (Fig. 1.9) also illuminates the idea of employing a historical *sideshadow* as heuristic for GIS landscape visualization analysis. J. Bonnett notes that whilst uses of virtual spaces (such as the ArcGIS and Google Earth platforms) are often used to "generate representations that cohere with our experience of actual space,"[39] the *sideshadow* representations comprise intentional and fantastic depictions utilizing virtual reality platforms to "interpolate objects not encountered in real space."[40] Bonnett states that:

> Objects that are normally distinguished -such as factual and counterfactual time-lines or macro and micro scales of history -are placed in proximity to enable users to assess their relationship. Metaphors —such as ghost images and tears or rips — are appropriated to represent and emphasize a relationship between one object and another.[41]

Ground-truthing surveys revealed that many of the access roads leading to the entrances of the proposed SSC ring tunnel still exist. Informal interviews with an individual who lived in the tunnel construction area during initial excavations confirmed that the model's routes and roads still exist on the abandoned site. The project would have provided numerous job opportunities for citizens of 13 cities whose locations intersected with the SSC project in positions ranging from SSC facility maintenance to project research positions. However, in 1993, construction on the SSC was halted after only 25% or 14 miles of ring tunnel was bored. Historical map comparison of the region with the SSC GIS model illustrates how little the main roadways in Ellis County have changed from the 1990s. The imprint of the SSC project on its landscape would have been enormous:

> [...] over 16,000 acres of land with a tunnel fourteen feet in diameter and 52 miles in circumference -more than twice the distance from Waxahachie to Dallas. It would require the construction of onsite office space, laboratories, and two cavernous subterranean facilities that would house particle detectors with weights comparable to battleships. It would comprise 4,728 magnets - most 17 meters long -that amounted to 41,500 tons of iron (roughly four Eiffel towers) and would require the supercooling effect of 2 million liters of liquid helium. Plus, the people . . . the total staff would eclipse 1,900.[42]

In comparison with the NASNAE plans, the SSC GIS model scale was fairly accurate with a minimal difference of 9.07% error. The 2D map representation and 3D model paint a geospatial portrait of the potential the SSC project may have had on transforming human and physical landscape of Waxahachie and Ellis County, Texas, and how it may have been pictured in *Einstein's Bridge*.

1.4 Conclusion

This chapter has considered GIS mappings of a single locale, parsed by the lenses of different spatialized narratives. The texts of a film, a novel and engineering plans, juxtaposed the phenomenological impressions of place, in GIS against Cartesian perspectives, to plot the unique historic trajectory traced by Waxahachie, Texas over the abstract grids and scales of space and time. In doing so, GIS illuminated how perceptions of landscape, identity and sense of place *shapeshift* depending on the eye of the beholder, by assuming different contingencies, meanings, sensibilities and nuances. In a sense, GIS also enabled artistic, humanistic and scientific narratives to find a confluent "platial" voice. Stephen Daniels and Patrick Bartlein state that "narrative has long been employed as a framework for expression in physical geography where 'telling the story' of places that range from a particular 'site' (e.g., a stream-channel cross section), to a landscape, to the whole Earth system, is accom-

[39]Bonnett, J., 2007. Charting a new aesthetics for history: 3D, scenarios, and the future of the historian's craft. *Histoire sociale/Social History*, 40(79), p. 183.

[40]Ibid.

[41]Ibid.

[42]Quirk, Ibid.

panied by explanations of how that story unfolded."[43] It is ironic then that GIS applications in historical geography, with a few notable exceptions, have not been used to tell more deeply multi-layered stories of places and plumb the wells of local histories. Many historical geography studies approach GIS as a type of "database visualizer," echoing the conservative methods of "traditional" historical geography, often critiqued for being "too closely tied to the apron strings of archival data sources and field observation."[44] Though traditional methods have produced "meticulous reconstructions of past patterns" the dynamic processes behind the creation of such patterns "in their cultural and political contexts" is at times relegated to the afterthoughts that frame a study's conclusion.[45] GIS contributed to the "spatial turn" in historical research and humanities scholarship in the 1980s and 1990s, and in "turn" influenced "GIScientists" to begin making "temporal runs."[46] However, many GIScientists still do not distinguish between Euclidean space and subjective concepts of place.[47] Thus, both historical geographers engaging GIS and GIScientists drawing on temporality could mutually benefit by exploring the various narrative frameworks employed in literary studies and in physical geography. As Betty Nowvskie notes, if we seek methods "capable of meeting more than the technical challenges of our massive geo-temporal datasets, we must develop design approaches that address recent theoretical merging's of background and foreground, space, and time."[48]

Acknowledgements Thanks to our UT Arlington geography professor, colleague and friend, Dr. Andy Milson for comments, suggestions and edits.

[43]Daniels, S., and Bartlein, P. J. 2017. Charting Time, *Annals of the American Association of Geographers*, 107:1, pp. 28–32, p. 29.

[44]Dennis, R., 1991. History, geography, and historical geography. *Social Science History*, 15:02, pp. 265–288, p. 266.

[45]Ibid.

[46]Yuan, M. 2014. Temporal GIS for Historical Research, *In Spatio-Temporal Narratives: Historical GIS and the Study of Global Trading Networks*, edited by A. Crespo Solana. Cambridge Scholars Publishing, pp. 45–55.

[47]D. Sui, and M. Goodchild, 2011. The convergence of GIS and social media: challenges for GIScience. *International Journal of Geographical Information Science*, 25 (11), pp. 1737–1748.

[48]Nowviskie, B. 2014. "Digital Humanities in the Anthropocene." *Nowviskie.org* (blog), 10 July 2014. http://nowviskie.org/2014/anthropocene/.

Digital Mapping and the Narrative Stratigraphy of Iceland

2

Emily Lethbridge

Abstract

The relationship between narrative and place in Iceland (as elsewhere) is an intricate and symbiotic one that is always in process. Digital mapping tools make it possible to take steps towards establishing chronologies of storied places. In addition, such tools aid in the interrogation and characterisation of the reciprocal dynamics of story and place in Iceland in ways not conceivable before. The ongoing Icelandic Saga Map (ISM) project attempts to link Iceland's rich medieval textual corpus with the country's geography, thus facilitating a better understanding of the various functions the landscape fulfils in the medieval sagas and in other works, as well as encouraging reflection on the role that landscape has played in the transmission and reception of these works over a time-period of a millennium or so. This chapter provides an overview of the medieval Icelandic textual corpus and a description of some technical aspects of the ISM project, followed by a discussion that focuses on the methodological challenges encountered and theoretical insights gained from the mapping process. Examples drawn from the most famous of all sagas, *Njáls saga*, provide a sense of the complexity of correspondences between textual representations of Icelandic landscape and its 'real-world' counterpart(s).

Keywords

Icelandic sagas · Digital mapping · Landscape · Place-names · Literary tourism

2.1 Introduction: Land of Rocks and Sagas

A 'land of rocks and sagas' is how Louis MacNeice described Iceland in the poem *Letter to Graham and Anne Shepard* (MacNeice 1937), a phrase that perfectly captures the juxtaposition—even fusing together—of geological strata with stories all around Iceland. Today, as in past centuries, visitors to Iceland are unfailingly struck by the physical appearance of the North Atlantic island's lava-strewn and glacier-crowned landscapes, and equally impressed by the way that female and male founding figures of Icelandic society—together with outlaws, ghosts, trolls—animate Icelandic topography and link the past with the present. The country is young in geological terms and historically speaking (Thordarson and Höskuldsson 2014; Karlsson 2000) but it is thickly layered with story at local and national scales. The relationship between

E. Lethbridge (✉)
Árni Magnússon Institute for Icelandic Research, Reykjavík, Iceland
e-mail: emily.lethbridge@arnastofnun.is

C. Travis et al. (eds.), *Historical Geography, GIScience and Textual Analysis*, Historical Geography and Geosciences, https://doi.org/10.1007/978-3-030-37569-0_2

narrative and place in Iceland (as elsewhere) is an intricate and symbiotic one that is always in process. It can be difficult to define this relationship precisely, but digital mapping tools make it possible to take steps towards establishing chronologies of storied places, as well as aiding the interrogation and characterisation of the reciprocal dynamics of story and place in ways not conceived before.

The ongoing Icelandic Saga Map (ISM) project attempts to link Iceland's rich medieval textual corpus with the country's geography, thus facilitating a better understanding of the various functions the landscape plays in the medieval sagas and in other works, as well as encouraging reflection on the role that landscape has played in the transmission and reception of these works over a time-period of a millennium or so. In this chapter, after an overview of the medieval Icelandic textual corpus and a description of some technical aspects of the ISM project, discussion will focus on the methodological challenges encountered and theoretical insights gained from the mapping process. Following this, some examples drawn from the most famous of all sagas, *Njáls saga*, will illustrate points made in the previous sections and give a sense of the complexity of correspondences between textual representations of Icelandic landscape and its 'real-world' counterpart(s).[1]

2.1.1 The Medieval Icelandic Textual Corpus

Icelanders were quick to adopt and adapt the technologies of book-making and writing with the Roman alphabet after Iceland became a literate society following the country's conversion to Christianity in the year 1000 C.E. (Gunnlaugsson 2010). In 1117–1118, the *Grágás* law-code (previously memorised and recited orally every year at the Alþingi or National Assembly) was written down in the Icelandic language, the first long text to be inscribed on parchment (Sandvik and Sigurðsson 2005). Together with the two bishoprics (Skálholt in the south, established 1056; Hólar in the north, established 1106), monastic houses became centres of learning and book production (Óskarsdóttir 2004) but literacy was also a tool taken up and utilised by the ruling secular elite as a means of promoting ideologies and consolidating political power (Clunies Ross 1997; Kristinsson 2003). From the late twelfth century onwards, Icelanders began to write vernacular accounts of their history from the settlement period and establishment of the early Icelandic Republic (870–930) to the authors' own times, a period which saw Iceland cede its independence to the Norwegian crown in 1262–1264 (Þorláksson 2005; Clunies Ross 2010). As well as these narratives, a wide range of other types of texts were produced including treatises on grammar, mythology and poetry, religious and legal works, annals and other historical works, and secular and sacred poetry (Clunies Ross 2010).

The best-known historical works covering the earliest period of Icelandic history, *Íslendingabók* ('The Book of Icelanders') and *Landnámabók* ('The Book of Settlements'), are attributed to the priest Ari fróði ('the learned') Þorgilsson (d.1148), though the subsequent transmission of *Landnámabók* in particular was complex, and this work only survives in several later redactions (Benediktsson 1986). While the emphasis of *Íslendingabók* is especially on the country's conversion to Christianity and the founding of the Church in Iceland, *Landnámabók* has a more secular focus (Whaley 2009). *Landnámabók* moves systematically around the inhabited periphery of the island giving details about the earliest settlers (genealogy, geographical origins) and their land-claims, and combining inventory-like topographical detail (place-names, boundaries) with anecdotal material about the settlers and their experiences in the new country (Vésteinsson and Friðriksson 2003).

Other narrative works—the sagas—told the stories of the earliest settlers and their descendants in greater detail. The most widely read saga genre is that of the *Íslendingasögur* ('Sagas of

[1]Some of the text in sections 2 and 5 has been adapted from an earlier publication (Lethbridge and Hartman 2016).

Icelanders', sometimes called 'Icelandic family sagas') (Ólason 2005). Extant narratives in this category number around 40 and are preserved in hundreds of medieval and post-medieval manuscripts, the oldest of which date to the mid-thirteenth century (Lethbridge 2014; McDonald Werronen 2018). They range in length from around 20 to 300 pages in printed editions or paperback translation and some are prosimetric in form: the prose is studded with verses uttered by characters to mark certain occasions or to serve as commentary on events (Harris 1997). The sources that the anonymous authors drew on belonged to the sphere of oral culture on the one hand (genealogical and legal material, anecdotes associated with specific places) and learned book culture on the other (hagiography, continental romance). Rooted in the landscapes of Iceland and set in the ninth to early eleventh centuries, the *Íslendingasögur* describe the initial claims of specific tracts of land and the building of farms on these sites, and more generally, the establishment of a new society and system of local and national politics and law (Ólason 2005). They include descriptions of movement around local areas and further afield in conjunction with agricultural practices (e.g. livestock management), politically and legally motivated travel (e.g. to local autumn and spring assemblies or to the summer Alþingi), or in the context of unfolding local disputes (e.g. ambushes, raids), which are a prominent theme. While the geographical focus of the *Íslendingasögur* is on Iceland, these narratives—as others, including the *Íslendingaþættir*, or short tales about early Icelanders (Rowe 2005)—also contain accounts of journeys abroad, whether visits to royal courts around Scandinavia and the British Isles, or beyond (Barraclough 2017).

Another saga category is the *samtíðarsögur* ('contemporary sagas'). These narratives are set in thirteenth-century Iceland and describe the warring politics and society—secular and ecclesiastical—that the descendants of *Íslendingasögur* protagonists (including the authors of *Íslendingasögur* and *samtíðarsögur* genres themselves) experienced. The best-known is the *Sturlunga saga* compilation, of which a long work called *Íslendinga saga* forms the backbone (Bragason 2005). While the authors of the *Íslendingasögur* are unknown, the identities of those writing *samtíðarsögur* (the category includes the *biskupasögur*, 'sagas of bishops of Iceland') are recorded: Sturla Þórðarson (1214–84), for example, was the author of *Íslendinga saga*. Yet other saga genres are the *konungasögur* ('sagas of kings'), which are biographies of Scandinavian kings (Jakobsson 2005); the *fornaldarsögur* ('legendary-heroic sagas'), which tell of heroes from mythical times and are set around Scandinavia (Tulinius 2005); the *riddarasögur* ('chivalric sagas'), romances set in Scandinavia and beyond (Glauser 2005); and the *heilagramanna sögur* ('sagas of holy men'), which are saints' lives (Cormack 2005).

2.2 The Icelandic Saga Map Project

The prominence and importance of physical setting in *Landnámabók* and in the sagas—in particular the *Íslendingasögur* and the *samtíðarsögur* (Callow 2006; Hoggart 2010; Jakobsson 2017)—make them a highly productive body of texts to consider from a spatial perspective. It has long been standard practice to include maps of locations in printed editions and translations (e.g. in all volumes of the widely used Íslenzk fornrit editions, first published in the 1930s). These static maps can only serve to roughly orientate the reader though, and are inevitably limited with regard to what they can display. The possibilities that GIS and other digital humanities tools offer here, in contrast, are almost boundless. The great potential that geo-referencing textual data has for generating new research questions and perspectives in the field of medieval Icelandic literature and Icelandic studies more widely is becoming evident with the development of the ISM project as well as other larger-scale projects such as Sagnagrunnur (a geographically mapped database of

Icelandic folk legends), Jarðabókin/Storied Lines (data pertaining to agricultural land use in medieval and early modern Iceland), and the Norse World project (data pertaining to medieval worldviews in East Norse texts).[2]

Initial work on the ISM project began in 2012 and from 2014 onwards, the project has received a number of grants enabling proper technical development and implementation.[3] The project website (http://sagamap.hi.is) displays texts of *Íslendingasögur*, *Sturlunga saga*, *Landnámabók* and certain other works (such as nineteenth-century travel writing on Iceland, discussed below) in which all named places have been geo-referenced and hyperlinked to the map next to the text panel. The user can click on place-names in the texts to see where they are located on the map, or, alternatively, can search for and click on places displayed on the map navigating from there back to the sagas or other texts that mention them (see Fig. 2.1). Multiple texts can be selected so that the places in them appear simultaneously on the map: their distribution reveals the degree of geographical overlap between texts, whether in the sense that the narrative stages of different sagas or other works intersect and cover the same regions or areas, or in the sense that the same place is mentioned in multiple works. Points on the map are colour-coded for text and a thick white border around a point denotes a place that is named in multiple texts (see Fig. 2.2). When a place is selected on the map or a place-name clicked on in a text, a pop-up box with information (place-type, what works mention it, latitude/longitude co-ordinates etc.) appears. The pop-up boxes also include summary information regarding debate or uncertainty about the location or identification of the place (mostly derived from notes in the Íslenzk fornrit editions) and links to other online resources (e.g. archaeological excavation reports). Elsewhere on the website, the user can find information about the project, the sagas, the text editions used for the interface, the backend technical specifications and development, and bibliographic resources useful for spatially oriented study of the sagas.

Linking sagas and other medieval Icelandic texts to the landscapes in which their action is set enables specialists and non-specialists alike to better understand how and where the events described in these narratives unfold around Iceland. It can be hard for those without a good grasp of Icelandic geography to appreciate the significance of the landscape stage and the implications that topography and the relative positioning of places may have for the dynamics or logistics of plot, for example. Amongst other details, the connotations of the distance between farms, or the routes that characters take when travelling, or the location of places where characters die and are buried in the landscape are often implicit but easy to miss if one is not familiar with the locale. Tracing the parabolas or axes of the sagas' action on a map while reading them, or being able to move around the settled perimeter of Iceland while reading *Landnámabók* (whose structure and presentation of settlement farms is carefully organised on geographical principles), for example, serves to underline the importance of geography and place-names in these medieval works—as source, structural element and sometimes even as a narrative actor that has agency and

[2] https://sagnagrunnur.com/is/, Gunnell (2005); http://jardabok.com/, Pálsson (2018); https://norseworld.nordiska.uu.se/, Petrulevich et al. (2019).

[3] Funding has been received from The Icelandic Research Council/RANNÍS (as part of a Rannsóknasjóður research grant awarded 2014–2016 for the project 'Time, space and narrative in the Icelandic sagas', PI Torfi Tulinius with co-PIs Emily Lethbridge and Gísli Sigurðsson; also as part of a Rannsóknasjóður Sigrúnar Ástrósar Sigurðardóttur og Haralds Sigurðssonar grant awarded 2015 for the project 'Landnámabók og kortasaga Íslands', PI Emily Lethbridge); The University of Iceland/Háskóli Íslands (as part of a Rannsóknasjóður research grant awarded in 2016 for the project 'Insight and imagination: Saga pilgrimages mapped and measured', PI Emily Lethbridge; also in 2015 as first prize for the annual University of Iceland applied science Hagnýtingarverðlaun competition); and the National Science Foundation (as part of a RIDIR grant awarded in 2016–2020 to the dataARC project, PI Colleen Strawhacker, grant numbers 1637076). The website is hosted by the University of Iceland/Árni Magnússon Institute for Icelandic Research. The principal programmer for the project is Trausti Dagsson though a proof-of-concept prototype was built by Logi Ragnarsson in 2012. Other project members are: Zachary Melton, Hjördís Erna Sigurðardóttir and Gísli Pálsson.

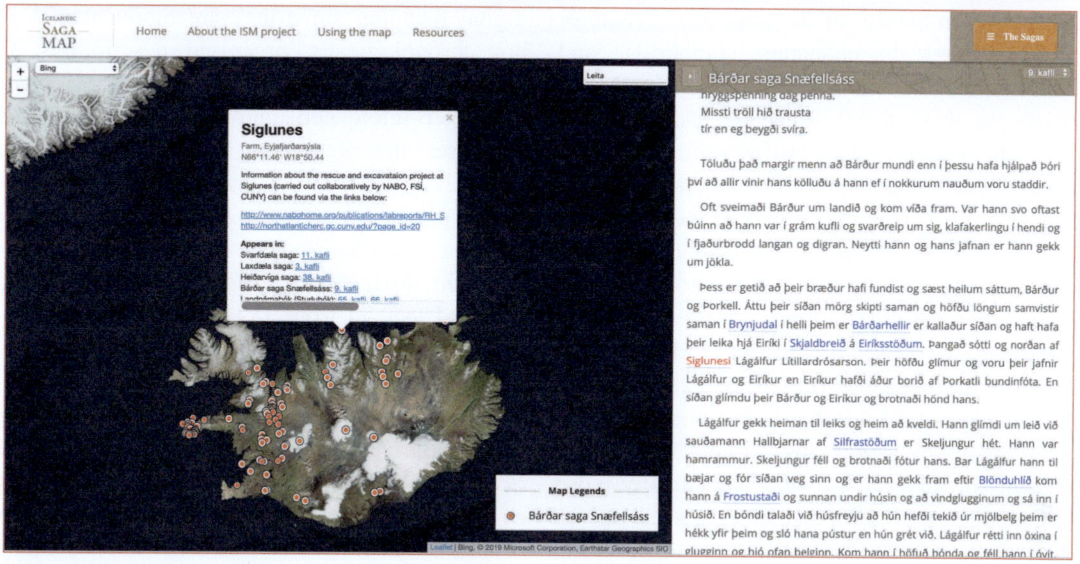

Fig. 2.1 *Bárðar saga Snæfellsáss* is the saga selected here for reading and the distribution of places named in it around the south, west and north of the country can be seen. The place-name 'Siglunes' has been clicked on in the text, bringing up its location on the map and hyperlinks to other texts (title/chapter) in which it also appears

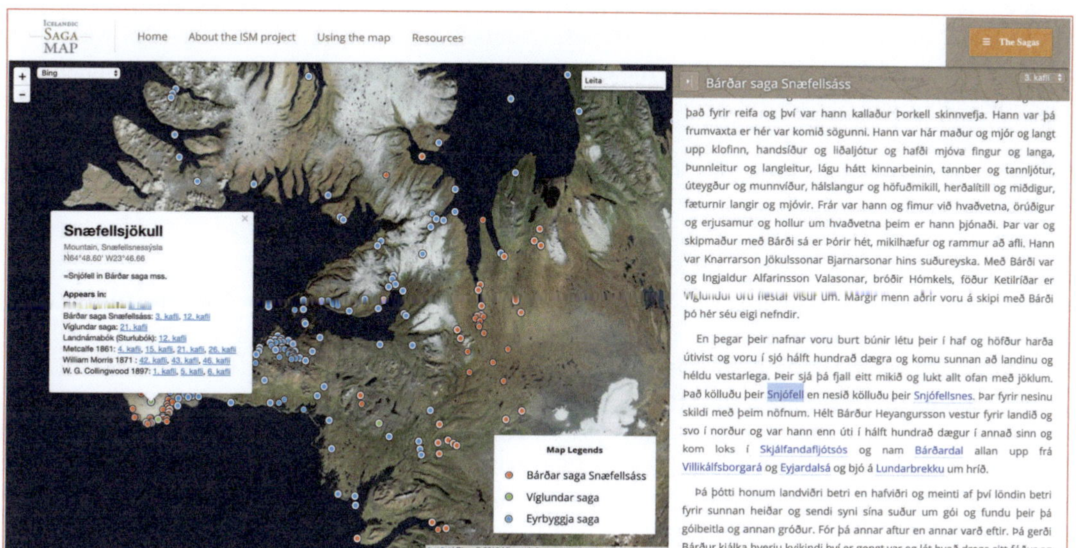

Fig. 2.2 As above in Fig. 2.1, *Bárðar saga Snæfellsáss* is the saga selected here for reading but locations of places named in two other local sagas, *Víglundar saga* (green points) and *Eyrbyggja saga* (blue points), are also displayed on the map alongside those named in *Bárðar saga* (orange points)

discernible influence on the course of events or decisions made by characters (Lethbridge and Hartman 2016; Lethbridge 2016; Kinniburgh 2018). As well as this, since the landscapes in which the sagas are set are also those in which they have been read and retold by Icelanders over many centuries, visualising the action of the sagas in a spatial framework gives modern readers a sense of how Icelanders experienced these stories in previous times while outdoors moving around the landscape, as well as reading indoors from manuscripts or editions (Lethbridge 2016).

The addition of non-Icelandic material to the ISM database such as nineteenth-century accounts of travel to Iceland—specifically those containing reports of visits to saga-sites—gives users of the website a further sense of how the sagas were accessed or read, and saga-landscapes experienced, by foreign visitors to the country.[4] These books and others belonging to the same genre are an interesting source for the landscape-based reception of the sagas and include many saga-related microtoponyms attached to small landscape features (rocks, mounds, caves, cliffs, hollows) that travellers were told about and shown by locals. Descriptions of saga-sites by 'saga pilgrims' furnish us with insights into the nature of the relationship between the sagas and geographic places in the nineteenth century, and ways in which the sagas were being presented to visitors by locals as an integral part of Iceland's cultural landscape. They arguably become part of the narrative stratigraphy of Iceland in their own right, sometimes even directly shaping the expectations and responses of later visitors to saga-sites and saga-landscapes (Helgason 2006; Aldred 2013; Lethbridge 2020).

As a corpus, these travel books lend themselves particularly well to geo-referencing and digital mapping since they are more often than not chronologically organised descriptions of visitors' itineraries. To date, four works have been geo-referenced and added to the ISM database: the Reverend Frederick Metcalfe's book *The Oxonian in Iceland; Or, Notes of Travel in that Island in the Summer of 1861, with Glances at Icelandic Folk-lore and Sagas*, published in 1861; William Morris's two journals of travel to Iceland from 1871 and 1873; and William Gershom Collingwood's and Jón Stefánsson's 1899 book *A Pilgrimage to the Saga-steads of Iceland.* Work on geo-referencing Sabine Baring-Gould's 1863 book *Iceland: Its Scenes and Sagas* is underway. Even on the basis of geo-referencing only these four works, a sense of how and where around the country saga-related narrative has accrued most densely can be gained from the visual display of the data. As more accounts are added to the ISM database, a diachronic, quantitative picture of the popularity of specific saga-sites and sagas over time will cumulatively be built up (where did saga tourists go, or not?, which saga-sites were most popular over time?), alongside a more qualitative and experiential picture of saga-sites through the visitors' eyes (how did different saga pilgrims describe the same saga-sites?) (cf. Belgum et al. 2019).

2.2.1 Logistical and Methodological Challenges

As is the case with many comparable projects, certain practical factors have had an impact on what we have been able to achieve thus far. One strength of the project is the relatively large number of medieval texts that have been geo-referenced to date (all of the *Íslendingasögur*, some *samtíðarsögur*, *Landnámabók*) but the difficulty of acquiring permission to republish texts under copyright, for example, has restricted the choice of editions (and, crucially, translations) freely available for use. It is for this reason that English translations of the sagas unfortunately are not included at present. However, as a first step towards finding a solution

[4]The 'Mapping the Lakes: A Literary GIS' project developed at the University of Lancaster by David Cooper, Ian Gregory, Sally Bushell and Zoe Bolton (https://www.lancaster.ac.uk/mappingthelakes/GIS%20Contacts.htm, Gregory and Cooper 2009; Cooper and Gregory 2010) was one early inspiration for this part of the ISM project.

here, all place-names in the database have been translated and will shortly be loaded onto the website. This place-name corpus is a valuable dataset in its own right, not least for those mining the sagas for detail pertaining to Iceland's environmental history. Icelandic texts of the sagas and of *Landnámabók* are those already available for reuse online on the *Netútgáfan/Snerpa* website[5]; the text of *Sturlunga saga* is Guðbrandur Vigfússon's 1878 out-of-copyright edition converted from pdf-form using *ABBYY FineReader OCR* software,[6] which supports Icelandic characters. Texts of nineteenth-century travel accounts (all out of copyright and available to download on websites such as Archive.org)[7] were also converted in the same way. The necessity of beginning somewhere and maximising resources available drove the decision to make use of editions already available online, but in certain respects, this is problematic from a text-critical perspective. The manuscript transmission of medieval Icelandic literature is complex and frequently, text editions conceal or suppress the textual variation that exists where more than one manuscript of the same work survive (see Quinn and Lethbridge 2012; Lethbridge and Óskarsdóttir 2018). Indicating variation with regard to place-names and topographical descriptions in manuscript witnesses of the medieval works uploaded would enable exciting analysis but hitherto, it has been beyond the scope of the ISM website to take differing manuscript traditions into account. The inclusion of this kind of textual variation is something that will be incorporated in the future though, with the aim of making the website a more useful philological tool.

Logistical aspects aside, the process of mapping the sagas and related material in and of itself has proved to be very illuminating and has highlighted crucial methodological and theoretical issues. The most interesting and challenging is arguably the question of what, actually, is being mapped (cf. Cooper et al. 2016). In terms of genre, the *Íslendingasögur* and *samtíðarsögur* are best described as historical fiction: they tell of events that are known to have taken place, and characters who are known from other sources to have been alive at certain points in time, but they elaborate and fill in gaps (or leave detail out) for literary purposes as well as to ideological and political ends. The saga authors' treatment of geography must also be assumed to be biased in certain respects, perhaps sometimes even involving modification of the topographical reality of their times or blending this with topographical invention (Nordal 1940; Schach 1955; Macrae-Gibson 1974–77; Wyatt 2004a). The impression given by the ISM user interface in its current iteration—with text and map laid out side-by-side, and aerial images of Iceland as the default map layer—is perhaps misleading or overly simplistic in this regard. The conclusion that an uncritical user might be tempted to draw is that geo-referencing place-names in the sagas is an unproblematic exercise, and that named places in the sagas can be directly correlated with those bearing the same name in the contemporary landscape. The very act of assigning GIS coordinates to saga places so that they appear on a map (whether digital or paper) involves conflating the sagas' toponymicon with the modern-day place-name record, thus implying a direct equivalence between the two.

The reality is more complex, however. The work entailed in geo-referencing places named in the sagas as part of the ISM project development highlighted the fact that, methodologically speaking, any attempt to map these works will inevitably merge literary mapping or cartography on the one hand (i.e. the mapping of imaginary literary realms or spaces, Reuschel and Hurni 2011; Piatti et al. 2009) with more 'historical' GIS cartographic practices on the other (Gregory and Ell 2007). The place-name record in Iceland is stable to a great degree and it can be assumed that most farms named in medieval sources were located on the same site as those with the same name today. There are nevertheless examples of place-names that have changed over time, or moved around, or been reallocated—sometimes

[5]https://www.snerpa.is/net/fornrit.htm.

[6]https://www.abbyy.com/en-au/ocr-sdk.

[7]https://archive.org/.

on the basis of knowledge of the sagas.[8] Some of the place-names recorded in the sagas have been lost over time, so there is no equivalent in the modern place-name corpus; others may have been invented by saga authors as a rhetorical means of increasing the authority of the narrative. Potentially, therefore, mapping the sagas will inevitably conflate fictional or semi-fictional places with 'actual' contemporary places: older topographies elide with present-day topography, the contemporary place-name record is superimposed onto partially fictionalised geographical spaces. The addition of nineteenth-century travel accounts which describe saga-sites—farms and larger landmarks as well as microtoponyms associated with saga characters or events in local tradition—adds another layer of complexity to this already multivalent real-and-imagined landscape.

2.2.2 Is This the Place…?

Emphasising the methodological complexities that underpin efforts to map the sagas is vital, not least because of the often simplistic and romantic nature of past scholarly and popular attitudes towards saga topography and the Icelandic landscape, as well as the ulterior motives of individuals who, today, have a vested interest in linking saga-sites with contemporary places in the landscape, whether for tourism-driven profit or other reasons (Þorláksson 2015). Archaeology in Iceland had its origins in the partly nationalist-fuelled nineteenth-century desire to prove the reliability of the sagas: saga narratives governed the selection of sites deemed to be worthy of excavation and often heavily influenced the interpretation of findings (Friðriksson 1994; Lucas 2008). Interrogating the 'fit' between topographical detail in the sagas and the observable or experienced contemporary 'reality', with the aim of asserting or rejecting the perceived reliability of individual narratives or aspects of them, has long been a theme in saga scholarship and local history publications. Discussion turning on geographical detail underpinned hypotheses set out the introductions to Íslenzk fornrit text editions about the identity of saga authors and the likely places of composition, for example. In more domestic, social contexts, too—that is the communal evening wake (*kvöldvaka*) held in households during the winter months up until the early twentieth century—whether or not certain events described in the sagas could have taken place where and how they were said to have was a perennial topic of debate.

Because the sagas sit on the boundary between history and fiction and because they are such a central part of Icelandic cultural heritage (Byock 1994; Wyatt 2004b), it is not surprising that the fit between sagas and contemporary landscapes should have given rise to extensive discussion and debate that continues to this day, even though few today would argue that the sagas—especially the *Íslendingasögur* but also the *samtíðarsögur*—present steadfastly reliable accounts of the historical events they describe (Þorláksson 2017). Nonetheless, critics and other interested parties will still go to considerable lengths to explain away differences between the contemporary landscape and saga landscapes in order to make them align better. Landscape change of one or another kind is often cited as the reason why an aspect of saga topography may not map perfectly onto modern-day terrain. Certainly, the extent to which the landscape has been altered over time as a result of environmental processes, natural and human, cannot be underestimated (Caseldine et al. 2005). But the possibility that there may be some deliberately creative treatment of topographical features (scaling up or down, conflating, inventing) that serves some narrative end is less often acknowledged.

To conclude this chapter and in order to illustrate the issues mentioned above, we will look to *Njáls saga* ('The Saga of Njáll'), the longest and possibly most 'national' of the *Íslendingasögur*, composed around 1280

[8] A good example here is the case of modern place-names in the valley of Hrafnkelsdalur in east Iceland, and their relationship with place-names in *Hrafnkels saga Freysgoða*. See, for example, discussion in Rafnsson.

(Sveinsson 1954). The climax of the saga is an infamous arson attack on the farm belonging to Njáll Þorgeirsson at Bergþórshvoll. Flosi Þórðarson is the leader of the party of burners who number about 100 men in all, with some 200 horses. The saga describes how the party makes a stop on the way to Bergþórshvoll: they tie up their horses and wait for evening to draw on in a hollow or small valley not far from the farm. This hollow is not named in any medieval manuscript text of the saga: it is only described in passing and its precise location is not specified—we do not know if it was thought to be on Bergþórshvoll land (this has been the assumption in later times) or somewhere further off. Arriving at Bergþórshvoll, they set the hall on fire and everyone dies except for Njáll's son-in-law Kári Sölmundarson, who is the only individual to escape from the blaze. We hear how after breaking out, he plunges himself in a nearby stream to extinguish his burning hair and clothes; after this, he rests in a ditch for a short while before running away under cover of the smoke. With the hall destroyed and no life inside, the burners ride back across the flat Landeyjar plains to Þríhyrningur, the distinctive triple-horned mountain about 22 kilometres to the northeast of Bergþórshvoll. Flosi proposes that the group remain there in hiding for three days, in order to watch movement in the area, before dispersing and going home; they do this. The rest of the saga narrates the aftermath of the burning: lengthy descriptions of legal proceedings are interspersed with violent vengeance killings led by Kári Sölmundarson in Iceland and beyond.

The burning of the farm at Bergþórshvoll was an historical event and it is mentioned in the Icelandic annals for the years 1009–1010. It is, however, impossible to ascertain how much of the saga's description of the burning reflected what may actually have happened. Instances where topographic details mentioned in the saga do not tally with the physical contours of Bergþórshvoll and the surrounding area as it is in modern times have been interrogated by academic critics (archaeologists, philologists and historians), locals and foreign visitors alike. The location and size of the hollow where Flosi and his men are said to have hidden before attacking the farm has been the subject of considerable discussion, as has the location and size of the area on Þríhyrningur where the burners hid before going home. Similarly, identifying the stream and the ditch that Kári rolled in and rested in after escaping the farmhouse has been a focus of enquiry. Nineteenth-century Icelandic printed sources (e.g. Sigurðsson 1856; Árnason and Vigfússon 1864; Kålund 1877; Vigfússon 1888; Jónsson 1896) mention a number of place-names attached to topographic features that, in local tradition, were believed to be associated with Flosi and the burners and with Kári. These include Flosalág ('Flosi's hollow') and Flosahóll ('Flosi's hill') on Bergþórshvoll land, along with Káratjörn ('Kári's pond'), Káradæla ('Kári's pond'), Káragröf/gróf ('Kári's ditch'); and Flosadalur ('Flosi's valley') and Flosahellir ('Flosi's cave') on Þríhyrningur.[9] Of these, only the place-names Káragröf/gróf and Flosadalur are given in the saga itself—and Flosadalur is only attested in one of the several medieval manuscripts that preserve texts of *Njáls saga*.[10] It is hard to date the other place-names precisely, though they are likely to be younger than the written saga, coined on the basis of knowledge of the saga and representing an articulation or reflex of a local need or wish to bind these dramatic narrative episodes even more tightly with the lived-in landscape. Discussion in these sources, as well as in numerous other twentieth-century publications (not least the introductory material and footnotes in editions of the saga such as Einar Ólafur Sveinsson's 1954 Íslenzk fornrit volume) mention the existence of these place-names and consider the plausibility of this or that hollow being 'the' hollow in the saga

[9]These place-names are also mentioned in the slightly younger place-name descriptions ('örnefnalýsingar') for Bergþórshvoll collected by the Place-name Institute of Iceland (Örnefnastofnun Íslands), now archived at the Árni Magnússon Institute for Icelandic Research in Reykjavík.

[10]The manuscript in question is AM 133 fol. (Kálfalækjarbók); the reading is at 79 recto. In English translation, the text reads "they ride up to the mountain and to a certain valley which has since been called Flosadalur."

where Flosi and his men hid, or where Kári rested, as well as other aspects of the saga's geography (Guðmundsson 1938; Johnson 1940: 18–21; Sveinsson 1954: lxxxiv-c, 325n3, 332n5, 332n6; Sveinsson 1977; Bjarnason 1999). The fact that the hollow known in later times as Flosalág could not conceal 100 men and 200 horses is rationalised by most critics (e.g. Sigurðsson 1856: 540) and even by a geologist (Einarsson 1965–67) as a consequence of landscape change (specifically, erosion caused by a nearby river). The possibility that topographical or other details in the narrative could have been worked up or even invented is not considered in efforts to explain discrepancies between text and landscape.

In 1861, *Njáls saga* became the first saga to be published in full in English translation and Bergþórshvoll became an iconic site on the saga-site circuit (Wawn 2000). In their accounts of travel in Iceland, Frederick Metcalfe and William Morris both describe visiting Bergþórshvoll and being shown places associated with the burning of the hall—specifically, the spot where Flosi and the burners hid ('Flossi-lóg' in Metcalfe's account; 'Flosi's Hollow' in Morris's) and the spot where Kári doused himself in the stream ('Kari-Tiörn' in Metcalfe; described but not named in Morris). Text and topography are interrogated by both men: issues with regard to lack of fit between saga and experienced landscape are articulated in both accounts. Metcalfe relates how the 'peasant' who showed him Flosalág remarked that 'In the memory of man, it was much deeper than it is now' (Metcalfe 1861: 370). Metcalfe himself expresses doubt this is the 'right' spot, arguing: 'as this place was in sight of the house, and the Saga tells us that the band kept out of sight, the above nomenclature must be erroneous, and the real place where they hid the horses must have been behind another knoll, somewhat nearer to the house' (Metcalfe 1861: 370). Morris states how he was shown 'the traditional places about the stead' by the farmer who 'going round the knolls … takes us first to a hollow close by the river side, and a few yards from the easternmost mound which they call Flosi's Hollow […] it is not big enough to hide a dozen men now (the Burners were a hundred) but they say the river has eaten up a big piece of it even in the memory of man' (Morris 1911: 44–45).[11] Morris's description of Káratjörn also communicates another discrepancy between experienced reality and the details of the saga narrative: 'There is no stream there now, but a rushy boggy piece of land marks what has been a pond, and there are traces of a brook that once flowed from it; a few hundred yards further on they show us a little mossy hollow under a grass bank which they say is the hollow Kári lay down to rest in after the stream' (Morris 1911: 45). Weather conditions when Metcalfe examines the spot provide the opportunity for him to yoke together the saga past and his present explicitly: 'Passing to the other, or north-west side of the house, I next visit a small pit in the swamp, which goes by the name of Kari-tiörn (the tarn of Kari) … The wind was probably in the same direction as it is now, viz. south-east; for, as the story relates, he was hidden by the smoke, so that the assailants outside did not detect his escape' (Metcalfe 1861: 370). Time is instantly compressed and thanks to the power of the landscape and the way it holds story, Metcalfe almost becomes a bystander at the burning of the farm some 850 years before his visit.

2.3 Conclusion

This climactic episode in *Njáls saga* cannot be divorced from its landscape context. It was envisaged as unfolding in the landscape from the start: the landscape played a vital role in the narrative mechanics of the episode (concealing characters, as well as being travelled through) and specific places and place-names were key to the subsequent transmission and reception of the episode for local Icelanders as well as foreign saga enthusiasts. In geo-referencing the named

[11]The direct echo between the two texts ('the memory of man') is striking: it might be coincidental but borrowings from older travel accounts are common and intertextuality is one of the features of this body of writing (Lethbridge 2020).

places in this part of *Njáls saga* and in Metcalfe's and Morris's descriptions of Bergþórshvoll, a mixed approach was taken. The saga's Bergþórshvoll was given the co-ordinates of the modern Bergþórshvoll farm complex: the precise locations of farm buildings may have moved around a little over the centuries but three typically Icelandic farm-mounds are the visible, accretive evidence of human habitation on the site, and excavations conducted there have revealed a dense and continuous stratigraphy of settlement from the earliest period of Icelandic history (Friðriksson 1994). Continuity is assumed—while the narrative about the fire as recounted in the saga may have been dramatised in various ways for literary purposes, its setting at Bergþórshvoll and the equivalence of the saga's Bergþórshvoll with today's Bergþórshvoll seems to be secure. The microtoponyms associated with Flosi and Kári are harder to pinpoint and display on the map, partly because of granularity but also because they are more likely to have moved about over time, according to contemporary needs or circumstances. One working solution here has been to give them the same co-ordinates as Bergþórshvoll so that their rough location could be indicated—as places pointed out to visitors intent on examining saga-sites, if nothing else—since they are part of the landscape- or place-based reception of the saga (see Fig. 2.3).

The most interesting aspect of the relationship between the sagas and places named in them is not the precise identification of those places in today's landscape, and the accompanying debate about how the sagas might or might not relate to historical reality. Rather, it is how perceived discrepancies or apparent incongruities between texts and landscapes invite us to interrogate the processes by which narrative (or narrativised) and experienced worlds are amalgamated and layered one on top of the other. Building up a chronologically deep and geographically distributed understanding of the role of the landscape in the transmission of the sagas following their written composition allows us to better appreciate the nature of the relationship between story and landscape or named places in it. This relationship is reciprocal and processual: the sagas have influenced perceptions and

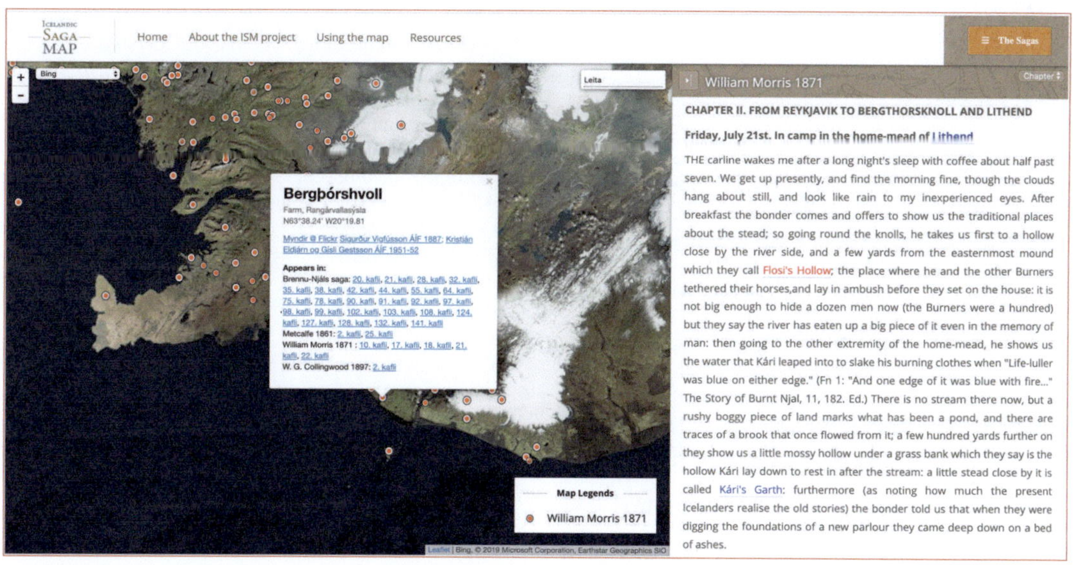

Fig. 2.3 The burning of the farm at Bergþórshvoll as described in Njáls saga was an historical event, and it is mentioned in the Icelandic annals for the years 1009–1010. William Morris's journal of travel in Iceland for the year 1871 is the text selected here for reading. The place-name 'Flosi's Hollow' has been clicked on in the text and since it is linked to the location of Bergþórshvoll, this is marked on the map

expectations of the landscape and geographical knowledge, as much as the landscape influenced the creation of the sagas.

Mapping the sagas and other medieval Icelandic works is a means of testing and highlighting connections between physical places and these rich and involved narratives, and exploring the reading experiences of Icelandic and non-Icelandic saga audiences over time at different scales, micro and macro. The map itself, in fact, might best be seen as a form or medium of reception: arguably, what we are mapping is not saga topographies but readers' topographies. Amongst the many possibilities offered by digital mapping, filtering data and overlaying different sources enable users to build up a more nuanced and textured picture of the narrative stratigraphy of Icelandic landscapes. The process of developing the ISM project up to this point has drawn attention to challenging methodological and theoretical issues, some of which are outlined above. Future development of the ISM project and database will continue to provide opportunities for reflection on the centuries-long tradition of engagement with saga landscapes in Iceland, and the continuous reconfigurations and rewritings of the country's narrative stratigraphy.

References

Aldred O (2013) A restless medieval? Archaeologies and saga-steads in the Viking Age North Atlantic. Postmedieval 4(2):136–149

Árnason J, Vigfússon G (1864) Íslenzkar þjóðsögur og æfintýri, vol 2. JC Hinrich, Leipzig

Baring-Gould S (1863) Iceland: its scenes and sagas. Smith, Elder & Co, London

Barraclough ER (2017) Travel. In: Jakobsson Á, Jakobsson S (eds) The Routledge research companion to the medieval Icelandic sagas. Routledge, London, New York, pp 210–217

Belgum K, Handley K, Bott R (2019) Mapping travel writing: a digital humanities project to visualise change in nineteenth-century published travel texts. Stud Travel Writ 23:306–324

Benediktsson J (1986) Íslendingabók. Landnámabók. Íslenzk fornrit I. Hið íslenzka fornritafélag, Reykjavík

Bjarnason B (1999) Njáluslóðir. Örnefni og staðfræði Njáls sögu. Mál og mynd, Reykjavík

Bragason Ú (2005) Sagas of contemporary history (*Sturlunga saga*): texts and research. In: McTurk R (ed) A companion to Old Norse-Icelandic literature and culture. Blackwell, Oxford, pp 427–446

Byock J (1994) Modern nationalism and the medieval sagas. In: Wawn A (ed) Northern antiquity: the post-medieval reception of edda and saga. Hisarlik, Enfield Lock, pp 163–187

Callow C (2006) Reconstructing the past in medieval Iceland. Early Mediev Eur 14(3):297–324

Caseldine C, Russell A, Harðardóttir J, Knudson Ó (eds) (2005) Iceland—modern processes and past environments. Elsevier, Amsterdam

Clunies Ross M (1997) Textual territory: the regional and genealogical dynamic of medieval Icelandic literary production. New Mediev Lit 1:9–30

Clunies Ross M (2010) The Cambridge introduction to the Old Norse-Icelandic saga. Cambridge University Press, Cambridge

Cooper D, Gregory IN (2010) Mapping the English Lake District: a literary GIS. Trans Inst Br Geogr 36:89–108. https://doi.org/10.1111/j.1475-5661.2010.00405.x

Collingwood WG, Stefánsson J (1899) A pilgrimage to the saga-steads of Iceland. W. Holmes, Ulverston

Cooper D, Donaldson C, Murrieta-Flores P (2016) Introduction: rethinking literary mapping. In: Cooper D, Donaldson C, Murrieta-Flores P (eds) Literary mapping in the digital age. Routledge, London, New York, pp 1–21

Cormack M (2005) Christian biography. In: McTurk R (ed) A companion to Old Norse-Icelandic literature and culture. Blackwell, Oxford, pp 27–42

Einarsson T (1965–67) Myndunarsaga Landeyja og nokkur atriði byggðarsögunnar. Saga 5:309–328

Friðriksson A (1994) Sagas and popular antiquarianism in Icelandic archaeology. Avebury, Aldershot

Glauser J (2005) Romance (translated *riddarasögur*). In: McTurk R (ed) A companion to Old Norse-Icelandic literature and culture. Blackwell, Oxford, pp 372–387

Gregory IN, Ell PS (eds) (2007) Historical GIS: technologies, methodologies and scholarship. Cambridge University Press, Cambridge

Gregory IN, Cooper D (2009) Thomas Gray, Samuel Taylor Coleridge and geographical information systems: a literary GIS of two Lake District tours. Int J Hum Arts Comput 3:61–84. https://doi.org/10.3366/ijhac.2009.0009

Guðmundsson B (1938) Staðþekking og áttamiðanir Njáluhöfundar. Andvari 63(1):68–88

Gunnell T (2005) Sagnagrunnur: a new database of Icelandic folk legends in print. Folk: Electron J Folk 45:151–62

Gunnlaugsson GM (2010) Reading and writing in medieval Iceland. In: Robinson PR (ed) Teaching writing, learning to write: proceedings of the XVIth colloquium of the Comité International de Paléographie Latine. King's College London, London, pp 155–162

Harris J (1997) The prosimetrum of Icelandic saga and some relatives. In: Harris J, Reichl K (eds) Prosimetrum: cross-cultural perspectives on narrative in prose and verse. Brewer, Cambridge, pp 131–164

Helgason JK (2006) Reading saga landscapes: the case of Samuel E. Waller. In: Eysteinsson Á (ed) The cultural reconstruction of places. University of Iceland Press, Reykjavík, pp 101–110

Hoggart, C (2010) A layered landscape: how the family sagas mapped medieval Iceland. Limina: J Hist Cult Stud 16:np. http://www.limina.arts.uwa.edu.au/__data/assets/

Icelandic Saga Map project http://sagamap.hi.is

Jakobsson Á (2005) Royal biography. In: McTurk R (ed) A companion to Old Norse-Icelandic literature and culture. Blackwell, Oxford, pp 388–402

Jakobsson S (2017) Space. In: Jakobsson Á, Jakobsson S (eds) The Routledge research companion to the medieval Icelandic sagas. Routledge, London, New York, pp 175–186

Jarðabókin/Storied Lines project. http://jardabok.com/

Johnson AJ (1940) Staðþekking Njáluhöfunda og Njála og Oddaverjar. Árbók Hins íslenzka fornleifafélags 47:1–35

Jónsson B (1896) Um nokkur vafasöm atriði í Íslendingasögum. Árbók Hins íslenska fornleifafélags 11:29–41

Karlsson G (2000) Iceland's 1100 years. The history of a marginal society. Hurst & Company, London

Kristinsson A (2003) Lords and literature: the Icelandic sagas as political and social instruments. Scand J Hist 28:1–17

Kinniburgh, MC (2018) Spatial reading: digital literary maps of the Icelandic outlaw sagas. Digit Mediev 11 (1):1–35. http://doi.org/10.16995/dm.66

Kålund K (1877) Bidrag til en historisk-topografisk Beskrivelse af Island. Syd- og Vest-Fjærdingerne. Gyldendal, Copenhagen

Lethbridge E (2014) "Hvorki glansar gull á mér / né glæstir stafir í línum", Some observations on Íslendingasögur manuscripts and the case of *Njáls saga*. Arkiv för nordisk filologi 129:55–89

Lethbridge E (2016) The Icelandic sagas and saga landscapes: writing, reading and retelling Íslendingasögur narratives. Gripla 27:51–92

Lethbridge E (2020) Americans in sagaland: Iceland travel books 1854–1914. In: Machan TW, Helgason JK (eds) From Iceland to the Americas: Vinland and historical imagination. Manchester University Press, Manchester, pp 137–159

Lethbridge E, Hartman S (2016) Inscribing environmental memory in the Icelandic sagas and the Icelandic Saga Map. PMLA 131(2):381–391

Lethbridge E, Óskarsdóttir S (eds) (2018) New studies in the manuscript tradition of *Njáls saga*. The historia mutila of *Njála*. Medieval Institute Publications, Kalamazoo

Lucas G (2008) Pálstóftir: a viking age shieling in Iceland. Nor Archaeol Rev 41(1):85–100. https://doi.org/10.1080/00293650802069193

MacNeice L (1937) Letter to Graham and Anne Shepard. In: Auden WH, MacNeice L (eds) Letters to Iceland. Faber & Faber, London

Macrae-Gibson OD (1974–77) The topography of Hrafnkels saga. Saga-Book Viking Soc 19:239–63

Mapping the Lakes project. https://www.lancaster.ac.uk/mappingthelakes/GIS%20Contacts.htm

McDonald Werronen S (2018) Late medieval and early modern Icelandic saga manuscripts. In: Kwakkel E (ed) Vernacular manuscript culture 1000–1500. Leiden University Press, Leiden, pp 187–217

Metcalfe F (1861) The Oxonian in Iceland; or, notes of travel in that island in the summer of 1860, with glances at Icelandic folk-lore and sagas. Longman, Green, and Roberts, London

Morris W (1911) Icelandic journals. In: Morris M (ed) The collected works of William Morris, vol 8. Longman and Green, London

Netútgáfan/Snerpa. https://www.snerpa.is/net/fornrit.htm

Nordal S (1940) Hrafnkatla. Ísafoldarprentsmiðja, Reykjavík

Norse World project. https://norseworld.nordiska.uu.se/

Ólason V (2005) Family sagas. In: McTurk R (ed) A companion to Old Norse-Icelandic literature and culture. Blackwell, Oxford, pp 101–118

Óskarsdóttir S (2004) The Church and written culture. In: Sigurðsson G, Ólason V (eds) The manuscripts of Iceland. Stofnun Árna Magnússonar á Íslandi, Reykjavík, pp 13–23

Pálsson G (2018) Storied lines: network perspectives on land use in early modern Iceland. Nor Archaeol Rev 51(1–2):112–141. https://doi.org/10.1080/00293652.2018.1468355

Petrulevich A, Backman A, Adams J (2019) Medieval macrospace through GIS: the Norse World project Approach. Cartog J 56(3). https://doi.org/10.1080/00087041.2019.1596341

Piatti B, Bär HR, Reuschel A-K, Hurni L, Cartwright W (2009) Mapping literature: towards a geography of fiction In: Lehn A, Gartner G, Cartwright W (eds) Cartography and art. Lecture notes in geoinformation and cartography. Springer, Berlin, Heidelberg, 16 pp. https://doi.org/10.1007/978-3-540-68569-2_15

Quinn J, Lethbridge E (eds) (2012) Creating the medieval saga: versions, variability and editorial interpretations of Old Norse saga literature. University of Southern Denmark Press, Odense

Rafnsson S (1990) Byggðaleifar í Hrafnkelsdal og á Brúardölum. Brot úr byggðasögu Íslands. Hið íslenska fornleifafélag, Reykjavík

Reuschel A-K, Hurni L (2011) Cartographies of fictional worlds. Cartogr J 48(4):218–223

Rowe E (2005) Short prose narratives (*þáttr*). In: McTurk R (ed) A companion to Old Norse-Icelandic literature and culture. Blackwell, Oxford, pp 462–478

Sagnagrunnur project. https://sagnagrunnur.com/is/

Sandvik G, Sigurðsson JV (2005) Laws. In: McTurk R (ed) A companion to Old Norse-Icelandic literature and culture. Blackwell, Oxford, pp 223–244

Schach P (1955) The anticipatory literary setting in the Old Icelandic family sagas. Scand Stud 27:1–13

Sigurðsson P (1856) Um forn örnefni, goðoraðskipan og fornmenjar í Rangárþingi. Safn til sögu Íslands og íslenzkra bókmenta að fornu og nýju 2:498–557

Sveinsson EÓ (ed) (1954) Brennu-Njáls saga. Íslenzk fornrit XII. Hið íslenzka fornritafélag, Reykjavík

Sveinsson EÓ (1977) Journey to the *Njála* country, 7th August 1973. Gripla 2:7–18

Thordarson T, Höskuldsson Á (2014). Iceland. Classic geology in Europe 3, 2nd edn. Dunedin Academic Press, Edinburgh, London

Tulinius TH (2005) Sagas of Icelandic prehistory (fornaldarsögur). In: McTurk R (ed) A companion to Old Norse-Icelandic literature and culture. Blackwell, Oxford, pp 447–461

Vésteinsson O, Friðriksson A (2003) Creating a past: a historiography of the settlement of Iceland. In: Barrett J (ed) Contact, continuity and collapse: the Norse colonisation of the North Atlantic. Turnhout, Brepols, pp 139–161

Vigfússon G (1878) Sturlunga saga including the Islendinga saga of lawman Sturla Thordsson and other works edited with prolegomena, appendices, tables, indices, and maps, 2 vols. Clarendon Press, Oxford

Vigfússon S (1888) Rannsóknir sögustaða, sem gerðar voru 1883 um Rangárvöllu og þar í grend, einkanlega í samanburði við Njáls sögu. Árbók Hins íslenzka fornleifafélags 7:35–62

Wawn A (2000) The vikings and the Victorians: inventing the old north in nineteenth-century Britain. Brewer, Cambridge

Whaley D (2009) A useful past: historical writing in medieval Iceland. In: Clunies Ross M (ed) Old Icelandic literature and society. Cambridge University Press, Cambridge, pp 161–202

Wyatt I (2004a) Narrative functions of landscape in the Old Icelandic family sagas. In: Hines J, Lane A, Redknap M (eds) Land, sea and home. Proceedings of a conference on viking-period settlement, Cardiff, July 2001. Maney, Leeds, pp 273–282

Wyatt I (2004b) The landscape of the Icelandic sagas: text, place and national identity. Landscapes 5:55–72

Þorláksson H (2005) Historical background: Iceland 870–1400. In: McTurk R (ed) A companion to Old Norse-Icelandic literature and culture. Blackwell, Oxford, pp 136–154

Þorláksson H (2015) Menningararfur á skiltum. Gömlum söguskilningi hafnað og lifir þó. In: Rastrick Ó, Hafstein VT (eds) Menningararfur á Íslandi: gagnrýni og greining. Háskólaútgáfan, Reykjavík, pp 161–192

Þorláksson H (2017) The bias and alleged impartiality of Sturla Þórðarson. In: Sigurðsson JV, Jakobsson S (eds) Sturla Þórðarson: skald, chieftain and lawman. Brill, Leiden, Boston, pp 200–211

Dead Men Tell Tales: History and Science at Duffy's Cut

3

William E. Watson, J. Francis Watson and Earl H. Schandelmeir

Abstract

The historical problems that the *Duffy's Cut Project* has sought to address concern the location of an 1832 mass grave of Irish immigrant railroaders and the cause of their deaths at Mile 59 of the Philadelphia and Columbia Railroad track in Chester County, Pennsylvania. Stories of the workers' death from a cholera outbreak were originally disseminated by a few contemporary newspapers: *The Village Record* (West Chester), *The National Gazette* and *Literary Messenger* (Philadelphia), and *The Philadelphia Inquirer*. Stories also abounded in railroad and local folklore, which told of Irish ghosts dancing on their graves close to the railroad fill being built by the deceased laborers at the time of their deaths. This chapter examines archival, geological, archeological, and anthropological perspectives, employed in tandem with GIScience and global positioning satellite techniques. These combined methods were used to inform the forensic excavation of the Mile 59 site, work that identifies the cause of the deaths of the immigrant Irish laborers as homicide, and not a cholera epidemic. The story unearthed by the *Duffy's Cut Project* illustrates the power of integrating historical geography, geoscience, and textual approaches to re-frame the historiographies of immigration and labor during the early years of the nineteenth-century America Industrial Revolution.

Keywords

Public history · Nineteenth-century immigrant labor · Railroad construction · Homicide · GPR/electrical resistivity · Excavation · Re-internment

3.1 Introduction

The *Duffy Cut Project* (DCP) began in 2003, and over the course of a decade, identified and excavated a mass grave, 30 miles West of Philadelphia, near Malvern, Pennsylvania. The grave was found to contain the remains of Irish Catholic immigrants, purported to be cholera

W. E. Watson (✉) · E. H. Schandelmeir
Immaculata University, Immaculata, PA, USA
e-mail: wwatson@immaculata.edu

E. H. Schandelmeir
e-mail: eschand@usa.net

J. Francis Watson
Christ Lutheran Church, Whiting, NJ, USA
e-mail: fmwats@aol.com

J. Francis Watson
Lutheran Archives in Philadelphia, Philadelphia, PA, USA

C. Travis et al. (eds.), *Historical Geography, GIScience and Textual Analysis*, Historical Geography and Geosciences, https://doi.org/10.1007/978-3-030-37569-0_3

victims, hired to construct a section of the Pennsylvania Railroad in 1832.[1] However, the DCP uncovered a darker, more sinister story about the cause of these workers' deaths—one that goes to the heart of the story of immigration in America. Starting with the discovery of a confidential and ignominious 1909 Pennsylvania Railroad file on the "incident," the project expanded into archival research on 1830s Pennsylvania Canal Commission reports, contemporary newspapers, immigration records into the Port of Philadelphia, and Irish property valuations. This research helped to establish a Pennsylvania State Historical Marker for the site in 2004, and permission was soon given to undertake an archeological survey of the site with the purpose of locating the exact site of the mass grave. Archival research was then bolstered by scientific research in geology, archeology, forensic anthropology, and forensic dentistry. Geological techniques were used to examine the construction of the 1832 railroad cut and fill, map the site, and examine its contents by means of electrical resistivity and ground-penetrating radar. One goal was to find *stopping voids* within construction site which might signal the presence of skeletal remains.[2] Between 2004 and 2012, archeological techniques were used to excavate early railroad and personal artifacts from the Irish workers' camp, and finally the skeletal remains of seven of rail workers were found within the fill. Anthropological techniques were used to examine the remains, with forensic examination showing that the seven excavated workers were killed by means of blunt force trauma and did not die from cholera, leading to the conclusion that murder, and not an epidemic was the cause of their collective deaths. Five of these skeletons were re-buried in West Laurel Hill Cemetery in Bala Cynwyd, Pennsylvania (PA), in 2012. Two skeletons were instead buried in Ireland, having been identified by comparing the estimated ages of the skeletons at the time of death with the ages of workers listed on the *John Stamp* passenger list. Eighteen-year-old John Ruddy was re-buried in 2013 at Holy Family Cemetery in Donegal, Republic of Ireland, while 29-year-old Catherine Burns was reburied in 2015, at St. Patrick's Cemetery in Tyrone, Northern Ireland.

3.2 Archival Research

Chief among the archival sources for the *Duffy's Cut* story was Pennsylvania Railroad (PRR) file number 004.01 "C" (Watson and Watson 2018). Entitled *History of 'Duffy's Cut' stone enclosure east of Malvern, Pa., which marks the burial place of 57 track laborers who were victims of the cholera epidemic of 1832*, it was meant to be a secret internal file that was created by, and for the sole use of, the PRR. This file is the only contextual account of the *Duffy's Cut* story that was produced by the railroad. The first document in the file states that "it is not desired to let it get out of the office." And with good reason; though the contents of the file were held in corporate secrecy, it contained geographical clues which would later come to undermine the "official" account of the incident. From the time the file was created to the 1968 merger of the PRR with the New York Central Railroad to form the Penn Central Railroad, the file was never out of the hands of PRR staff and executives. The cover of the file is marked "Office File, Vice President in Charge of Operation, Philadelphia, PA," and dates from the term of office of its principle creator. The file consists of 32 pages from cover to cover and contains 22 separate documents. The documents in the PRR file date from 1889 to 1932 and include letters from railroaders and reminiscences of others that shed light on different aspects of the *Duffy's Cut* narrative. The file, furthermore, details the work of marking

[1]Railroad contractor Philip Duffy hired 57 Irish immigrants to construct a rail line through the region's densely wooded hills and ravines. Workers sailed to Philadelphia from Derry, and were drawn from the Ulster counties of Donegal, Tyrone, and Derry to work in Pennsylvania's growing railroad industry.

[2]In constructing a railway, cut and fill is the process by which the amount of material taken from the excavation for the rail bed, roughly matches the amount of fill needed to construct the rail embankments, thus minimizing the cost of construction labor.

the Duffy's Cut site by late nineteenth-century and early twentieth-century railroaders.

The PRR file was created, starting in 1909, by two railroad supervisors who were then serving in Paoli, Pennsylvania: Martin W. Clement (1881–1966) and George R. Sinnickson (1874–1934). While both Clement and Sinnickson solicited information on the events at Duffy's Cut from locals and railroad employees, Clement was the real architect and later guardian of the file. After rising to Vice President, Clement became the 11th President of the PRR in 1935 (Tripician 1970). His face made the cover of the March 16, 1936 issue of *Time* magazine, and 30 years later his obituary appeared in its September 9, 1966 issue. The survival of the PRR file can be credited to Clement's personal assistant, and later PRR Director of Personnel Administration, Joseph F. Tripician (1898–1977), who not only preserved the file after the PRR was merged with the NY Central Railroad in 1968, but also redacted it. Tripician had been present whenever the file was viewed by visitors to Clement's office. In addition, Tripician numbered the pages of his old boss' file and corrected several dates that had been inaccurately added to the file over the years. Most importantly, he ensured that the continuing existence of the only official railroad report on Duffy's Cut.

Clement's file is significant for a number of reasons. First, it became the official railroad record of what happened at Duffy's Cut; it also detailed the construction of railroader memorials at the site, including a wooden memorial fence that was first put in place in 1873, and was repaired subsequently until it was finally replaced in 1909 by a stone memorial wall commissioned by Martin Clement himself. The file indicated the exact area of the burial of the dead at Mile 59, under the original 1832 railroad fill, while the arrival in America of the Irish laborers in the summer of 1832 is mentioned in the file as well. Furthermore, the file definitively set the number of the dead at Duffy's Cut at 57, corresponding closely to the only contemporary internal railroad account of the deaths at Duffy's Cut, the 1833 *Mitchell Letter*, which stated that a number as high as 60 had died at the site. The number of fatalities was 57, which was officially accepted by the PRR, based on Clement's research. Finally, Clement's file is important in that it states unequivocally that the deaths of the Irish laborers occurred at a very particular period, during the cholera outbreak in Chester County in August of 1832. While the deaths at Duffy's Cut did indeed occur during the 1832 cholera epidemic, subsequent forensic investigation has shown that all seven of the individuals recovered at the site were murdered (Watson and Watson 2018; Watson 2009a, b, 2010a, b; Hughes 2010). Clement recognized the fact that the track workers who died building that section of the railroad were "Irish immigrants," and that they were hired by a railroad contractor named Duffy to labor at Mile 59, located between Malvern and Frazer.

In his file, Clement furthermore detailed his efforts to memorialize the Irish work crew who died at Duffy's Cut. In 1909, he had the stone memorial wall erected at the Duffy's Cut site, made of original 1832-era stone sills—the stone blocks on which the original tracks were placed. This stone wall replaced the original wooden fence that had been erected in 1873 by Irish-American railroaders as a memorial to the Irish laborers who died there 41 years earlier. Clement also stated in the file that he had wished to place a sign on the site, indicating who was buried there: Clement noted that his request for a sign was turned down because of its projected cost. It is ironic that the man who created the PRR file, and who erected the still-standing stone wall at Duffy's Cut, and who even wanted to place a public sign at the spot that told the story of who was buried there would later insist that his file be kept a secret. Throughout his tenure as railroad president, Clement kept the Duffy's Cut file in his office, and he only allowed invited guests to view the file privately there. He, furthermore, swore all viewers of the file to secrecy.

Included in the file are three newspaper articles on Duffy's Cut, two of which date from 1889, while one is dated 1909. These articles were efforts by two local Chester County historians who sought to bring the story of Duffy's Cut to the light of day. The first of the two 1889 pieces, both of which were penned by Julius F. Sachse, entitled *The Legend about Duffy's Cut on the Pennsylvania Railroad Between Malvern and Frazer*, related the experience of a local Chester County man whose reminiscences helped in the discovery of the location of the shanty and the burial site. In this piece, Sachse used a local source (an "old resident") to describe the site of a shanty where the laborers lived as they worked along Mile 59. A thorough search in the valley just south of the original 1832 rail line resulted in the discovery of the shanty site as it was described by Sachse's "old resident." Subsequent archeological investigation of this shanty site has yielded some very significant finds that link this location with the Irish laborers who lived and died there in the summer of 1832, and who were buried nearby. Based on his conversations with his local source, Sachse furthermore described the burial site of the laborers, which also assisted the archeological team in the work of excavating the graves located there (Watson and Watson 2018; Watson et al. 2006). Sachse's second piece, *The Last Relic of the Pennsylvania Railroad Originally Called the Philadelphia and Columbia Railway*, presented details of the building of Mile 59 and the construction of Duffy's Cut and Duffy's Fill, or Duffy's Bank as it was also known. Both of Sachse's articles originally appeared in the pages of the *Village Record* newspaper of West Chester, PA, and they both assisted the archeological investigation at the Duffy's Cut site.

The third newspaper article in the PRR file is from the *Daily Local News* of November 30, 1909. Entitled *That Small Burial Mound Near Malvern*, it was written by Chester County clergyman and historian Rev. Alden W. Quimby. This short article detailed the location of the mass grave at Duffy's Cut in light of later track construction that moved the roadbed slightly northward, and over at least a portion of the Duffy's Cut mass grave site. Quimby wrote: "Many years later, when the course of the tracks was changed to avoid severe curves, the unmarked graves, unknown to the constructors, were covered over by the new roadbed." Quimby concluded: "almost every minute ponderous trains roll and rumble over the real resting place of the cholera victims." Geophysical work by Dr. Timothy Bechtel at the stone wall has indicated that Alden Quimby's reflections are likely on target, as a large anomaly has been located below the stone wall that curves toward, and moves under, the current track bed.

Among a variety of letters from Irish-American railroaders, who worked at the Duffy's Cut site in the late nineteenth century and early twentieth century, one particular letter in the PRR file made a very significant contribution to the discovery of the mass grave at Duffy's Cut. This letter, written by Chester County born Irish-American railroader George Dougherty (1845–1924), assisted in locating the graves of the first seven bodies (Watson and Watson 2018; Watson 2010a, b). Dougherty was a life-time employee of the PRR and, except for a stint during the Civil War working in the Union Army Railroad Construction Corps, he spent his whole career with the PRR working in Chester County. Dougherty's father, John, was from County Donegal, as were many of the laborers who died at Duffy's Cut. John Dougherty was also a contemporary of the Duffy's Cut laborers. Significantly, the father told his son where he had heard the bodies of the Irish laborers were buried at Duffy's Cut. Dougherty wrote: "I heard my father say that they were buried where they were making the fill" (*PRR Duffy's Cut file*, 1909). This indeed proved correct. The fill was the earthen railroad bridge that the Irish laborers were constructing when they were murdered. In the later history of the railroad, the practice of burying laborers who died while working on a rail line under the tracks they were building became commonplace, and the burial under the fill at Duffy's Cut appears to be the first case of such a practice in North America. The *Dougherty Letter*, and its indication that the bodies of the dead Irish workers were buried "where they were making the fill," became the key archival source for helping to locate the first seven burials at Duffy's Cut.

While the *PRR Duffy's Cut* file states that the railroad contractor at Mile 59 was named Duffy, there are no other details about this man's life or his later career with the railroad in the file itself. A thorough search of a variety of archival materials has enabled Duffy's life-story to be pieced together from birth to death. Duffy's naturalization paperwork is found in the United States National Archives. Duffy filed his petition for naturalization on October 1, 1813. This petition states that Duffy arrived in America on June 18, 1798. This was the year of the United Irishmen Rebellion in Ireland, and like many of his fellow Irishmen and women, Duffy came to America for a new life. Arriving as a 15-year-old young man, Duffy worked his way up from laborer to contractor. His work history can be traced through the PA Canal Commission records located in the PA State Archives in Harrisburg, PA. From these sources it is clear that Duffy became a trusted and dependable railroad contractor, who had a reputation for completing very difficult tasks. What appears to be Duffy's first contract with the Philadelphia and Columbia (P&C) railroad is on February 1829, where he was a partner with his brother-in-law, James Smith, on "section 16." Duffy received his own contract for Mile 60 of the P&C in June of 1829. While Duffy worked on that mile of the railroad, he was recorded in the 1830 US census as living in Willistown Township in Chester County, PA. At that time, he was renting a property and had 23 men between the ages of 20 and 40 living with him. These were all foreign-born men and were undoubtedly the "sturdy looking band of the sons of Erin" who were said in the local newspapers to be working with Duffy in what was described as the "Herculean task" of working on Mile 60 of the railroad.[3]

After completing Mile 60, on May 18, 1831, Duffy was awarded the contract for what became the most expensive and most deadly section of the P&C, Mile 59 (Fig. 3.1). Charged with completing this mile of railroad by April 1, 1832, the section of the track became known as "Duffy's Cut." A very experimental mile of rail line, its construction required both the digging out of a "cut" and the filling in of a large "chasm" the size of a modern American football field to make an earthen railroad bridge (which became known as Duffy's Fill or Duffy's Bank). This was because no money had been allocated in the state budget for a bridge at Mile 59. The work at Mile 59 proved so problematic that by the second half of June, Duffy found himself more than two-and-a-half months behind his contracted completion date. The *PRR Duffy's Cut* file states that he then hired another work crew of recently arrived Irish immigrants to complete the job. The PRR file specified that "in the early summer of 1832" Duffy "employed a large number of Irishmen, who had but lately arrived on these shores." A thorough search of the shipping records at the National Archives in Philadelphia resulted in the discovery of the only ship that brought a full complement of laborers into the port of Philadelphia in the summer of 1832, the *John Stamp*, a barque that sailed from Derry in April and landed in Philadelphia on June 23, 1832.

3.3 The John Stamp

The captain of the *John Stamp,* John Young, was Irish and mostly sailed from the port of Cork. However, in this case, Captain Young sailed the ship from Derry in April of 1832 with 160 passengers, 100 of whom were male, and 60 were female. The ship arrived to American shores on June 23, and was required to dock at the Lazaretto Quarantine Station in Tinicum, Delaware County, a few miles south of Philadelphia, before it was allowed to proceed to the Port of Philadelphia. The ship's arrival coincided with a global cholera pandemic which started in 1829 in the Bay of Bengal and moved westward across Central Asia into Russia in 1830 and then into western Europe and Great Britain in 1831. It jumped across the Atlantic in 1832, arriving first in Canada and making its way south following the Hudson Valley into New York City in June. From there, it came to Philadelphia in early July (Watson and Watson 2018). Anyone who was ill

[3]Attributed variously to *The Village Record, The National Gazette, Literary Messenger,* and *The Philadelphia Inquirer.*

Fig. 3.1 **Main image**: Duffy's Cut looking west toward, Frazer, Pennsylvania, USA. Insets, clockwise from top right: Topographical map showing location of the shanty area on its left, south of the old fill and current railroad line; Philip Duffy's signature on Mile 59 contract, and Pennsylvania State Historical Marker for Duffy's Cut, listing cholera as the cause of death

was offloaded and placed in a hospital. According to Lazaretto records, no one on the *John Stamp* was ill with cholera when the ship arrived in Pennsylvania. No one was offloaded, and no one was placed in hospital. All the laborers were healthy when they arrived in Philadelphia (Watson and Watson 2018). Duffy likely recruited the workers when they arrived, probably having advertised for railroad work throughout Donegal, Tyrone, and Derry beforehand. Given the segregated, sectarian nature of labor practices in Ulster during the period of the Protestant Ascendancy, Catholic manual laborers in 1832 would have had very few opportunities outside of potato farming for absentee landlords. Catholic political emancipation championed by Daniel O'Connell in 1829 did little to alleviate the harsh economic segregation of the northern Irish economy (Miller 1985; Kenny 2000). Things did not improve for Catholic laborers in Ireland until the time of Charles Stewart Parnell's Irish National Land League in 1879–1882 (Ranelagh 2012). Captain Young's record of his passengers is quite detailed. In addition to the names of passengers, he listed their county of origin, their age in April 1832, their occupation, whether they traveled together or alone, and finally, what they brought with them as personal items. Passengers brought with them all their worldly belongings; items such as chests, boxes, trunks, bundles, barrels, bags, sacks, and "packages." One 22-year-old laborer, George Quigley, who hailed from County Donegal, actually brought three trunks with him. In addition, there were 18 children aboard the *John Stamp*. Its passengers hailed from a variety of northern counties: 70

from Tyrone; 50 from Donegal; 27 from Derry; 6 from Fermanagh; 5 from Leitrim; and 1 from Monaghan. A total of 39 passengers who were listed as laborers came from counties Donegal, Derry, Tyrone, and Leitrim. In terms of age, most of the laborers were in their 20s. The youngest laborers were 15 years old; one was 70. As Duffy had been utilizing Irish immigrant laborers at Mile 60 of the railroad, he supplemented their ranks with these 39 men from the *John Stamp*. A detailed genealogical search was made of the passengers from the ship and revealed that these laborers disappeared from history after their arrival in America. Among the names of the laborers was an 18-year-old named John Ruddy, from County Donegal (Hughes 2010; Sullivan 2013). He brought with him a box. His was the first body recovered at Duffy's Cut in 2009 (Watson and Watson 2018; Tucker 2010). The last person recovered at the base of the 1832 railroad fill was a widow, 29-year-old Catherine Burns, from County Tyrone. She traveled with her 70-year-old father-in-law, John Burns, who was listed as a laborer. What is known from the *PRR Duffy's Cut* file is that all 57 laborers died within eight weeks of their arrival. Martin Clement believed that the Irish laborers died from cholera, as was reported in the 1832 newspapers.

A highly significant archival source for the Duffy's Cut investigation is the *Mitchell Letter*, written in the spring of 1833. Penned by Duffy's immediate railroad supervisor, the P&C Superintendent of the Eastern Division of the railway, William B. Mitchell, this letter to the PA Canal Commission details the reason why Mile 59 was not, as of then, completed. This letter is the first official report by the railroad detailing the deaths of the Irish laborers at Mile 59. It is also important in that it contains the first internal admission by the railroad that a number as high as 60 men had died at Duffy's Cut. Located in the PA Canal Commission records in the PA State Archives, this letter was written to James Clarke who led the PA Canal Commission. Mitchell was born in Northern Ireland of Scots-Irish ancestry. Originally working as a teacher after his arrival in America, Mitchell helped in the original survey of the P&C rail line and rose to become Superintendent of the Eastern Division of the railroad, on which Duffy's Cut was located. As such, Mitchell reported back to his superiors the reason for the year-long delayed completion of Mile 59. Mitchell wrote: "Philip Duffy, the contractor on section 59 [...] has progressed with his contract nearly to completion—he may finish in a month." Mitchell made clear to his superiors that the unfinished work on Mile 59 was the reason that the rail line was not finished by its scheduled completion date: "his is the only contract on 10 miles that is not completed" (Watson and Watson 2018). Mitchell continued with some of the details of Duffy's inability to finish his mile of railroad within his contracted completion time. Mitchell stated that Duffy "complains that he is in great difficulty and distress [...] this man (Duffy) has been rather unfortunate during the last fall, nearly one half of his men died of cholera." An average railroad work crew in 1832 was around 120 men who worked along a whole mile of railroad. This number of half a work crew corresponds well to the number of 57 men who the PRR later admitted died there, as per Clement's internal file. Mitchell continued: "but it must be admitted that he is perfect mast[er] of the art of complaining, with or without cause." As is also clear in the Mitchell Letter, Duffy's only real concern with not finishing Mile 59 on time was a financial one: "his estimates are altogether inadequate to pay his expenses." Duffy claimed to be behind his expenses by $3,382.40. Mitchell encouraged the PA Canal Commission to pay Duffy an additional $500 to help him complete his work, but the Canal Commission refused.

Half of Duffy's work crew—as many as 60 men (despite PRR claims)—died in August of 1832, and were buried under the track bed. Nonetheless, the railroad was only interested in completing the work at Mile 59. Not atypical of the attitude to workers during these years of the Industrial Revolution, Mitchell's letter is seen today as a rather insensitive assessment of the deaths of Duffy's Irish-born work crew. The fact remains that the railroad did nothing to acknowledge these deaths. The cover of the Mitchell letter plainly states: "File this away—no action." And indeed no action was taken to acknowledge what happened at Mile 59 on the part of the railroad, and no effort was made by Duffy to contact their families in Ireland. While the railroad quietly covered up the deaths of the Duffy's Cut 57, and the newspapers downplayed the number of the dead throughout the nineteenth century (stating that a number as low as 6 died there), Duffy himself never publicly spoke of the deaths that occurred at Mile 59. It was only after his death in 1871 that later Irish-born railroad work crews were free to speak and remember the deaths of 57 Irish laborers that occurred at Mile 59 in August of 1832.

After the murder of the Duffy's Cut laborers, Philip Duffy continued on to a very lucrative career with the railroad. After the events at Mile 59, local Chester County newspaper accounts reveal that in the spring of 1833 Duffy advertised for local laborers to complete the job, which was duly done in the same year. Duffy received many more contracts with the P&C and PRR, and in what was perhaps his last railroad contract on the PRR, signed on October 8, 1849, Duffy was commissioned to re-work the track so as "to avoid the inclined plane" near the Philadelphia Main Line. Beyond the railroad, Duffy also received contracts for other work projects in the city of Philadelphia, including expanding roadways in his Port Richmond neighborhood. Duffy even served as a Commissioner for the Gunner's Run Improvement Company that built the ill-fated Aramingo Canal, and he was a Commissioner for the newly incorporated district of Richmond within the county of Philadelphia. Duffy died a rich man, and on his death certificate his occupation was listed as a "gentleman." His will left a substantial sum of money to his children, none of whom ever married, and who were eventually buried with their father. The records of Duffy's death and burial are found in his death certificate, located in the Philadelphia City Archives, as well as in the parish register of St. Anne's Catholic Church. He was buried in his family plot in the cemetery of that church, in Port Richmond (Watson and Watson 2018; Watson 2016a, b; Farr 2018). In contrast, the 57 Irish workers who died at Duffy's Cut remained in unmarked graves for 180 years with scarcely more than a secret railroad file to attest to their passing.

3.4 Scientific Research

Duffy's Cut is located at Amtrak mile 22.3, west of Philadelphia. During the years of the operations of the Philadelphia and Columbia Railroad (1834–1857), it was designated as Mile 59. However, that system's miles were measured west to east from the Susquehanna River in Columbia to the Schuylkill River in Philadelphia. The Duffy's Cut site today sits divided on Amtrak-owned property and privately-owned property. The fill started by Duffy's ill-fated 1832 crew, and finished by his 1833–1834 crew, spans about 300 feet east to west across the valley, bridging natural high spots. It sits along the route mapped out by Major John Wilson and his son William Hassell Wilson in their 1827 survey for the Pennsylvania Canal Commission. In the 1830s, it was about 40 feet in height from the valley floor, while the current, much-expanded fill, is 125 feet in height. The fill was made higher and wider by the Pennsylvania Railroad circa 1870–1873 and again circa 1909–1911 in order to accommodate heavier locomotives. The newer fill is also significantly wider than the old fill, extending about 100 feet northward from the old fill to the westbound tracks of the contemporary active train line. The workers' shanty was located in 2005–2006 with assistance from railroad historian John Hankey, ground-penetrating radar specialist Ron LaBarca, the Battlefield Restoration and Archaeological Volunteer Organization, and the Greater Philadelphia Search and Rescue team (Watson and Watson 2018; Watson 2014).

Hundreds of artifacts from the workers' shanty and the early railroad were excavated at Duffy's Cut between 2004 and 2009, in a 30' × 30' ashen burned field. Located about 30 feet south of the eastern end of the 1832 fill, the field contained ash uniformly distributed about three to four inches below the surface (Fig. 3.2). The ash was evidence of the fire described in the

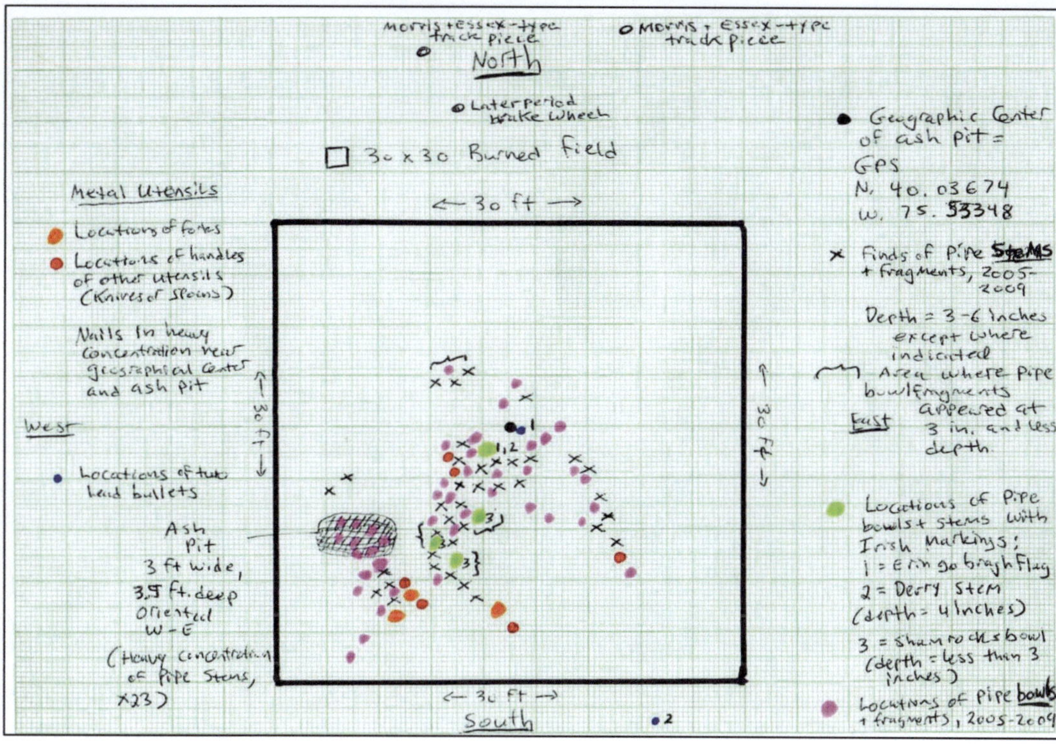

Fig. 3.2 Map of artifact distribution in relation to the shanty area south of the fill

November 7, 1832 *Village Record* newspaper. There were also several artifacts unearthed showing evidence of the fire, including three clay pipe stems with melted pewter on them, and a nail with a button fused to it. Eighty-nine percent of the artifacts excavated at Duffy's Cut were found within the burned field.

Artifacts included several dozen clay pipe bowls and pipe stems with various designs on them. One pipe bowl had an *Erin go bragh* flag, shamrocks, and a round tower on it, with the words around the flag reading "Flag of Ireland." Another pipe bowl had a Brian Boru harp and shamrocks on it, and a pipe stem was stamped "Derry," the port of departure of the ship *John Stamp* that brought 47 of the 57 workers to America (Fig. 3.3). It was the only pipe stem with a city imprinted on it. Fifteen glass buttons were excavated (14 white and 1 dark blue), several dozen glass and pottery fragments, several small-caliber lead bullets, and a number of iron implements including a cooking pot and ladle, knives and forks, a horseshoe, a fire starter, a chest hasp, a pick axe head, a shovel head and handle, a stone chipper, a hook and a looped implement for placing the stone sleepers in the ground, over 40 nails that could have been used in making the kind of lean-to shanties that housed the workers, as well as many unidentified pieces from other metal items. The glass and pottery dates from the 1830s through the 1870s, the latter time frame corresponding to the time of Patrick Doyle's work crew at the site (Watson and Watson 2018; Tucker 2010).

The cooking pot and ladle were the first artifacts located at Duffy's Cut in 2006, in a depression toward to the eastern end of the fill and about 50 feet south of it. The depression was caused by a spring which runs when it rains (Fig. 3.4). The heavily rusted pot still had charcoal fragments beneath it, indicating that it had been buried in haste. The ladle, broken in half, was excavated approximately four feet north-east of where the pot was excavated (Watson et al.

Fig. 3.3 *Erin go bragh* flag pipe bowl, and Derry pipe stem excavated at Duffy's cut site

Fig. 3.4 View of original railroad fill from the north. **Left inset**: Earliest strap rail of the Morris & Essex type excavated. **Right inset**: Earliest stone spike excavated

2006). There were also several rare early railroad artifacts excavated at the site, closer to the fill, which were only used for a discrete period of time in the 1830s by the railroad. Many of the railroading artifacts were identified by John Hankey of the B&O Railroad Museum and Kurt Bell of the Railroad Museum of Pennsylvania. Two pieces of cast iron strap rail of the Morris and Essex type were excavated, one found within the fill at a depth of 15 in., and one that was found at a depth of 8 in. on the valley floor, four feet south of the eastern end of the fill (Fig. 3.4, left inset). A thick square iron post holder was excavated from within the eastern end of the fill at a depth of 7 in. This would have been used to hold a wooden post for a canopy over an early railroad car. An 1860s-era iron brake wheel was also found in the fill, perhaps falling to its location as a result of a mid-nineteenth century railroad mishap at Duffy's Cut. Later iron spikes were found throughout the upper portions of fill, and the remnants of a barrel were excavated containing 40 rusted but intact spikes at the western end of the fill, just east of the stone culvert (Watson et al. 2006; O'Hanlon 2007; Grilli 2005; McRorie 2004).

Seven early experimental stone spikes were excavated in the western end of the valley, near the stone culvert (Fig. 3.4, right inset). The railroad experimented at the first stages with different kinds of spikes to affix the tracks to the stone sleepers. The use of stone spikes was short lived because they frequently cracked, negating their effectiveness. By the time the rail line opened in 1834, iron spikes came into use and have been used ever since. Several generations of early iron spikes have been excavated from within the fill all along the line of the fill. The stone spikes were made of granite, the same stone used for the sleepers. The spikes were probably being chipped in the same area where the sleepers were being made, and the heavy iron chipper was found 15 feet north of where the stone spikes were found. The culvert was built over the Valley Creek in 1831 by Irish contactors McCartney and McCartney, prior to Duffy's contract for Mile 59. Remnants of the blacksmith's smithy were found circa 2005 near the culvert along the stream (Watson and Watson 2018). The artifacts are housed in the Immaculata Duffy's Cut Museum, in the Gabriele Library, at Immaculata University. The 1832 railroad crew lived and worked in the valley. The 1870s local railroad crew working on track expansion resided in Garrett's Siding, near Fraser, and not in the valley. The *Erin go bragh* flag pipe bowl with the words "Flag of Ireland," shamrocks and a round tower on the reverse, may indeed be the oldest examples of Irish nationalist material culture found in North America (Watson and Watson 2018).

The old fill is apparent below the new fill on the eastern end of the valley, but disappears halfway across the valley, and appears to have been purposely removed at a later time by another work crew. A gully created by runoff from the tracks above divides the entire fill in half and marks the western end of the old fill line. As it was theorized in the PRR file on Duffy's Cut that the Irish railroaders were buried within the older railroad fill, the Duffy's Cut team needed to be able to discern where the manmade fill was distinguished from the natural hilly landscape in the valley. Geologist Tim Bechtel joined the Duffy's Cut team in 2008 from the University of Pennsylvania and provided necessary geographical information about the natural and man-made components of the valley's structure. Bechtel utilized a variety of techniques to obtain information about the composition of the old fill from ground penetrating radar, electrical resistivity, and magnetometry. He created a subsurface map of the entire area, and found that the old fill was created by using the soil and rocks from the "cut" above and backfilling it into the valley east to west toward the 1831 stone culvert built over the Valley Creek in the western end of the valley (Bechtel 2014; Gibbons 2013).

Bechtel identified an 1826 report for the Pennsylvania Society for the Promotion of Internal Improvements that discussed how railroad cut and fills were created in the 1830s. Architect William Strickland was sent by the society to examine British rail line construction at Lancashire, Leeds, and Newcastle which employed "solid earth embankments" and stone sleepers to affix the tracks to the embankments

(Bechtel 2014). According to Bechtel, most of the rock of the "cut" was removed by pick axes and shovels to make the track line as straight and flat as possible, with black powder being used only on the hardest rock. The debris was then put into wheelbarrows which were dumped into horse-drawn wagons and taken along wooden tracks to the spot where the fill was to be constructed. The soil and rocks "were hand-pushed against bumpers that tipped their loads onto the advancing embankment" on an "angle of repose" which needed "little grading to achieve" (Bechtel 2014). Understanding how the fill was constructed was an essential step in discerning the probable locations of the railroader graves. Based on the information in the Dougherty letter in the PRR file, the men "were buried where they were making the fill," which was understood to mean within the original fill rather than the area above where the fill was grubbed out of the landscape (Bechtel 2014).

With assistance from the Chester County Emerald Society, the Duffy's Cut team received permissions from the Chester County district attorney and coroner to handle excavated human remains within the county. The homeowners agreed to the work, and the East Whiteland Township commissioners approved the plan. The Pennsylvania Historical Commission certified that the 1832 burials at the spot did not constitute a cemetery in any legal sense in the Commonwealth of Pennsylvania, so that excavations could occur at the site. Testing of any skeletal remains would be done by a team of scientific specialists who joined the Duffy's Cut team, including physical anthropologist Janet Monge and archeologist Samantha Cox (both from the University of Pennsylvania), forensic dentist Matthew Patterson (who had expertise in identifying American servicemen's remains on the basis of surviving teeth), and from Chester County Deputy coroner and forensic dentist Norman Goodman. West Laurel Hill Cemetery in Bala Cynwyd, Pennsylvania, came forward to donate grave space to rebury the remains once the tests were complete. Immaculata University students were trained by Cox to assist in the dig, and the university agreed to provide a Celtic cross for re-buried remains at West Laurel Hill (Watson 2014).

Bechtel employed both electrical resistivity equipment and ground-penetrating radar in the effort to locate human graves in the 1832 fill (Fig. 3.5). The objective was to find a subsurface "stopping void" in the fill, wherein a solid object, such as a skeleton buried long ago, dissolves over time and collapses down to the bottom, leaving a void of looser soil on top of the collapsed object. Bechtel possessed experience in locating graves with such methods at a variety of sites including at a French and Indian War (1754–1763) battlefield in Western Pennsylvania (Bechtel 2014). After narrowing down the possible areas of interest throughout the valley in 2008, Bechtel surveyed the eastern end of the surviving 1832 site in December and placed his findings on a 3D spatial grid for the archeological team in March 2009. His survey located a stopping void within the fill, about 18 in. below the surface at the base of the fill and extending down several inches below the valley floor. Informed by these results, the Duffy's Cut excavation crew located the first set of remains, designated SK001, on March 20, 2009. Forensic analysis by Monge and Cox indicated that the remains were from the nineteenth century, and were those of an 18-year-old male about five feet, four inches in height, who was heavily muscled from manual labor and who had died of a perimortem (time of death) blow to his head. The skeleton also had a shoe buckle and a pants clasp with it. Matthew Patterson X-rayed SK001's skull and found that the person had an M-1 *agenesis*, which is a rare dental anomaly in which the top-right front molar is absent from birth (Fig. 3.6). SK001's remains also had an outline of a coffin, with some pieces of wood and 141 coffin nails. The remains were oriented with head in the west and feet in the east, following Catholic burial customs (Cox 2014; Barry 2013; Hughes 2010).

Contemporary newspapers and railroad sources provided a consensus of information in terms of the number of victims and the cause of death. The official sources stated a number between 57 and 60, indicating that the Irish workers had died of cholera during a widespread epidemic that

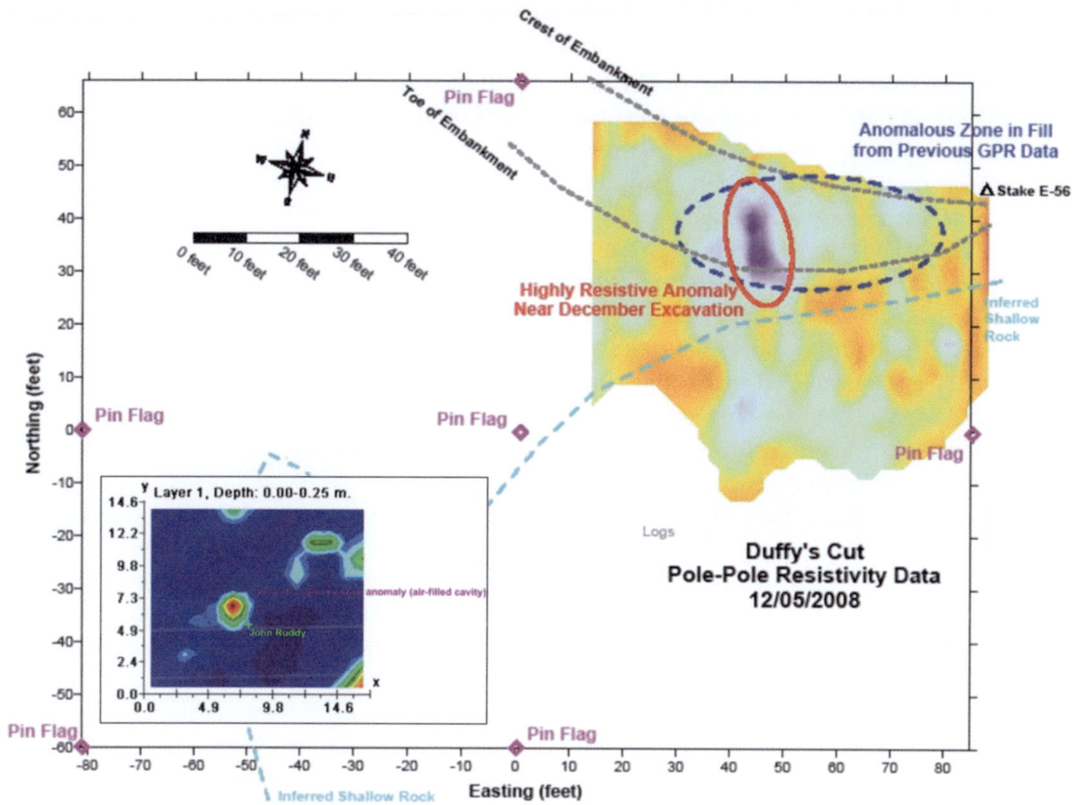

Fig. 3.5 **Main image**: Tim Bechtel's GPR/electrical resistivity image, showing location of John Ruddy. **Inset image**: Tim Bechtel's next GPR/electrical resistivity image showing the location of other *stopping voids* in proximity of a large tulip poplar, under which were four more sets of skeletal remains

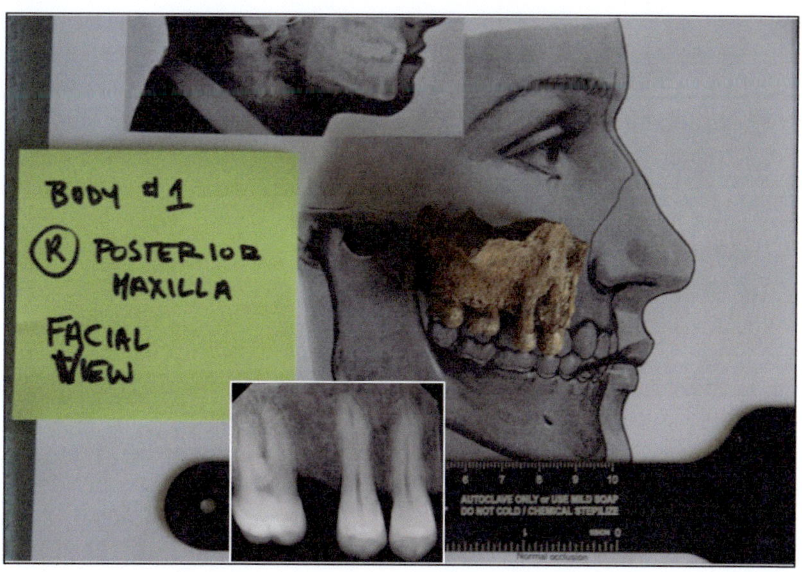

Fig. 3.6 View of John Ruddy's (AKA SK001) jaw showing M-1 *agenesis* (Credit. Matt Patterson)

took a minimum 900 victims in the Delaware Valley in the summer and fall of 1832 (Watson 2009a, b). Duffy's other workers would have been gouging out the landscape for the "cut" above the valley along the track line and doing the customary grubbing and leveling work needed to make the tracks as straight and flat as possible. Internal railroad sources such as the 1833 *Mitchell letter* noted that Duffy lost half of his crew to cholera in 1832 at the site. The Pennsylvania Railroad file specifically noted the number of 57 and maintained that number long after it would have been not only possible, but expedient to alter the internal official record, as in the November 7, 1832 *Village Record* newspaper report where the railroad claimed only eight or nine workers died. However, our archeological evidence indicates violence as the cause of the deaths at Duffy's Cut. Emerald Society police officers assisting our investigation concurred with historical researchers that as cholera outbreaks in the nineteenth century generally killed about half of the people it infected, the railroad's insistence that all 57 men in the valley died of cholera instead implies that something more than a just an outbreak occurred at Duffy's Cut. If cholera was involved, it indicates that perhaps the quarantining of the infected workers went horribly wrong and resulted in mass murder (Watson and Watson 2018).

Bechtel discovered several more *stopping voids* using the same methods at the end of July 2009 near and under an 80-foot tall tulip poplar tree situated within the fill, and about 4.5 feet northwest of the remains of SK001. The anomalies at the location were in proximity to the tree's roots, between 12 and 18 in. below the surface of the fill. Under Cox's supervision, the team excavated two more skeletons (SK002 and SK003) on August 10 and 11, which were also buried in coffins and oriented west–east, and then found the outline of a fourth skeleton with only small fragments of the bones left in the grave (SK004). The remains of SK002 in particular showed signs of extreme perimortem violence. The right sides of the skeletons buried at this location had portions of the limbs on their right sides (parts of arms or legs) washed away from water flowing down from the tracks above over the 177 years since their burial. A fourth grave (SK005) was located in mid-August, but the roots of the tulip poplar entwined the skeleton so that it could not be fully excavated until the tree was removed (which would not be for another three years). Two more sets of remains (SK006 and SK007) were located in early July 2010, four feet to the west of the tulip poplar. They were both buried about 18–20 in. below the surface of the fill as it intersected with the valley floor and were buried in coffins, oriented west–east. These remains also had portions of the limbs on their right sides washed away. The remains closest to the tulip poplar, SK006, were excavated on August 4, and proved to belong to a taller adult male who stood about six feet in height (Fig. 3.7). This skeleton experienced the most extreme perimortem violence of any skeleton excavated, including an axe blow to the right side of the cranium and a small-caliber bullet fired at close range into the top of the skull, which remained lodged in the bottom of the skull.

SK007 was excavated on August 5 and proved to be an adult female (by examination of pelvis and palate) who stood five feet, three inches in height, and who was estimated by Monge and Cox to be around 30 years of age. SK007's remains included 110 coffin nails as well as fragments of wood. The skull also showed signs of perimortem violence. Women were occasionally employed as cooks and laundresses at railroad work sites, especially if they were related to the workers (Watson 2014). SK005 was excavated after the removal of the tulip poplar in two stages in spring and summer, 2011. The skeleton was excavated from the roots in an extended dig that lasted from August 8, 2011 to February 20, 2012. Oriented west–east like the others, SK005 proved to be an adult male who was buried with a sheathed barlow knife on his right side and two pewter buttons which may have been from a haversack. The excavation presented a problem in that the tree roots were entwined within the skeletal remains, and so a non-conventional dig plan was necessary. Only when the stump was removed in early 2012, it was possible for the last bones and coffin nails to

Fig. 3.7 **Left**: Skeleton of SK006 with inset showing massive perimortem violence to skull. **Right**: Skeleton and skull of Catherine Burns (AKA SK007)

be excavated. More teeth were retrieved from SK005 than the other skeletons because the root ball encased the skull. A bullet fragment was also found with SK005.

All the excavated skulls showed signs of perimortem violence, and none of the remains had defensive wounds, indicating murder after being incapacitated, possibly by being tied up. The coffins each had an apparent over-abundance of nails—between 82 and 141 per coffin. The consensus among the Duffy's Cut historians and the Emerald Society police officers was that the large number of nails might be an evidence of an effort to conceal violence done to the workers. The Sachse narrative that appeared in the *Village Record* newspaper, and was repeated in the Pennsylvania Railroad file, stated that some of the workers escaped from the valley and were forced to return when doors were closed to them. It is probable that these workers were not simply forced back into the valley, but that they were killed off-site, perhaps by members of the East Whiteland Horse Company, resentful of the intrusion of a foreign-born labor force perceived as taking away lower-end manual jobs from the local workforce. The eight or nine victims reported by the November 7, 1832 *Village Record* edition most likely refers to the seven skeletons found in coffins in the eastern end of the fill. The Duffy's Cut team believes that they were returned to the valley in sealed coffins to be buried within the fill by their colleagues who were at this time still likely uninfected by cholera (which burial practice later became standard railroad procedure for workers who died at the work site).

If healthy workers had opened the coffins and seen the violent ends their colleagues had faced, "disorders and riots" might have resulted, of the type suggested in an 1829 *Village Record* article which warned violence might occur in Chester County at Duffy's Mile 60. This fear was based on the fact that violence had occurred elsewhere along the line between workers and local residents due to property damage that occurred during the course of normal railroad construction (Watson and Watson 2018). One or two sets of skeletons may have been washed away by the flow of water, or have yet to be located in the fill, to accommodate the number of 8 or 9 victims in the newspaper account. The railroad could not dispute the fact that those workers in coffins had died, because it is likely that the local populace saw the coffins being brought back into the valley. The sheltered nature of the valley made possible a more comprehensive quarantine to occur after that time, as more workers became ill and were incapacitated. The workers who had fled the valley were most likely not infected with cholera. Based upon the examination of the remains themselves, Janet Monge has stated that "if they had cholera, that is not what killed them," and "it really looks like a massacre" (Snow 2010; Browne 2014).

Identifying some of the workers is possible by the time frame mentioned by Sachse in his account, in which he stated that the workers had recently arrived in the USA. As noted earlier, the *John Stamp* carried 47 of the workers from Derry and were paired at the site with the ten "alien, non-naturalized laborers" living in Duffy's rented house in Willistown, Chester County about a mile south of the rail line. The *John Stamp* was the only ship that arrived in Philadelphia in that time frame (June 23), and Derry is the only city on any of the clay pipe stems or any other artifact found at Duffy's Cut. As previously noted, the names of the *John Stamp* laborers who vanished from the historical record after arrival were ascertained using standard genealogical protocols. The ship passenger list recorded the origin and age of the workers. Monge ascertained the gender and the approximate ages of the skeletons excavated at Duffy's Cut, and with the first (SK001) and last (SK007) it became possible to link a name on the ship list with each skeleton.

SK001 was a male, approximately 18 years of age, and SK007 was a female, approximately 30 years of age. The 18-year-old male has been identified as John Ruddy of Donegal, a laborer, passenger number 4 on the *John Stamp* passenger list, who came with one box. The approximately 30-year-old female has been identified as Catherine Burns of Tyrone, aged 29, passenger number 120, a widow, who came with her father-in-law John Burns, who had no possessions noted on the passenger list (O'Hanlon 2009; Valania 2010). The skeletal remains were examined by Monge and Cox at the University of Pennsylvania, and Patterson obtained DNA from living Ruddys in Donegal to try to link up at least one set of skeletal remains with Irish and Irish diaspora lineages. Some bones from each skeleton (such as John Ruddy's jaw fragment with the M1 *agenesis*) were retained for research purposes, but re-interment of most of the remains occurred in three stages. Five sets of remains were re-interred at West Laurel Hill Cemetery in March, 2012; Ruddy was re-interred at Holy Family Cemetery in Ardara, Donegal, Republic of Ireland, in March, 2013; a portion of Catherine Burns' remains was re-interred at St. Patrick's Cemetery in Clonoe, Tyrone, Northern Ireland, in July, 2015 (O'Hanlon 2013; Watson 2014; Holmes 2012; Matheson 2013; Browne 2014; Glenza 2015, Wylie 2015; Philip 2015).

3.5 Conclusion

The Duffy's Cut story is a tragic tapestry woven from various narrative strands: archival and epistolary research, GIScience techniques (ground penetrating radar/electrical resistivity imaging) archeological and forensic anthropological methods and journalistic texts. However, the story is not yet complete, as further remains were located by Bechtel under the stone monument in 2011, and further core sampling needs to be conducted. DNA research on the remains still needs to be conducted as well (Barter 2017). There is also another known Irish immigrant laborer mass grave site at Downingtown, 11 miles west of Duffy's Cut in Chester County, PA, which remains to be examined. It is connected to Duffy's Cut by one worker who fled the valley to another Irish immigrant work crew at railroad Mile 48 under contractor Peter Connor. The work crew also died during the epidemic and were buried in a potter's field north of the track line on private property, which was later incorporated into Northwood Cemetery. The two sites were connected in 1832 by mass casualty epidemics. They were also associated when Doyle erected a wooden fence at Duffy's Cut in the 1870s after locating skeletal remains and when a cemetery incorporated the potter's field in Downingtown. In 1909, efforts to commemorate both sites stories were undertaken by Clement at Duffy's Cut and by the local historian Charles H. Pennypacker in his *History of Downingtown* (Watson and Watson 2018; Watson 2019). There are dozens of similar sites across the East Coast of the USA where anonymous Irish laborers who died building rail lines and canals were buried in mass graves in the nineteenth century. Sites stretched from the Erie Canal in New York to the New Orleans Canal in Louisiana. There are perhaps 20,000 Irish immigrants dead in these mass graves of the American Industrial Revolution, and Duffy's Cut is unique by the fact that some of the names of the victims were recovered along with their skeletal remains. Only the Funk's Grove mass grave in Illinois, dating from the 1850s, might be compared to Duffy's Cut, where the names of Irish railroaders who died in a later epidemic are recorded.

References

Barry D (2013) With shovels and science, a grim story is told. New York Times. http://www.nytimes.com/2013/03/25/us/secrets-of-duffys-cut-yield-to-shovel-and-science.html?pagewanted=all&_r=0. Accessed 25 Mar 2013

Barter P (2017) Duffy's cut secrets unearthed. Sunday Times. https://www.thetimes.co.uk/edition/ireland/duffys-cut-secrets-unearthed-zvtlsdj3v. Accessed 9 July 2017

Bechtel T (2014) The engineering geology of Duffy's cut. In: Annual conference of the Eastern American studies association, LaSalle University, Philadelphia. https://www.researchgate.net/publication/263579838_The_Engineering_Geology_of_the_1832_Duffy's_Cut_Railroad_Embankment_Malvern_PA. Accessed 28 Mar 2014

Browne B (2014) Blood on the tracks. Irish Examiner. http://www.irishexaminer.com/lifestyle/features/blood-on-the-tracks-were-50-irish-immigrants-murdered-in-philadelphia-in-1832-268069.html. Accessed 19 May 2014

Chariton F (2010) 1832 cholera outbreak in Philadelphia and Duffy's cut. Catholic Historical Research Center of the Archdiocese of Philadelphia. http://www.pahrc.net/1832-cholera-outbreak-in-philadelphia-and-duffys-cut. Accessed 15 Sept 2010

Cox SL (2014) Duffy's cut: a brief report of the bioarchaeological findings from a massacre of Irish immigrant railroad laborers from 1832. Am J Phys Anthropol 153

Farr S (2018) New discovery in Duffy's cut 1832 tragedy. The Philadelphia Inquirer, 16 May 2018, p B2

Gibbons K (2013) The application of forensic geophysics to Duffy's cut (MP59) archaeological site in Malvern, PA. https://www.sas.upenn.edu/lps/sites/www.sas.upenn.edu.lps/files/GibbonsCKelsey2013.pdf

Glenza J (2015) Reburial of woman in native Ireland highlights 183-year-old mystery. The Guardian. http://www.theguardian.com/us-news/2015/jul/19/reburial-woman-ireland-murder-mystery-duffy-s-cut. Accessed 19 July 2015

Grilli A (2005) Unexplained events fuel interest in Duffy's cut. Suburban Advertiser, 7 April 2005

Hage H (1839) Report relative to the survey of a rail road from Chambersburg to Pittsburg, and a survey of the raystown branch of the Juniata River, with estimates of the cost of the work, Harrisburg, PA

Holmes KE (2012) Long-forgotten dead of Duffy's cut get proper rites. Philadelphia Inquirer, 10 March 2012

Hughes S (2010) Bones beneath the tracks. Pennsylvania Gazette 34–43. http://www.upenn.edu/gazette/1110/feature1_1.html

Kenny K (2000) The American Irish: a history. Pearson Education Limited, Essex

Matheson K (2013) 19th-century PA immigrant reburied in Ireland. AP wire. https://www.cnsnews.com/news/article/19th-century-pa-immigrant-reburied-ireland. Accessed 28 Feb 2013

McRorie J (2004) New artifacts unearthed at Duffy's cut. Daily Local News, 15 Nov 2004

Miller KA (1985) Emigrants and exiles: Ireland and the Irish exodus to North America. Oxford University Press, Oxford

BBC News (2015) Duffy's cut: catherine burns is buried in clonoe. http://www.bbc.com/news/uk-northern-ireland-33586042. Accessed 19 July 2015

Noble work in caring for destitute patients in the almshouse here during the cholera epidemic of 1832 (Undated) printed source of unknown provenance in archives of the daughters of charity, De Paul Provincial House, Albany, New York

O'Hanlon R (2007) D-day at Duffy's cut. Irish Echo, 8 August 2007

O'Hanlon R (2009) Sacred Duty: Duffy's cut find has already yielded a name. Irish Echo, 12. 1 April 2009

O'Hanlon R (2013) Journey home: full circle as John Ruddy's remains back in Ireland. Irish Echo, 13 February 2013

Philip A (2015) Unearthing a deadly secret: were 57 Irish workers murdered in 1832 Pennsylvania? Washington Post. http://www.washingtonpost.com/news/morning-

mix/wp/2015/10/8/unearthing-a-deadly-secret-were-57-irish-workers-murdered-in-1832-Pennsylvania/?tid=sm_fb. Accessed 8 Oct 2015
Ranelagh JO (2012) A short history of Ireland, 3rd edn. Cambridge University Press, Cambridge
Snow M (2010) Wolf Blitzer's "Situation room." CNN. http://www.cnn.com/video/?/video/crime/2010/08/21/snow.csi.1832.cnn
Sullivan M (2013) Revisiting the ghosts of Duffy's cut. Hist Irel 21(3):6–7
Sweeney E (2007) The londonderry sentinel and the Philadelphia railway mystery. Londonderry Sentinel, 29 August 2007
Sweeney E (2009) Grisly find in Philadelphia. Londonderry Sentinel http://www.londonderrysentinel.co.uk/news/Grisly-find-in-USA.5130445.jp. Accessed 1 Apr 2009
Tauxe RV (1998) Cholera. In: Evans AS, Brachman PS (eds) Bacterial infections of humans, 3rd edn. Plenum Medical Book Company, New York
The statutes at large of Pennsylvania, Harrisburg, PA (1911) Laws passed session, 1832–1833. http://www.rootsweb.com/~usgenweb/pa/pafiles.htm
Tripician JF (1960) The role played by the Pennsylvania railroad in the 'main line' area of Philadelphia. Unpublished paper, Narberth, PA
Tripician JF (1970) Idiosyncrasies of people in high places. Unpublished paper, Narberth, PA
Tucker A (2010) Ireland's forgotten sons. The Smithsonian 41(1):14–18. http://www.smithsonianmag.com/ist/?next=/history/irelands-forgotten-sons-recovered-two-centuries-later-9194680
University of Pennsylvania Museum of Archaeology and Anthropology (2012) Dead men of Duffy's cut. https://www.penn.museum/research/projects-researchers/physical-anthropology-section/144-dead-men-of-duffy-s-cut
Valania J (2010) Murder in the time of cholera. Philadelphia Weekly, 18 August 2010
Via N (2005) 173-year-old main line mystery. Main Line Life, 1 June 2005
Watson WE (2009) First human remains recovered at Duffy's cut. The Keystone 42(3):43–44
Watson WE (2009) The sisters of charity, the 1832 cholera epidemic in Philadelphia and Duffy's cut. U.S. Cathol Hist 27(4):1–16
Watson JF (2010a) George Dougherty: a section foreman looks at Duffy's cut. The Keystone 43(1):81–82
Watson JF (2010) Update: Murder at Duffy's cut. The Keystone 43(4):32–33
Watson JF (2013) Duffy's cut burial in Ireland. The Keystone 46(3):41, 44
Watson WE (2014) History and memory at Duffy's cut. Railr Hist 211:76–87
Watson WE (2015) Duffy's cut. The Encyclopedia of Greater Philadelphia. http://philadelphiaencyclopedia.org/archive/duffys-cut
Watson JF (2016a) Philip Duffy and Duffy's cut among the Irish-American railroading community. Seanchas Ard Mhacha (J Armagh Dioc Hist Soc) 26(1):151–167
Watson JF (2016) Who was the real Philip Duffy? Railr Hist 214:82–89
Watson WE (2017) The Irish and Ireland. The Encyclopedia of Greater Philadelphia. N.p. http://philadelphiaencyclopedia.org/archive/irish-the-and-ireland
Watson WE (2019) Addendum to Duffy's cut. Irish Ed xxxviii(1):7, 10
Watson WE, Halus EJ (2015) Irish Americans: the history and culture of a people. ABC-Clio Publishers, Santa Barbara
Watson WE, Watson JF (2018) Massacre at Duffy's cut: tragedy and conspiracy on the Pennsylvania railroad. The History Press, Charleston
Watson WE, Watson JF, Ahtes JH, Schandelmeier EH (2006) The ghosts of Duffy's cut: the Irish who died building America's most dangerous stretch of railroad. Praeger Publishers, Westport
Weir C (2007) Mass grave may house immigrants. Belfast Telegraph. http://www.belfasttelegraph.co.uk/northwest-edition/daily/article2884216.ece. Accessed 22 Aug 2007
Whyte BJ (2010) Duffy's cut victim to be given a new face. Irish Central. http://www.irishcentral.com/news/Duffys-Cut-victim-to-be-given-a-face-103779529.html. Accessed 25 Sept 2010
Wilson WH (1896) The Columbia-Philadelphia railroad and its successor. York, PA, 1985. Originally entitled reminiscences of a railroad engineer
Wilson JG (1965) A brief history of east whiteland township. Frazer Press, Frazer
Wylie C (2015) Remains of woman murdered 183 years ago brought home. Irish Times. http://www.irishtimes.com/news/ireland/irish-news/remains-of-woman-murdered-183-years-ago-brought-home-1.2290182. Accessed 20 July 2015
The cholera in Philadelphia, Archives of the daughters of charity, De Paul Provincial House, Albany, New York. Record Group 11-01, Box 1, Folder 03-02, Item no. 1. The sisters of charity subsequently changed the name of the order to the daughters of charity

“Please Mention the Green Book:” The *Negro Motorist Green Book* as Critical GIS

4

Ethan Bottone

Abstract

The Green Book, a Jim Crow era travel guide created by African-Americans for African-Americans, has received much recent popular and academic scrutiny. Consisting of almost 30 editions published between 1936 and 1966, the *Green Book* features thousands of addresses for businesses that catered to African-Americans during a period of institutionalized discrimination and segregation. Use of the guide allowed for safe travel by black travelers through hostile areas of the United States as it provided escape from harassment and potential violence instigated by unwelcoming shopkeepers and patrons. As a tool of resistance developed to spatially subvert white supremacy, the many editions of the *Green Book* provide a kind of road map that can reveal black geographies previously forgotten by hegemonic knowledge structures. However, despite this recognized social and historical importance, few studies have investigated the spatial data contained within the pages of the guidebook, or more broadly, the spaces of black geographies. This chapter seeks to fill this gap by understanding how the text of the *Green Book* can be read through the epistemologies of black geographies and critical geographic information science (GIS). Simultaneously, it provides insights into the geography of African-American travel patterns during an era of state-sponsored discrimination. This study embraces technological advances since the time of the *Green Book's* publication to visually map spatial data published during the Jim Crow era to demonstrate how the study of black geographies may benefit from the use of critical GIS and texts such as the *Green Book.* Using a case study of New Orleans, Louisiana (USA), the author shows how the *Green Book* can be read to reveal how shifts in American racial politics, from overt segregation in the 1930s to racial liberalization in the 1960s, led to shifts in the spaces associated with African-American travel. By comparing the spatial data of the *Green Book* to historical census data, trends in urban neighborhood composition can explain how and why African-American travel patterns shifted within the case city. Furthermore, such mapping reveals the complex networks of spaces developed by black Americans to live within a segregationist society while actively resisting discrimination through the construction of counter-public spaces. Finally, this chapter demonstrates how historical texts, including guidebooks, can be used to provide insights into the historical geography of a largely understudied population, African-American travelers.

E. Bottone (✉)
Department of Geography, University of Tennessee, Knoxville, TN, USA
e-mail: ebottone@vols.utk.edu

C. Travis et al. (eds.), *Historical Geography, GIScience and Textual Analysis*, Historical Geography and Geosciences, https://doi.org/10.1007/978-3-030-37569-0_4

Keywords
Green book · Black geographies · Critical GIS · African-American tourism · Black mobility · Counter-public spaces · New Orleans

4.1 Introduction

During the period that lasted roughly from 1876 to 1964, African-Americans in the United States experienced intense levels of discrimination, were segregated from the white population, intentionally left economically disadvantaged, and routinely threatened with violence or physically attacked in outright manners. This discrimination affected all parts of the country, but was arguably most prevalent in the American South, where laws enacted shortly after Reconstruction codified and institutionalized much of the hatred felt toward black Americans. These so-called Jim Crow laws reached into all aspects of life, from the prohibition of interracial marriages to the creation of a racial *apartheid* in the region's restaurants, bathrooms, and on its public transportation. Lasting officially for almost a century, Jim Crow ended with the passage of the 1964 *Civil Rights Act* that made racial discrimination illegal and opened the gates to a potentially more just and equal future. However, the effects of Jim Crow laws are undoubtedly still felt today, with African-Americans typically occupying lower socio-economic classes and being incarcerated at much higher rates than other populations, signifying just how disruptive government-sponsored discrimination was (and, as some have argued, continues to be).[1] Travel was one facet of life where the restrictions of Jim Crow were felt more than others, especially on long-distance journeys and vacations. Black travelers, potentially unfamiliar with the local dimensions of Jim Crow practices, were forced to contend with dangers in places identified as *Sundown Towns* (localities where African-Americans were told not to stay after sundown), harassment by hostile police, or violence from local residents and business owners as they moved from town to town.[2] African-American tourists also had to contend with the spatial realities of segregation and had to search hard to find accommodations willing to accept them for a night's rest, a fuel refill and bathroom break, or a quick meal. In her book, *The Warmth of Other Suns* (2010) Isabel Wilkerson describes perhaps one of the more poignant and disturbing experiences an African-American faced while traveling in her account of Robert Joseph Pershing Foster's trek from Louisiana to California in 1953. During his journey Foster was forced to drive non-stop across three western states as he was unable to find a hotel that would welcome him.[3] Reaching a breaking point in Arizona, Foster, who had been rejected from at least three other local motels, asked an owner:

> I'm looking for a room […] now, if it's your policy not to rent to colored people, let me know now so I don't keep getting insulted […] It's a shame that they would do a person like this […] I'm no robber. I've got not weapons. I'm not a thief. I'm a medical doctor. I'm a captain [in the army] I have money to pay for my services [….] Now if you don't rent to colored people, let me know so I can go on to California. This is inhuman.[4]

Examples such as Foster's showcase the undignified and potentially dangerous situations created by Jim Crow laws for black travelers. However, African-Americans were not helpless under this institutionalized racism. As has been documented throughout the Civil Rights Movement, African-Americans resisted subjugation and oppression through a variety of means, both formal and informal.[5] This resistance even extended to

[1]Alexander, M. 2010. *The New Jim Crow: Mass Incarceration in the Age of Colorblindness*. New York: The New Press.

[2]Loewen, J.W. 2005. *Sundown Towns: A Hidden Dimension of American Racism*. New York: The New Press.

[3]Wilkerson, I. 2010. *The Warmth of Other Suns: The Epic Story of America's Great Migration*. New York: Vintage Books.

[4]Ibid., 208–209.

[5]For examples of formal resistance, see McKnight, G.D. 1998. *The Last Crusade: Martin Luther King, Jr., the FBI, and the 1968 Poor People's Campaign*. Boulder: Westview Press; and Arsenault, R. 2006. *Freedom Riders: 1961 and the Struggle for Racial Justice*. Oxford: Oxford University Press. For examples of informal resistance, see Alderman, D.H. and Inwood, J. 2016. Mobility as Antiracism Work: The "Hard Driving" of NASCAR's Wendell Scott. *Annals of the American Association of Geographers*, 106:3, 597–611.

travelers as they developed the means to literally navigate through, and sometimes around, a white-dominated world. One such method used to escape the indignities and dangers of the Jim Crow highway was developed by an African-American postal worker named Victor Hugo Green. His creation, dubbed the *Negro Motorist Green Book*, or *Green Book* for short, was a travel guide that listed locations by street, city, and state across the United States that were willing to accept the business of African-American travelers (Fig. 4.1).

While traditional travel accommodations such as lodging, restaurants, and service stations were listed, other establishments including beauty parlors, drug stores, and taverns were provided as well, giving a sense of the broader black communities, or counter-public spaces, that existed during the Jim Crow era.[6] Within each edition of the *Green Book*, published from 1936 to 1966, street addresses in major cities of each state were provided for listed businesses. As the guide grew in popularity, and tourism became more popular among middle-class African-Americans, the *Green Book* began listing several international travel accommodations as well (such as the Bahamas) and changed its official name to the *Negro Travelers Green Book* in 1952. With the end of formal segregation and discrimination in 1964, the *Green Book*'s purpose quickly became moot and publication ceased two years later. Despite its quick end, the guide provided many African-American tourists the confidence needed to travel to unfamiliar locations and allowed for the successful planning of both business and pleasure trips. This is not to say that the *Green Book* was a panacea for the discrimination of Jim Crow travel. Wilkerson offers the caveat that "the books were often out of date by the time they were printed [and] the accuracy of their entries based on the fortunes of "hoteliers" who may have only been renters themselves [allowed] for the possibility that [the traveler] might arrive at a place in the guidebook only to find that the proprietor had been gone for years."[7] However, even with such disappointments, "the mere presence of the guidebooks [...] gave a sense of order and dignity to the dispiriting prospect of driving cross-country" and created what can be thought of as a modern-day underground railroad that gave safe passage to African-Americans traveling through the hostile landscape of Jim Crow America.[8]

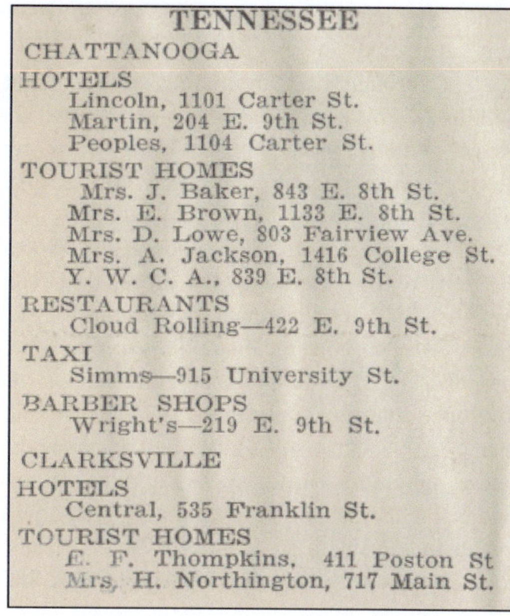
TENNESSEE
CHATTANOOGA
HOTELS
Lincoln, 1101 Carter St.
Martin, 204 E. 9th St.
Peoples, 1104 Carter St.
TOURIST HOMES
Mrs. J. Baker, 843 E. 8th St.
Mrs. E. Brown, 1133 E. 8th St.
Mrs. D. Lowe, 803 Fairview Ave.
Mrs. A. Jackson, 1416 College St.
Y. W. C. A., 839 E. 8th St.
RESTAURANTS
Cloud Rolling—422 E. 9th St.
TAXI
Simms—915 University St.
BARBER SHOPS
Wright's—219 E. 9th St.
CLARKSVILLE
HOTELS
Central, 535 Franklin St.
TOURIST HOMES
E. F. Thompkins, 411 Poston St
Mrs. H. Northington, 717 Main St.

Fig. 4.1 Example of *Green Book* listings found in the 1939 edition of the guidebook (*The New York Public Library Digital Collections*) (The New York Public Library Digital Collections. Retrieved https://digitalcollections.nypl.org/collections/the-green-book#/?tab=about&scroll=3)

Though the guide never left the memories of those who used it, the public lost sight of the *Green Book* and only "rediscovered" the text within the last decade. Featured in several podcasts, several news stories, and even an Academy Award winning film, the *Green Book* has also captured the imagination of a part of America that is attempting to come to terms with the

[6]Inwood, J.F.J. 2011. Constructing African American Urban Space in Atlanta, Georgia. *The Geographical Review* 101:2, 147–163.

[7]Wilkerson, *Warmth of Other Suns*, 204.

[8]Ibid., 204.

country's racialized past and present.[9] Furthermore, academics are beginning to engage with the text through critical theories of race and mobility.[10] This chapter seeks to add to the literature building around the *Green Book* by viewing the guidebook through a critical GIS framework influenced by critical race theorists and black geographies.[11] By applying a critical GIS lens and methodology to the spatial data contained within the numerous editions of the *Green Book*, a fuller understanding of black life and resistance to white supremacy can be developed that previous studies, which made no attempt to visualize the spaces described in the guide, ignored. Through the application of modern GIS technology, this chapter will demonstrate, using a case study of New Orleans, Louisiana, how these vibrant black spaces and places can be mapped and their stories, both oppressive and liberating, told.

[9]Hall, D. (Producer). 2016, February 23. Episode 201: The Green Book. *99% Invisible* [Audio Podcast]. Retrieved from https://99percentinvisible.org; Staples, B. 2019, January 25. The Green Book's Black History: Lessons from the Jim Crow-era Travel Guide for African-American Elites. *The New York Times*. Retrieved from https://www.nytimes.com/2019/01/25/opinion/green-book-black-travel.html.

[10]Hall, M.R. 2014. The Negro Traveller's Guide to a Jim Crow South: Negotiating Racialized Landscapes During a Dark Period in the United States Cultural History, 1936–1967. *Postcolonial Studies* 17:3, 307–319; Mitchell, J.T. and Collins, L. 2014. *The Green Book*: "Safe Spaces" from Place to Place. *The Geography Teacher* 11:1, 29–36; Alderman, D.H. and Inwood, J. 2016. Civil Rights as Geospatial Work: Rethinking African-American Resistance. In *Race, Ethnicity, and Place in a Changing America, 3rd Edition*, eds. J.W. Frazier, E.L. Tettey-Fio, and N.F. Henry, 177–185. Albany: SUNY Press.

[11]Schuurman, N. 1999. Critical GIS: Theorizing an Emerging Science. *Cartographica* 36: 1–108; Pavlovskaya, M. 2006. Theorizing with GIS: A Tool for Critical Geographies? *Environment and Planning A* 38: 2003–2020.

4.2 Mapping Race

Attempts to visualize the distribution of racial and ethnic groups and to locate racialized landscapes are not new.[12] However, much of this work has been criticized for being positivistic, reducing the lives of people of color "to essential measurable 'facts.'"[13] Such studies, while providing important statistical information about the conditions in these areas, often affirm hegemonic narratives and harden "spatial binaries, in turn suggesting that some bodies belong, some bodies do not belong, and some bodies are out of place."[14] As such, "race becomes attached to place in detrimental ways because local conditions reify and naturalize identity-difference: [B]lack women live in 'bad'/black neighborhoods, have unhealthy children, restricted employment opportunities and resources, longer work days, and so on."[15]

By not recognizing the social construction of space, championed by Henri Lefebvre, positivist scholars are decontextualizing the placement of black bodies in the United States from the violent and oppressive histories that created the racialized spaces they are found in.[16] A white supremacist society, founded on institutions that actively placed people of color in the margins and peripheries of society, has created the American landscapes that geographers now seek to map and visualize. Furthermore, these mappings that put marginalized populations in their

[12]Block, J.P., Scribner, R.A., and DeSalvo, K.B. 2004. Fast Food, Race/Ethnicity, and Income: A Geographic Analysis. *American Journal of Preventative Medicine* 27:3, 211–217; Sharma, M. 2017. Income Divide and Race/Ethnicity in Tennessee Metropolises. *International Journal of Geospatial and Environmental Research* 4:1, 1–23.

[13]McKittrick, K. and Woods, C. 2007. "No One Knows the Mysteries at the Bottom of the Ocean." In *Black Geographies and the Politics of Place*, eds. K. McKittrick and C. Woods. Toronto: Between the Lines, 6.

[14]McKittrick, K. 2007. *Demonic Grounds: Black Women and the Cartographies of Struggle*. Minneapolis: University of Minnesota Press, xv.

[15]Ibid., 12–13.

[16]Lefebvre, H. 1991. *The Production of Space*. Malden: Blackwell.

place hide the work that many have performed to struggle and resist against white supremacy, which itself significantly contributes to the production of space.[17] Through presentations of what is seen as wrong with society, such as high crime rates, low education levels, and health disparities, "a black sense of place" and "black geographic knowledges" that combat against the harsh realities of a racialized existence are "undermined by hegemonic spatial practices" that include traditional forms of mapping and visualization.[18]

With the theoretical development of critical GIS, however, scholars can use this framework to avoid falling into the positivist trap. Born out of social constructionist critiques on the rise of geographic information sciences in the mid-1990s, critical GIS "[investigates] the impacts of GIS upon participation, power relations, and existing inequalities in access to spatial data and technologies, and [theorizes] how these impacts occur."[19] Proponents of critical GIS studies seek to include forms of data that may have been previously thought to be useless (i.e., qualitative data) or excluded purposefully as they were deemed subversive to dominant discourses.[20] Specifically, critical GIS seeks to uncover, question, and reverse hegemonic power relations surrounding spatial data creation, analysis, and visualization and the use of those practices to exclude certain populations from prevailing ideas of who and what counts as data. As such, we can view the positivistic mapping of race in place, which is usually conducted using top-down derived methods and data sets, as antithetical to the goals of critical GIS.

To answer the calls of critical race scholars who decry the use of traditional mapping techniques in research, this chapter intertwines the epistemologies of critical GIS with black geographies.[21] Informed by critical race theory, black geographies recognizes the subaltern experiences of black Americans, which are produced both by marginalizing forces from above and through the work of those being oppressed as they create spaces in which to survive and resist.[22] Specifically, "black geographies seek to highlight black agency in the production of space and black geographic experiences in the articulation of black geographic visions of society."[23] In this way, the experiences and spaces of black life are brought into view in spite of their hegemonic exclusion and are properly centered to bring into view a more nuanced socio-historical context of spatial construction within the United States.

With the recent attention swirling around the *Green Book*, now is the time to use this text as a jumping off point to further showcase how the epistemology of black geographies can engage with critical GIS to produce visualizations of black life, movement, and spatial construction during a time of intense hegemonic geographic segregation and discrimination. By using the spatial data found within the pages of the guidebook, produced personally by the African-Americans who made use of the text, this chapter will add to the literature about mapping racialized landscapes through the combination of epistemologies and the visualization of geospatial intelligence produced by typically

[17]McKittrick and Woods, "No One Knows the Mysteries", 6.

[18]Ibid., 7.

[19]Elwood, S. 2008. Volunteered Geographic Information: Future Research Directions Motivated by Critical, Participatory, and Feminist GIS. *GeoJournal* 72:3/4, 177.

[20]Kwan, M-P. 2002. Is GIS for Women? Reflections on the Critical Discourse in the 1990s. *Gender, Place and Culture* 9:3, 271–279; Gilbert, M.R. and Masucci, M. 2006. The Implications of Including Women's Daily Lives in a Feminist GIScience. *Transactions in GIS* 10:5, 751–761; Pavlovskaya, Theorizing with GIS.

[21]McKittrick, *Demonic Grounds*; McKittrick and Woods, "No One Knows the Mysteries"; Pacheco, D. and Velez, V.N. 2009. Maps, Mapmaking, and Critical Pedagogy: Exploring GIS and Maps as a Teaching Tool for Social Change. *Seattle Journal for Social Justice* 8:1, 273–302; Tate IV, W.F. and Hogrebe, M. 2011. From Visuals to Vision: Using GIS to Inform Civic Dialogue About African American Males. *Race Ethnicity and Education* 14:1, 51–71.

[22]McKittrick, *Demonic Grounds*; McKittrick and Woods, *Black Geographies*.

[23]Allen, D., Lawhon, M. and Pierce, J. 2018. Placing Race: On the Resonance of Place with Black Geographies. *Progress in Human Geography*: 2.

excluded individuals. Furthermore, historical landscapes of African-American travel will be revealed, contributing to the growing study of black travel and movement.

4.3 Black Mobilities

One essential aspect of daily life, mobility, has been studied extensively in terms of its relationship to white-supremacy and African-American resistance. To maintain the economic and racial status quo developed during the rise of slavery, the dominant white hegemony and its resultant white supremacy created a racial divide that sought to subjugate black Americans and leave them disadvantaged, including in access to mobility and the resources it provided. By limiting travel, the white supremacist society was effectively limiting work opportunities, access to food and other resources, chances to increase cultural capital through leisure travel, and the gathering of black leaders who could potentially form resistance against the hegemony. Despite numerous obstacles to travel, black Americans did successfully move throughout this period of (attempted) forced immobility. Isabel Wilkerson's epic narrative of the Great Migration proves just that, as she follows several characters through their movements in and out of the Jim Crow South.[24]

Recent work has shown how nuanced and sophisticated these acts of *resistant mobility* became during the Jim Crow period. On a macro-scale, both Franz and Seiler discuss how the rise of private car travel allowed for African-Americans to escape racial segregation and violence routinely experienced on public transportation, while simultaneously enabling access to more economic opportunities.[25] Similarly, Mark Foster details how prosperous black citizens were especially able to travel "in the face of Jim Crow," through the creation of their very own resorts, athletic tournaments, and recreational spheres.[26] On a micro-scale, Derek Alderman and Joshua Inwood explore the tactics of Wendell Scott, one of the first African-American drivers to participate in all-white NASCAR races.[27] Scott's geographic and social mobility, and the mobility of other African-Americans during this period, can be viewed as antiracism mobility work, as their practices were "actively fashioned and employed to transform their geographic immobility (or controlled mobility) into movement that subvert[ed] white supremacy and embodi[ed] antiracism."[28] This antiracism mobility work helped black Americans to resist against the power structures of white supremacy and fight for opportunities of work, pleasure, and resources that accompany the ability to move.

Contrary to the sense of motion that mobility conveys, understanding the movements of African-Americans can also reveal the settlement of black spaces and places. Through the paradigm of critical mobilities and its underlying concept that spatial mobility is racialized, we can recognize the resistance tactics that allowed for the thriving of black communities that catered to travel for leisure or work. For instance, Foster's work on prosperous black travelers exposes this side of mobility when he discusses the development of "black resort areas [that] sprouted up near eastern cities with relatively large black populations."[29] Despite being permanent locations, these resorts are essential parts of the geographic mobility complex due to their status as destinations for black travelers. Katie Algeo provides another example of this relationship between settlement and movement in her study of an African-American hotel established at the edge of Mammoth Caves, one of the nineteenth

[24]Wilkerson, *Warmth of Other Suns*.

[25]Franz, K. 2004. 'The Open Road': Automobility and Racial Uplift in the Interwar Years. In *Technology and the African-American Experience: Needs and Opportunities for Study*, ed. B. Sinclair. Cambridge: The MIT Press, 131-154; Seiler, C. 2006. "So That We as a Race Might Have Something Authentic to Travel By": African American Automobility and Cold-War Liberalism. *American Quarterly* 58:4, 1091–1117.

[26]Foster, M.S. 1999. In the Face of "Jim Crow": Prosperous Blacks and Vacations, Travel and Outdoor Leisure, 1890–1945. *The Journal of Negro History* 84: 130.

[27]Alderman and Inwood, *Mobility as Antiracism Work*.

[28]Ibid., 602.

[29]Foster, In the Face of "Jim Crow", 136.

century's premier tourist destinations.[30] While she recognizes the limitations set by Kentucky's Jim Crow society, Algeo narrates the agency that black hoteliers and travelers used to construct a settled place that was welcoming within a largely white space.[31] Because of these established connections, we can understand mobility in a similar light, using tourism data not only as a source to investigate black movement but also to learn more about the settled *travelscapes* this resistant mobility created, especially in urban settings that have been largely unstudied in the literature on tourism.

Places for accommodating the *resistant mobility* of black travel also allowed for the flourishing of more formal black resistance, as African-Americans used the spaces they created, under the laws of Jim Crow, to gather and strategize about how to further resist white supremacy. For instance, Inwood describes how black entertainers would use their mobility, provided through musical tours of the country located in night clubs like those listed in the *Green Book*, to exchange information between African-American communities across the Mason-Dixon Line.[32] Simultaneously, these accommodating spaces facilitated travel as the hotels, restaurants, barber shops, beauty parlors, and tourist homes established to harbor black Americans from the harshness of Jim Crow created a network that allowed for economic and cultural capital gains, a goal that was in direct violation of the purpose of the Jim Crow laws. And the *Green Book*, published by African-Americans for African-Americans, organized these locations for easy access by black travelers as they moved across America's roads.

The *Green Book*, representing this network that allowed for safe travel and resistance, was created from the geospatial work originally performed by its editor, Victor Hugo Green. Initially developed through his contacts with the U.S. Postal Service, who detailed what businesses in New York City accommodated African-Americans, this network expanded as the scope of the guide grew to include the rest of the United States (and eventually several international destinations as well). This expanded network used the readers themselves as agents of data collection, as they would read the *Green Book*, see a call for new listings, and write to the offices of Victor Green to detail new businesses they discovered were willing and able to facilitate the safe travel of African-Americans. Green also hired agents during this period who would write to different cities to locate new listings as well. These examples provide evidence that African-Americans not only conducted antiracism mobility work, as previously described, but performed valuable geospatial work as well. The data collected by Green's agents and the users of the guide were spatial in nature, as they contained locational information stored as addresses. As such, the work performed to collect and organize the spatial information can be viewed as geospatial work and a form of spatial counter-mapping, as argued by Alderman and Inwood.[33] Through a combined lens of black geographies and critical GIS, we can see how this geographic data, used to resist and survive in spaces constructed through hegemonic discourses of white supremacy, can be visualized to further understand spaces of black mobility and the wider socio-spatial production of the Jim Crow South.

4.4 Reading New Orleans Through the *Green Book*

An example of how the text and spatial data found in the *Green Book* can be read to map the network of safe spaces and places that facilitated black travel and resistance follows using the city of New Orleans, Louisiana as a case study. New Orleans was chosen as a representative city because of its long history as a destination for tourism as well as its racialized past before and

[30]Algeo, K. 2013. Underground Tourists/Tourists Underground: African American Tourism to Mammoth Cave. *Tourism Geographies* 15:3, 380–404.

[31]Ibid.

[32]Inwood, J. 2014. The Great Migration. In *North American Odyssey: Historical Geographies for the Twenty-first Century*, eds. C.E. Colten and G.L. Buckley. Lanham: Rowman & Littlefield, 103–115.

[33]Alderman and Inwood, *Civil Rights as Geospatial Work*.

during the Jim Crow era.[34] With the *Big Easy's* population during the peak of institutionalized discrimination only about a third African-American, white supremacy easily shaped the landscape and society of the city. The tourism industry was especially affected by this dynamic, since "in Jim Crow New Orleans, tourism sites and guides were developed to either erase traces of black culture or present blacks as subservient to whites."[35] The location of black touristic spaces, in terms of hotels, restaurants, and other businesses that catered to travelers, were separated socially and spatially as "white civic leaders structured the cityscape to reflect their vision of the past and to reinforce their [white supremacist] values in the present."[36] Through this racialized construction of space, New Orleans, as a place, and its associated tourism landscapes were developed to support white supremacy and maintain the immobility and subjugation of black Americans that originally began with the enslavement of Africans.[37]

However, black residents and travelers in New Orleans were not without agency against this institutionalized racism. The largest slave revolt in American history occurred on the plantations of the German Coast in 1811, when near 500 enslaved individuals marched on New Orleans until they were brutally suppressed and executed.[38] After abolition, when Jim Crow laws were enacted in the city to preserve the racial status quo, African-Americans actively protested the segregation of public street cars by boarding those set aside for whites and refusing to leave until physically removed or arrested.[39] Black New Orleanians used their culture as well to resist white supremacy, with the birth of jazz leading to the rise of popular night clubs that contributed to the growth of a black middle class.[40] Despite the imposition of restrictions on black spaces, movement, and other aspects of social life, African-Americans in New Orleans were able to survive, resist, and prosper within the Crescent City.

With the introduction of automobiles, and the adaptation of Jim Crow laws to segregate facilities associated these vehicles, black residents of, and travelers to, New Orleans worked to once more subvert white supremacy. Included in this antiracism mobility work is the *Green Book*, which features New Orleans-based businesses that welcomed black patrons. Both African-American agents working for Victor Hugo Green and black travelers patronizing these businesses contributed to the guide, as demonstrated by calls for volunteered information such as this section from the introduction to the 1938 edition: "There are thousands of places that the public doesn't know about and aren't listed. Perhaps you might know of some? If so send in their names and addresses and the kind of business, so that we might pass it along to the rest of your fellow Motorists."[41] Each subsequent publishing of the *Green Book* included such solicitations, and later editions even asked for assistance in ensuring spatial data quality, especially in regard to the creation of safe spaces and black counter-public spaces: "If in applying for accommodations you are refused, kindly notify us about same, giving us the reasons, we shall contact this particular place and remove their listing."[42] Through these actions that disrupted

[34]Stanonis, A.J. 2006. *Creating the Big Easy: New Orleans and the Emergence of Modern Tourism, 1918–1945*. Athens: University of Georgia Press; Gotham, K.F. 2007. *Authentic New Orleans: Tourism, Culture, and Race in the Big Easy*. New York: New York University Press.

[35]Stanonis, *Creating the Big Easy*, 196.

[36]Ibid., 213.

[37]Campanella, R. 2017. *Cityscapes of New Orleans*. Baton Rouge: Louisiana State University Press, 23–31.

[38]Paquette, R.L. 2009. "A Horde of Brigands?" The Great Louisiana Slave Revolt of 1811 Reconsidered. *Historical Reflections* 35:1, 72–96.

[39]Blassingame, J.W. 1973. *Black New Orleans, 1860-1880*. Chicago: The University of Chicago Press, 189–190.

[40]Inwood, The Great Migration, 111; Stanonis, *Creating the Big Easy*, 198–212.

[41]Green, V.H., ed. 1938. *The Negro Motorist Green Book, 1938 Edition*. New York: Victor H. Green & Co, Introduction.

[42]Green, V.H., ed. 1954. *The Negro Traveler's Green Book: 1954*. New York: Victor H. Green & Co, 5.

the established knowledge production industry, founded on white supremacy, African-Americans, including those in New Orleans, developed a form of geospatial intelligence that was parlayed into a black geographic knowledge which allowed for black travel and resistance.

In publications spanning from 1938 to 1966 (all but the first edition), 735 listings were printed in the *Green Book* for New Orleans, divided between 80 different businesses. Locations represented in the guidebook for the city include those regularly associated with the travel industry, like hotels, tourist homes, and gas stations, as well as other businesses that served the wider black community, such as beauty schools, drug stores, and barber shops. This finding is characteristic of other large cities featured within the *Green Book*, revealing the wider implications of the guide's spatial data to reveal the abundant agency and success that African-Americans had in resisting white supremacy. By far, the most popular type of business listed in the *Green Book* for New Orleans was hotels, which comprised 287 of the total listings; restaurants were the second largest category, with 140 entries counted. The fact that hotels were the most listed type of business exposes several facets of black life and travel during the Jim Crow era. First, and probably most reasonably, travelers on long vacations need somewhere to stay; African-American travelers to New Orleans were no different, and because hotels were in ready supply, we can infer that black travelers traveled both long distances and long durations to the *Big Easy*, similarly to whites. Second, as a result of the long trips taken to New Orleans, African-Americans not familiar with the city needed safe spaces to sleep and relax, and these hotels provided escape from the harsh realities of legalized discrimination. *Green Book* listed restaurants, gas stations, and taxicabs provided similar relief, but due to the vulnerable and residential nature of hotels, their importance in facilitating black resistance and resilience shines through.

Because addresses are stored in the *Green Book* from the time of its contemporary publication, it is possible to use modern technology to visualize the counter-public spaces originally mapped and presented by this representation of black geographies. To showcase the network of these spaces, the 735 addresses listed for the city of New Orleans were geocoded and loaded into a modern computer-based geographic information system. Information such as the year the entry was listed, the type of business, the business's name, and notes about the entry were recorded along with the address. To further visualize how these locations represented safe spaces for African-Americans, as well as physical manifestations of white supremacist segregation tactics, the geocoded addresses were overlaid onto historical census data displaying black population levels within census tracts. For the purposes of this study, two maps were created at separate time periods (1940 and 1960) to easily display the changes experienced in black travel to New Orleans.

Figure 4.2 depicts a map of establishments listed in the 1940 edition of the *Green Book* compared to the percentage of people of color residing in census tracts of Orleans Parish. It is easy to see the clear segregation that existed in the spaces of white and black residents. We can interpret from the placement of *Green Book* sites that the travel industry was highly segregated as well. All but two businesses listed in the 1940 guidebook are located in tracts that are inhabited by a majority of people of color. Several locations are right on the edge of these majority–minority tracts, but nevertheless are segregated from white spaces. Despite this visible segregation, we can also see that a concentration of businesses is seen in the same area, in the center of the map and along a north–south axis to the east of the cluster. This clustering represents the development of a tourism industry for African-Americans by African-Americans created by businesses that catered to different aspects of the tourist experience. The visualization of such a concentrated district reveals the resilience and prosperity of black New Orleans residents in the face of a harsh Jim Crow reality.

Geocoded entries from the 1960 edition of the *Green Book* are shown in Fig. 4.3. Similarly to the 1940 map, a large number of the 1960 *Green Book*-listed businesses are located in majority-people of color census tracts. However,

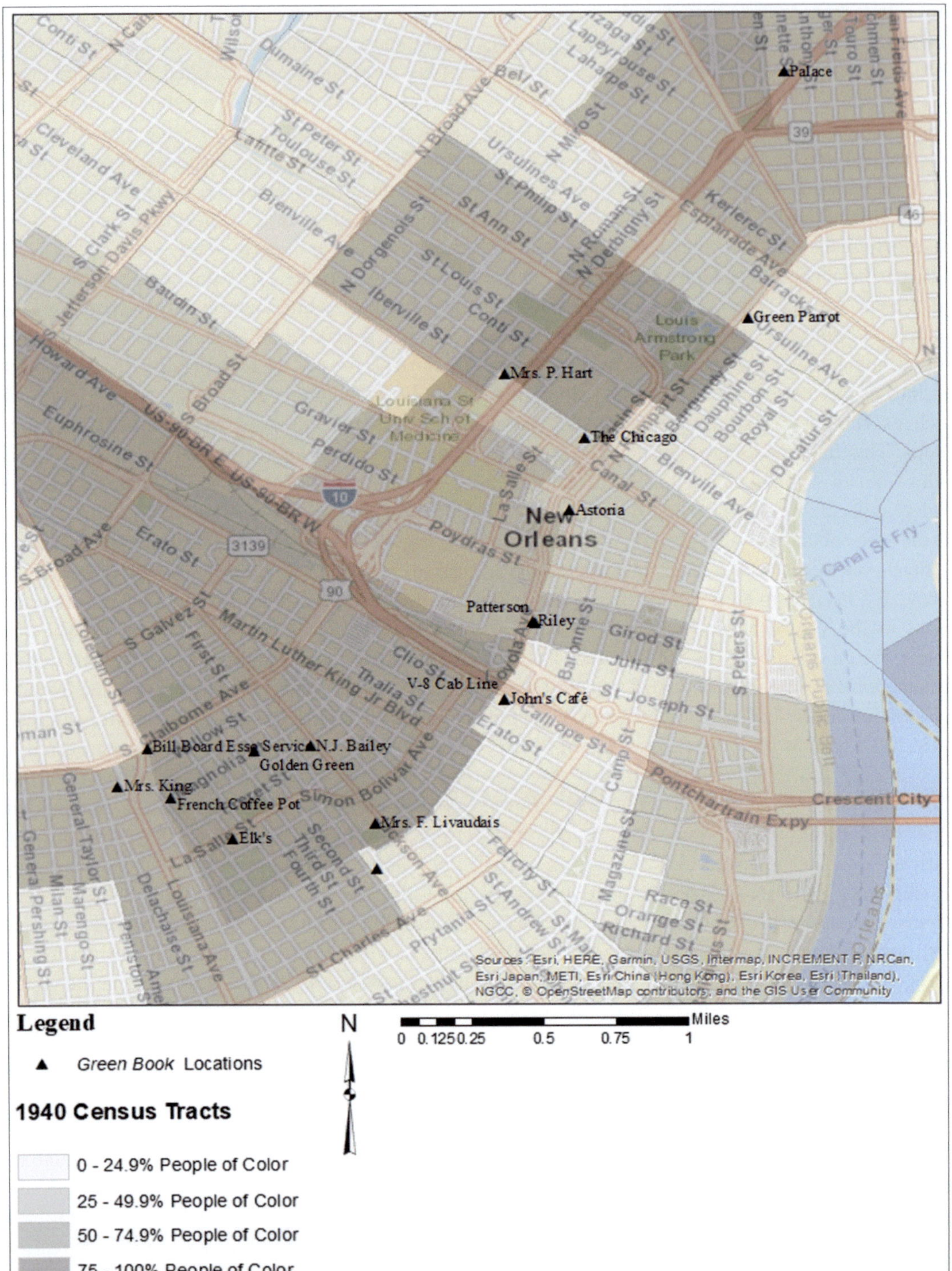

Fig. 4.2 Locations of New Orleans businesses listed in the 1940 edition of the *Green Book*. Map by author

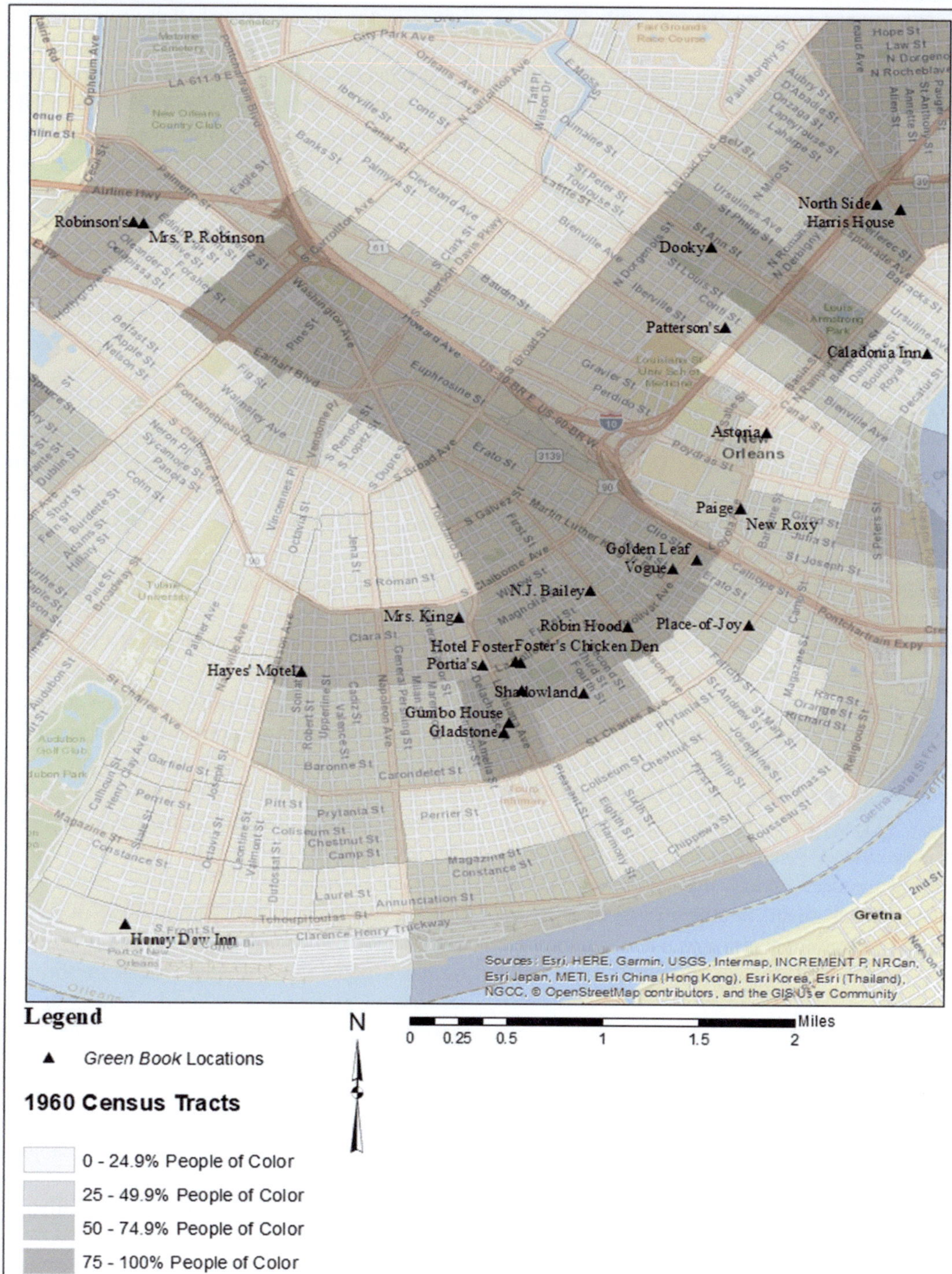

Fig. 4.3 Map of New Orleans businesses listed in the 1960 edition of the *Green Book*. Map by author

differences exist between the two editions, as in 1960 there were three businesses sited in majority-white areas. These three points represent a restaurant, the Honey Dew Inn, and two hotels, the Astoria and the Caladonia Inn. The appearance of these establishments deep in white census tracts provides insight into the changing geography of New Orleans travel, revealing a possible liberalization of racial attitudes by some white residents. It is not known at this time if these three businesses were white-owned and accepted black travelers or were African-American owned and moved into areas with more accepting white neighbors. Either way, this development demonstrates changing attitudes that led to new geographies of black spaces in New Orleans and the United States. As the civil rights movement advanced, helped along by the mobility afforded through the *Green Book* and other antiracism mobility work, urban geographies changed with increasing integration. The citizen-driven nature of the *Green Book*'s spatial data added to this integration, since black patrons potentially submitted these three addresses, thus traveling into white neighborhoods, or saw the addresses in the guide and later visited these locations. Despite this potential liberalization, it appears that the tourism industry and the black community of New Orleans as a whole remained largely segregated in the 20 years between publications, thus revealing the deep divides that characterized (and continues to do so in) Southern cities affected by legalized and informal types of white supremacy.

4.5 Conclusion

Through this chapter, I assert that scholars can learn much about the changing black geographies of travel and urban landscapes by reading the textual and spatial data of the *Green Book*. By understanding the guidebook through a framework of black geographies and critical GIS, we can gain a more nuanced understanding of the resistance and resilience that African-Americans displayed in the face of the great adversity thrust upon them by the white leaders in the nineteenth and twentieth century. In the case of New Orleans specifically, a reading of the *Green Book*'s volunteered spatial data revealed the presence of a thriving and concentrated travel-related industry that existed separately from the white-led tourism center focused on the French Quarter. When the data is compared temporally, patterns in racialized travel and settlement can be seen, as in the appearance of several black-accommodating businesses in majority-white areas of New Orleans in the 1960 edition of the guide. Even with this apparent liberalization, segregation still dominated the geographies of black life in the *Big Easy*. Through a critical GIS lens though, we can understand that this segregation did not hamper the life and travel of many African-Americans, as they were able to create their own knowledges of spaces and places, such as the *Green Book*, that subverted white supremacy and its control of many geospatial intelligences. Furthermore, the work of this chapter has shown how positivist mapping of race can be transcended by the recognition and engagement of black geographies. The *Green Book* consists of information entirely created by African-Americans in resistance to the hegemonic construction of space by white supremacists. As such, it "brings into focus networks and relations of power, resistance, histories, and the everyday, rather than locations that are simply subjugated, perpetually ghettoized, or ungeographic."[43] This chapter's reading and mapping of the *Green Book* illustrates that despite the spatial segregation imposed on the black population of New Orleans, a thriving network of *resistant mobility* developed in *and* outside segregationist spaces, revealing the wider agency that black travelers and entrepreneurs had during the Jim Crow era. Instead of being hemmed in and "ghettoized" by the spatial construction of racialized landscapes, this mapping of the *Green Book* reveals the counter-production of space by African-Americans that facilitated safe travel and business. An epistemological combination of black geographies and critical GIS helps us to not map race *in* place, but to contribute to the spatial liberation of black

[43]McKittrick and Woods, "No One Knows the Mysteries", 7.

Americans and recognize the place-making abilities of marginalized populations.

While this chapter provides a short example of the types of research that can be done by geographically reading the *Green Book*, future work remains. For instance, examining how landscapes associated with black travel have changed since the time of the *Green Book* can reveal how urban spaces have been transformed by processes such as urban renewal and gentrification, not captured by the guidebook. Comparing not only addresses between the publication period and present day, but also street names, business names, and the photos of advertisements could lead to much insight about the landscapes of changing black geographies. Comparisons between cities during the same period of time as well could lead to interesting discoveries about the local dimensions of segregation, racial liberalization, and black resistance. These studies all make use of the volunteered geographic information found within the pages of the *Green Book*, spatial data and knowledges contributed by those seeking to create a better life for themselves in defiance of an institutionalized system of hatred and discrimination. Using this data keeps African-Americans voices heard and recognizes their contributions to the place-making, resistance, and a resilient culture.

References

Alderman DH, Inwood J (2016a) Mobility as antiracism work: the "Hard Driving" of NASCAR's wendell scott. Ann Am Assoc Geogr 106(3):597–611

Alderman DH, Inwood J (2016b) Civil rights as geospatial work: rethinking African-American resistance. In: Frazier JW, Tettey-Fio EL, Henry NF (eds) Race, ethnicity, and place in a changing America, 3rd edn. SUNY Press, Albany, pp 177–185

Alexander M (2010) The new Jim crow: mass incarceration in the age of colorblindness. The New Press, New York

Algeo K (2013) Underground tourists/tourists underground: African American tourism to mammoth cave. Tour Geogr 15(3):380–404

Allen D, Lawhon M, Pierce J (2018) Placing race: on the resonance of place with black geographies. Prog Hum Geogr 2

Arsenault R (2006) Freedom riders: 1961 and the struggle for racial justice. Oxford University Press, Oxford

Blassingame JW (1973) Black New Orleans, 1860–1880. The University of Chicago Press, Chicago, pp 189–190

Block JP, Scribner RA, DeSalvo KB (2004) Fast food, race/ethnicity, and income: a geographic analysis. Am J Preventat Med 27(3):211–217

Campanella R (2017) Cityscapes of New Orleans. Louisiana State University Press, Baton Rouge, pp 23–31

Elwood S (2008) Volunteered geographic information: future research directions motivated by critical, participatory, and feminist GIS. GeoJournal 72(3/4):177

Foster MS (1999) In the face of "Jim Crow": Prosperous blacks and vacations, travel and outdoor leisure, 1890–1945. J Negro Hist 84:130

Franz K (2004) 'The open road': Automobility and racial uplift in the interwar years. In: Sinclair B (ed) Technology and the African-American experience: Needs and opportunities for study. The MIT Press, Cambridge, pp 131–154

Gilbert MR, Masucci M (2006) The implications of including women's daily lives in a feminist GIScience. Trans GIS 10(5):751–761

Gotham KF (2007) Authentic New Orleans: Tourism, culture, and race in the big easy. New York University Press, New York

Green VH (ed) (1938) The negro motorist green book, 1938. Green & Co, New York

Green VH (ed) (1954) The negro traveler's green book. McKittrick and Woods, No One Knows the Mysteries, New York

Hall D (2016) Episode 201: The green book. 99% invisible [audio podcast]. Retrieved https://99percentinvisible.org

Hall MR (2014) The negro traveller's guide to a jim crow south: negotiating racialized landscapes during a dark period in the united states cultural history, 1936–1967. Postcolon Stud 17(3):307–319

Inwood JFJ (2011) Constructing African American urban space in Atlanta Georgia. Geogr Rev 101(2):147–163

Inwood J (2014). The great migration. In: Colten CE, Buckley GL (eds) North American odyssey: Historical geographies for the twenty-first century, Rowman & Littlefield, Lanham, pp 103–115

Kwan M-P (2002) Is GIS for women? reflections on the critical discourse in the 1990s. Gend Place Cult 9 (3):271–279

Lefebvre H (1991) The production of space. Blackwell, Malden

Loewen JW (2005) Sundown towns: a hidden dimension of American racism. The New Press, New York

McKittrick K, Woods C (eds) (2007) No one knows the mysteries at the bottom of the ocean. In: Black geographies and the politics of place. Between the Lines, Toronto, p 6

McKittrick K (2007) Demonic grounds: black women and the cartographies of struggle. University of Minnesota Press, Minneapolis

McKnight GD (1998) The last crusade: Martin Luther King, Jr., the FBI, and the 1968 poor people's campaign. Westview Press, Boulder

Mitchell JT, Collins L (2014) The green book: "safe spaces" from place to place. Geogr Teach 11(1):29–36

Pacheco D, Velez VN (2009) Maps, mapmaking, and critical pedagogy: exploring GIS and maps as a teaching tool for social change. Seattle J Soc Justice 8 (1):273–302

Paquette RL (2009) A horde of brigands? The great louisiana slave revolt of 1811 reconsidered. Hist Reflect 35(1):72–96

Pavlovskaya M (2006) Theorizing with GIS: a tool for critical geographies? Environ Plan A 38:2003–2020

Schuurman N (1999) Critical GIS: theorizing an emerging science. Cartographica 36:1–108

Sharma M (2017) Income divide and race/ethnicity in tennessee metropolises. Int J Geospatial Environ Res 4 (1):1–23

Seiler C (2006) So that we as a race might have something authentic to travel by: African American automobility and cold-war liberalism. Am Q 58(4):1091–1117

Stanonis AJ (2006) Creating the big easy: New Orleans and the emergence of modern tourism. University of Georgia Press, Athens, pp 1918–1945

Staples B (2019) The green book's black history: lessons from the jim crow-era travel guide for African-American elites. The New York Times. Retrieved https://www.nytimes.com/2019/01/25/opinion/green-book-black-travel.html

Tate WF IV, Hogrebe M (2011) From visuals to vision: using gis to inform civic dialogue about African American males. Race Ethn Educ 14(1):51–71

The New York Public Library Digital Collections. https://digitalcollections.nypl.org/collections/the-green-book#/?tab=about&scroll=3

Wilkerson I (2010) The warmth of other suns: the epic story of america's great migration. Vintage Books, New York

Part II

Cultures, Networks and Mobilities

Queer Cartographies: Urban Redevelopment and the Changing Sexual Geography of Postwar San Francisco

5

Damon Scott

Abstract

This chapter reflects on methodological and historiographic insights gained using GIS to reconstruct the impact of urban renewal on "hangouts for homosexuals" on the San Francisco waterfront during the 1950s and early 1960s. These stigmatized places—which entered the popular-spatial imaginary through contemporaneous local press coverage of bar raids—were initially documented as historical sites by LGBTQ community archivists who culled basic information about the physical location, approximate years of operation, and business type from oral history transcripts, newspaper clippings, early gay bar directories, and nightlife columns in the city's first gay periodicals dating to the early 1960s. In a collaborative project with the GLBT Historical Society of Northern California, the author mapped these and similar sites in order to "sketch out" temporal and spatial changes in the sexual geography of the city. The pragmatic and technical challenges of mapping the waterfront sites, in particular, drew the author deep into the urban planning and redevelopment archives of the city in a search for basic information about the physical landscape prior to the transformation of the area from a low-rise, mixed-use, wholesaling, and maritime trade-oriented district into a freeway-oriented gateway to an expanding downtown financial district. Charting this transformation demonstrates the productive possibilities of triangulating a historical-geographic fix on the queer past from personal recollections, contemporaneous published sources, and historical urban planning documents in pursuit of a new spatial imaginary of "gay community" formation. Informed by—but ultimately not constrained by poststructural critiques of GIScience—this chapter illustrates how the representational artifices and associative power of GIS can serve as instruments for unraveling teleological narratives that conflate urban development and modern LGBTQ subjectivities.

Keywords

LGBTQ history · Urban redevelopment · Cadastral GIS platforms · San Francisco waterfront · Gay and lesbian bars · Critical GIS · Queer GIS · Spatial citizenship · Homonationalism

5.1 Introduction

The on-line GIS platform for property information management (PIM) of the San Francisco Planning Department includes a map layer of

D. Scott (✉)
Departments of Geography and Global & Intercultural Studies, Miami University, Oxford, OH, USA
e-mail: scottd2@miamiOH.edu

C. Travis et al. (eds.), *Historical Geography, GIScience and Textual Analysis*, Historical Geography and Geosciences, https://doi.org/10.1007/978-3-030-37569-0_5

rainbow-shaded parcels called out for their "association with LGBTQ history." The PIM—used principally to manage and share publically all tax, permitting, and zoning decisions linked to each individual parcel of real property within the City and County of San Francisco—flags sites of LGBTQ historical interest documented by public historian Donna Graves and architectural historian Shayne Watson as part of a city-wide historical preservation and landmarking initiative (Graves and Watson 2016). The map layer includes all sites documented in their encyclopedic theme study "Citywide Historic Context Statement for LGBTQ History in San Francisco" with discrete addresses and is annotated with short descriptions and page references linked to the study. Near the central waterfront of the city, three different parcels—the former Army and Navy YMCA on the Embarcadero, Sydney G. Walton Square, and the ten-story Southern Pacific Building at the foot of Market Street—are all shaded rainbow-bright to call out their historical association with San Francisco's LGBTQ past (Fig. 5.1). The inclusion of these sites in this on-line cadastral GIS platform is a study on how contemporary urban planners are beginning to at least representationally acknowledge the part queer people have played in the city's past and provide a model for a queerly informed spatial citizenship (De Luca et al. 2015).

The waterfront YMCA—like other branches of the philanthropic organization across the country that offered short-term lodging, athletic, and social facilities to single, traveling men (Gustav-Wrathall 1998)—was a reliable destination for men seeking to secure sex with other men during the first half of the twentieth century. The second site in the inset, Sydney G. Walton Square is named for a local banker who helped shepherded the area through the federal urban renewal process as a member of the San Francisco Redevelopment Agency and occupies an entire city block in the heart of the Golden Gateway development. On what is now the southeast corner of this privately owned greenspace, Charlotte Coleman operated a lesbian bar, "The Front," from 1958 until the underlying land was acquired, cleared, and then sold to private developers by the city in 1962 (Strait 1962b; Boyd 2003). In grossly exaggerating the footprint of Coleman's bar by swathing the entire city block in prismatic hues, the PIM layer overcompensates for the problematic, long-standing practice of urban planners marginalizing queer subjects (Sandercock 1998; Frisch 2002). This queer case of *rainbow-spreading*, however, is not the outcome of an insurgent, queer mapping intervention. It is the outcome of a geo-coding algorithm that used a defunct street address to interpolate a bygone lesbian bar in the core of a superblock development that replaced not only the building that housed that bar but also all structures on that block as well as roughly 20 additional, adjacent blocks.

Like the outsized footprint of Coleman's bar, the C-shaped building at the base of Market Street is also poorly registered as the site of a former queer bar on the PIM map layer. The Southern Pacific Building, built in 1917 as the corporate headquarters of the western rail company, never housed the Ensign Club, a waterfront "place where 'sexual perverts were permitted regularly to congregate'" (Liquor Licenses of 2 S. F. Spots Are Threatened 1959). The class-A office building did, however, adopt the Ensign's high-profile address—"1 Market"—after the block of low-rise waterfront buildings in which the waterfront bar complex was located was razed in 1965. To accurately represent the historic footprint of the Ensign Club in prismatic hues, the San Francisco Planning Department first would need to incorporate a preurban redevelopment map of the foot of Market Street outlining the city lots acquired by eminent domain to construct a park-like buffer between the Embarcadero freeway and a modern office building and hotel complex. Razed in the mid-1960s to make way for the eastern portion of what became Justin Herman Plaza, the Ensign Club was an underground sex club that operated in the basement below the Ensign Café bar-restaurant and a second-floor cocktail lounge on a prominent corner building in a bygone low-rise block of commercial establishments at the corner of Market and the Embarcadero.

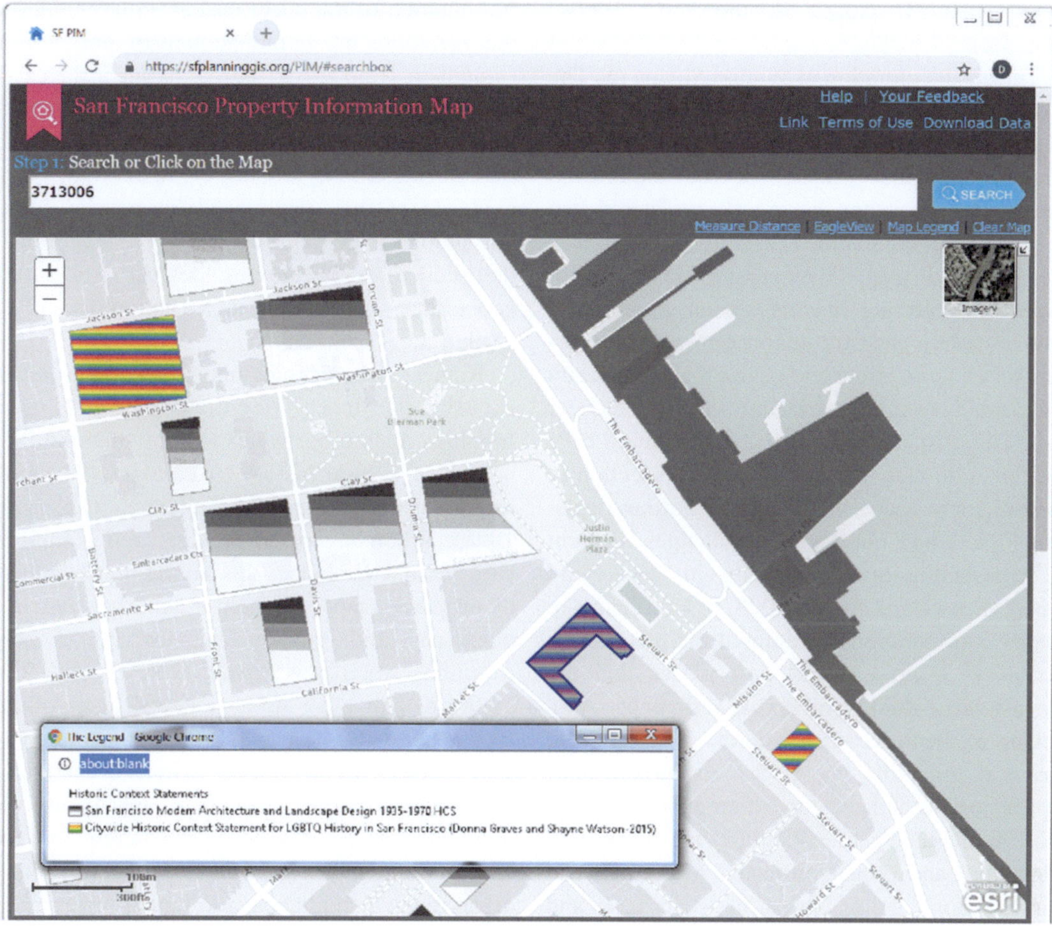

Fig. 5.1 A map view of the foot of Market Street area on the San Francisco Planning Department's Property Information Map (PIM) interface with the "Preservation" map layer active and the South Pacific Building selected (Accessed November 17, 2018)

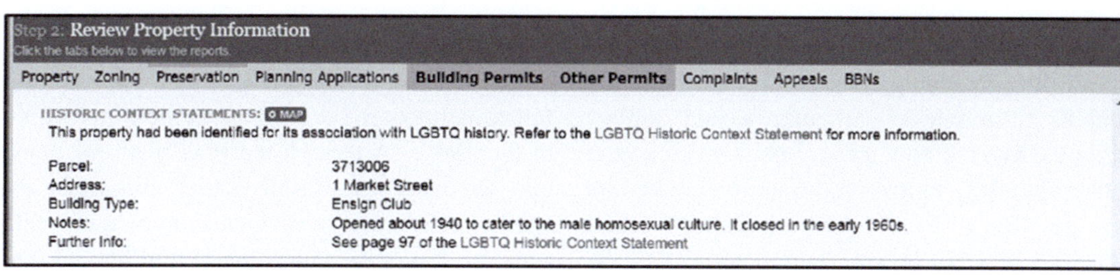

Fig. 5.2 Annotation associated with "Ensign Club" on the San Francisco Planning Department's Property Information Map (Accessed November 17, 2018)

The waterfront block had been situated at the principal gateway to the city and benefited from a steady stream of ferry commuters, new arrivals, and maritime workers before improved motorized and rail access to the city turned the area into a marginal zone. The Planning Department's PIM platform includes scant attribute information about the Ensign Café, noting simply that "It closed in the early 1960s" (Fig. 5.2).

The queer coloration of Walton Square (as The Front) and the Southern Pacific building (as the Ensign Club) might be dismissed as minor mapping errors—the output of an automated geo-coding algorithm that poorly registered two anachronistic commercial establishments on a modern parcel map. The anachronisms and slippages of projecting bygone sexual geographies onto a modern cadastral map of property ownership raises more issues, however, than a correctly registered map could ever resolve about the queer past. The misplaced, prismatic map symbol calls out for more than a simple address correction in the geo-spatial dataset—it raises "profound epistemological and ontological challenges inherent in representing queer identities in space [and time]" (Brown and Knopp 2008, p. 43). There are deeper considerations here about the incommensurability of present and past places and the *longue durée* fluidity of sexual difference. As a linking element, the street address is an unstable means of orienting oneself in past urban landscapes just as indices for differentiating sexual subjectivities and gender identities are historically contingent and, when projected backward, can draw tenuous equivalences between past and present socio-sexual domains of experience (Domosh 1997; Ahmed 2006).

Following DeLaurentis' discursive exercise of queer map "reading" (2018), deconstructing the elisions and slippages embedded in the PIM's spatial representations of gay spaces opens new potentials for situating the power of cartographic symbols to privilege particular sexual subject positions. The rainbow map symbol—a clear nod to the pride flag, an invented tradition created by the "gay Betsy Ross", Gilbert Baker in San Francisco in 1978 (Haag 2017)—projects post-gay liberation social identities onto a queer past and lays claim to a progressive narrative of the modern global expansion of LGBTQ civil rights with San Francisco as the point of origin. A postcolonial *queer of color* map critique of the PIM platform might interpret this spectral illumination of the queer past as a placeholder in the semiotics of homonationalism and homocapitalism (Rao 2014, 2015), as an instance of grounding an emergent global teleological homonormativity in a particular neoliberal formation of LGBTQ urbanity (Oswin 2015), or as evidence of the urgent need for a "new queer idioms of the geopolitical (p. 166)" that "refuse the seductions of homing devices that suture geopolitics to…region, area, nation (p. 159)" (Anjali and Geeta 2016).

Rather than deconstructing the map layer, exposing its silences, or critiquing its ideological framings, the author wants to show that the PIM is part of a long history of cartographic representations of urban landholdings (Cosgrove 1998). When registered together and combined with textual evidence, this cartographic archive of property maps can help reach back into the queer past and provide a window into a more extensive waterfront "queer world." The PIM platform, like older urban renewal maps of the area generated as a part of the administration of state eminent domain power to transfer land from one private owner to another, are themselves imbricated in the ongoing queer heritage of the city. Both the PIM map layer and mid-century urban renewal land acquisition maps have served as municipal instruments for managing and growing real property valuations and mitigating urban blight. Both have also perpetuated silences about the role of a state-directed redevelopment scheme in profoundly reshaping the sexual geography of the waterfront with the wholesale transformation of the physical landscape and the functional patterns of land use at the water's edge in the 1960s. Although the PIM platform draws into the map sexual and gender minorities and symbolically celebrates inclusionary queer citizenship, it is important to point out that the map layer is nestled under the "Preservation" tab—which summarizes the protective status of each building in the city based primarily on its architectural merit, rather than its social significance. In the context of the city's stratospheric real estate values, this "preservation" framing of

LGBTQ sites is less a celebration of the city's queer past than an inventory of places at risk of heteronormative adaptive reuse or redevelopment (Cramer 2014; Lucchesi 2015; Brass 2016; Perhan 2017). In San Francisco, as in many other cities, LGBTQ neighborhoods and commercial establishment are becoming less vital socially (Doan and Higgins 2011; Ghaziani 2015). These changes are in part due to new rounds of capital investment in the districts where primarily white, property-seeking gay "urban pioneers" once staked a claim to undervalued urban lands along the rent gap "frontier" (Knopp 1992; Smith and DeFilippis 1999).

While this study resonates with Tolia-Kelly's call for a valuing "otherworldly" cultural geographic research beyond "the usual moral geographies of the discipline" (2010, p. 358), it also shares an affinity with recent work that uses GIScientific tools to pose historical-geographic questions about the queer past. A growing body of literature on historical urban geographies of sexualities have taken up Matt Houlbrook's call (2001) to "situate geographies of sexual encounter, identity and experience within wider urban social relations (p. 498)" and to relate changing discursive and material geographies of the city to "the historical production of categories of sexual deviance (p. 501)." Notable examples are two recent special journals—one on the historical geography of moral regulation and another that investigates the absences and possibilities of works that constitute a "Historical Geographies of Sexualities?" (Legg and Brown 2013; Podmore and Brown 2015). Many of these works draw on close readings of archival sources generated by and about queer subjects to "place" sexuality and gender at the center of revised spatialized histories of social life, regulatory control, and urban development. Since the "spatial turn" in history, however, few geographers have produced cartographic representations of the past in order to theorize the historical-geographic production of sexual subjectivities or to chart the historical and chorographic contingencies of sexuality-inflected social relations (for exceptions, see Podmore 2006, 2013, 2015). Recent work that deploys GIScientific methods to generate new geo-visualizations and empirically based spatial histories of past queer worlds positions those outcomes within a self-reflective, processual "queer" critique of GIScience (Brown and Knopp 2008; Cockayne and Richardson 2017; Gieseking 2017; Elwood and Leszczynski 2018; Gieseking 2018). The recent growth of what Wilson (2015) has indexed as "Queer GIS" is one of multiple subfields that have emerged out of critical GIS. This intellectual genealogy is useful in tracing perhaps the beginnings of a geo-spatial turn away from human geography's critique of the epistemological failings of ontological reasoning in GIScience and a willingness to critically engage with the neogeography of "big data" by directly intervening in the ways it is used materially as an "ontic device" (Leszczynski 2009).

5.2 A Shared Desire to Map San Francisco's Queer Past

Informed by critical GIS's processual approach to geo-spatial research and inspired by the self-reflexive "desire for history" of Allan Bérubé (2011), this section situates the Planning Department's map error within a longer history of efforts—including my own—to cartographically index and interpret the queer past. There is a long history of mapping LGBTQ + historical sites in San Francisco that predates the "Citywide LGBTQ Historic Context Statement" (cf. Castells 1983; Godfrey 1988; Rubin 2000; Roque Ramirez 2001; Boyd 2003; Scott 2008, 2015).[1] In the early 1980s, a group of gay men and lesbians began to systematically collect, document, and interpret the queer past of San Francisco Bay

[1]In addition to these published examples, an unknown number of additional, equally well-researched maps exist in LGBTQ community archival collections. For example, see the undated "Partial List of Sites in the Mid-Market Street Area of Historical Interest to Gay/Lesbian/Bisexual/Transgender Communities" created by the Center for the History of Sexual Diversity and stored at the Gay and Lesbian Historical Society of Northern California in San Francisco.

Area and, in 1985, co-founded a community-based archive and public history organization. Still very actively engaged in queer history projects over 30 years later, the Gay, Lesbian, Bisexual and Transgender Historical Society of Northern California (GLBTHSNC) maintains an expansive collection of periodicals, organizational records, oral histories, visual media, ephemera, and material culture that documents the LGBTQ life and history in the San Francisco Bay Area. The Society has also taken a leading role in initiatives to ensure that San Francisco remains a beacon for sexual and gender minorities through cultural heritage programming, planning, and advocacy.

Several of the original founders of the organization, while collecting and organizing complete sets of serials, guidebooks, and ephemera, began also compiling a list of gay bars, restaurants, bathhouses, and businesses in San Francisco and in the Bay Area. The compilation of LGBTQ sites assembled by Len Evans and Eric Garber and managed by Willie Walker consisted of a set of word processing files each containing a single table and used by Garber and Walker to organize and update entries by city or business type.[2] Each table had columns for the names, address, dates of operation, and source notes. Some entries had multiple names as businesses changed identity or ownership. In some cases, businesses moved and the table contained multiple entries to keep track of the different addresses. The dates of operation were incomplete and imprecise—owning to a complex set of challenges in locating queer social spaces in time and space. Some of the entries in the database were well documented in multiple, written sources produced by different authors who recorded the name, location, opening date, type of place, and narrative details about its social significance.

As an intern with the GLBT Historical Society of Northern California (GLBTHSNC), the author merged the "sites" listings compiled by Evans, Garber, and Walker into a single, searchable and user-friendly "Sites Database" which used to map nearly 1,400 extant and historic places. With the technical assistance of a public interest GIS consulting firm, the author used ArcView to transform the sites database into the first city-wide map of the city's LGBTQ past in the form of a large format, printed poster in July 2000. Based on archival, text-based references to past places initially produced by queer subjects and later cataloged by queer community historians, this poster was a forerunner of later qualitative GIS "sketch maps" constructed by projecting individual queer perceptions of space derived from interviews onto a geographical referenced base map (i.e. Boschmann and Cubbon 2014). Like the Planning Department, the author ran into the same problem of address mapping what he called "queer sites" on the modern street line network. The procedure returned a handful of errors. A significant number of rejected "queer sites" was located at or near the foot of Market Street and near the Ferry Building in the 1950s. In particular, a cluster of 1950s-era waterfront bars generated error messages by the ArcView geo-coding algorithm, which led to manually approximate their location with handpicked data points.

While several of these sites have figured prominently in the scholarship of gay community formation, little has been known until now about the historical and geographic context in which these waterfront bars operated. For the author the rejected addresses—which included one Market and the Ensign Club—marked the start of deep personal and professional investment in trying to disentangle the urban planning history and queer social history of the San Francisco waterfront by piecing together a wide variety of sources. The reason for the incongruity was the redevelopment of a large section of the waterfront and adjacent wholesaling district bring in the early 1960s. Not only did the redevelopment project dissolve older parcel boundaries but it also involved the vacation of some street segments within a 20-block urban renewal project area, known as "Embarcadero-Lower Market Urban Renewal

[2]Referred to as the "Sites Database," this annotated FileMaker database of over 3,000 entries is used as a research tool by archivists and visiting scholars at the GLBT Historical Society.

Project, Area, E-1." The technical challenges of producing a more accurate geo-visualization of queer sites on the San Francisco waterfront—as well as other areas of the city—prompted the author to seek out and assemble a much archive of the sources capable of pinpointing the timing and extent of urban landscape change in postwar San Francisco. By drawing on variety of sources that contain mappable social, spatial, and temporal details, intertextual analysis when combined with geo-visualization tools has helped to more accurately and precisely locate queer places in time and space.

5.3 On Texts and Contextualization

Prior to 1961, the best historical sources of male homosexuality on the San Francisco waterfront are oral history transcripts, press clippings of legal actions against bar operators, maritime labor union newsletters, and a crime exposés in a national men's pulp magazine. In October 1961, "community" organizing among gay bar operators, employees, and patrons produced regular printed accounts of the gay male social networks and commercial establishments in the city. These sources have been useful in expanding annotations in the "Sites Database" to include details about the changing sexual character individual places in the city. In addition, the first and last mention of specific queer sites serves as a rough approximation of the years of operation of individual places, which has helped reconstruct cartographic snapshots of queer geographies of the past.

Before the publication of "gay guidebooks" and community newspapers—which were supported by advertising revenues and produced for a growing audience of gay consumers—informal directories and address listings are circulated among queer men who shared their knowledge of places where they could make contacts with other likeminded men (Meeker 2006). Both informal and commercial gay bar directories were synchronic and explicitly locational, making them valuable authorities in fixing queer life in particular moments and to past places. The earlier privately circulating, handwritten listings that were jotted down in the back of an address book, incorporated in private correspondence, or recorded in personal diaries often include a mix of named bars, public cruising spots, and sexually permissive commercial areas. Commercial guidebooks produced with the cooperation of bar operators who sought to attract more patrons by promoting their business with an entry or at times a sponsored advertisement do not appear with regular frequency until the early 1960s.

In contrast, memoirs, oral history transcripts, and public history projects draw on personal and collective past experiences to retrospectively conjure up bygone queer life worlds. Over the past three decades, the GLBTHSNC has collected over 500 oral histories of queer people in the region, some conducted as part of book monograph projects and others as public history projects, like the "Uncles Project" and "Nephews Projects." Nearly half of the oral histories are accessible on-line as searchable PDF transcripts. Keyword searches of nearly 10,000 pages of available transcribed interviews for gay bars on or near the waterfront in the 1950 and 1960s garnered a numerous short references and a handful of more detailed descriptions of the look, feel, and operation of the place. These texts sometimes provide passing mention or thick descriptions based on personal experiences or word-of-mouth exchanges with other men. In some cases, these provide clues about physical location of past queer places—but more often offer impressionistic recollections of setting in which meaningful events occurred decades ago. Sexual subjectivities, however, are both personal and shared, making it difficult to make broader claims about sexual subcultures in the past from limited historical sources. Generalizing from individual experiences recorded in oral histories, memoirs, and diaries to trace the contours of emergent socio-sexual identities likely flattens the complexities and multiplicities of sexual expressions and relations in the past. On the other hand, hewing too narrowly to the particular risks missing the broad trends and historical patterns in how those expressions and relations have changed over time.

Urban newspaper coverage of bar raids, street sweeps, and arrests targeting "sex deviates" often include some specific, mappable information about the queer past. When they do, these texts—although embedded in moral discourses about social disorder and sexual danger—provide the best fixes on queer places in time and space. Reconstructing the sexual geography of the past—and building a narrative interpretation of geographic change over time—from these sources is challenging. When used as historical evidence, these sources mark episodic instances of public exposure and can be useful in charting alternating periods of permissiveness and intense scrutiny in the policing of sexual and gender difference. On their own, however, they do not provide much help in mapping the shifting terrain of past queer places. Do crackdowns, for example, mark the "discovery" of new—or expanding—queer presence or an ongoing effort to manage and contain queer subjects within a prescribed district? Do "drives on deviates" dislodge queer people from one part of the city targeted for "clean-up" or do they temporarily push them underground? Do they also unintentionally popularize queer commercial spots and cruising zones by making them legible to a broader public?

Beginning in the early 1960s, published gay community newspapers, newsletters, and "bar rags" constitute an archive of "invaluable intergenerational memory" (Holmes 2015). Holmes, in his methodological and historiographical essay on black gay social history, shows how when treated as "an archive of experience" gossip columns "produced a critical mass of politically and historically legible black gay subjects in Washington, and around the country, for the first time." In a similar vein, one of the most useful sources for reconstructing the shifting micro-geographies of queer life in San Francisco in the early 1960s is the "Roving Report," a gossip column in The Citizen's News (Oct 1961–Dec 1967), a gay "bar rag." Published by the League for Citizen Education to rally gay bar patrons to vote for an openly gay candidate for the San Francisco Board of Supervisors, the editor promulgated a "community" politic, imploring gay bar patrons to do "your part in the preservation of your personal liberties" (Strait 1962a). He explicitly asked readers to act as a "community of interest" in cultivating "greater mutual assistance" in the areas of "legal aid, job placement, interests in social responsibility, and the growing awareness of political potency." The camp-inflected, peripatetic narrative provides a twice-monthly rundown of openings, closings, and law enforcement actions against commercial establishments and cruising spots. Often the businesses referenced in the report are also promoted in paid advertisements in the same issue, making it an imperfect source for documenting spaces of queer sociability by privileging places that operated within an emerging political economy of relatively open, gay "community-path of recessoriented" drinking establishments.

In general, the relatively stable-built environment of streets, parks, and buildings has been helpful in reconstructing the movement of homosexual men to and through the city, as they built social networks and a new commercial geography of "gay" bars, bathhouses, and restaurants post World War II in San Francisco. For commercial establishments, location is the common key that links the textual sources together. When systematically surveyed, Polk's San Francisco Business Directories document changes in address, business name, business operator(s), and sometimes business type. They are useful in cleaning up typographical errors; narrowing down opening and closing dates; and inferring changes in the location, management, or operation of a commercial establishment. In some cases, changes from one operator follow soon after press accounts of liquor violations or police raids. This is particularly true for incorporated drinking establishments operated by a group of investors who hold the liquor license. The most complete set of San Francisco business directories—spanning the years 1850–1981—are accessible on-line as scanned, searchable PDFs (accessible as part of the San Francisco Public Library's contribution to archive.org). In the 1940s and 1950s, there are some gaps in annual coverage—due to some combined dual-year editions and some skipped years. In addition, with the 1952 edition, the publishers added a new "Street Directory" section in which all the alphabetically ordered entries are reorganized and reprinted by street name and number. This new format greatly facilitates tracking changes in operation because there are often inconsistencies in whether businesses are listed by name or business operator.

With complete and corrected business addresses in hand, the best sources of geo-spatial information to reconstruct the preurban renewal street grid were Sanborn Fire Insurance maps and land acquisition maps produced as part of San Francisco Redevelopment Agencies (SFRA's) efforts to assemble land for clearance and sale to private developers. The agency disposed of the first tract of cleared land in 1962 to Perini-San Francisco Associates for construction of the Golden Gateway Center, a complex of modern high-rise apartment towers. The final tract was sold to group of investors in 1967 based on their proposal to construct the "Embarcadero Center," a linear five-block-long office tower and hotel complex connecting the foot of Market to the financial district. In between the residential and commercial portions of the urban renewal project area, a set of new Embarcadero freeway ramps fed a massive new multi-story parking garage topped by a plaza and the Maritime Center office tower.

5.4 Toward a New Queer Spatial History of the San Francisco Waterfront

The nature, extent, duration, and social significance of sexual geography of the waterfront are not well understood. Two compelling entry points are historian Allan Bérubé's unpublished book manuscript and narrated slideshow about the multi-racial, queer-positive Marine Cooks and Steward's Union (2016) and Justin Spring's treatment of Samuel Steward's detailed records on his sexual experiences on the San Francisco waterfront in the summer of 1952 and 1953 (2010). In the popular press and in city planning reports, this area was depicted as a congested wholesale fruit and produce market. As rational land-use planning practices and modern urban design projects emerged in the post World War II period, the low-rent hotel and nighttime entertainment districts surrounding downtown San Francisco were targeted first in slum clearance and anti-vice campaigns and then as urban renewal project areas. These queer spaces were situated among and adjacent to produce wholesaling and maritime industry land uses located between the central business district and the waterfront piers at the foot of Market Street. The anachronistic errors in the PIM map layer and the textual evidence of an extensive prerenewal queer world on the waterfront point to a causal connection between urban redevelopment and the political mobilization of a new collective "gay community" at precisely the time and place when land acquisitions were most intense.

The urban landscape itself has paradoxically been both an obstacle to and an opportunity for posing and resolving more compelling questions about the nature and underlying causes of shifting spatial patterns of queer life in the city. Integrating property transactions, commercial histories, administrative filings, and building records linked to particular parcels, business operators, or addresses with the "Sites Database" has brought the historical geography of the queer waterfront to greater focus. San Francisco Redevelopment Agency's administrative records produced as part of its successful effort to transform the city blocks bracketed by the Embarcadero Freeway into a new automobile-oriented "Golden Gateway" constitute a rich archive of source material for reconstructing the queer waterfront. Ironically, documents produced to rationalize the planned reconfiguration of waterfront land uses have made it possible to map the very queer places that were razed by the functionalist reconfiguration of waterfront land uses.

The best base map for locating queer waterfront sites during the 1950s and 1960s is the land acquisition map that accompanied the formal urban redevelopment plan for Urban Renewal Project Area E-1 (San Francisco Redevelopment Agency 1959). With its prerenewal lot line, this

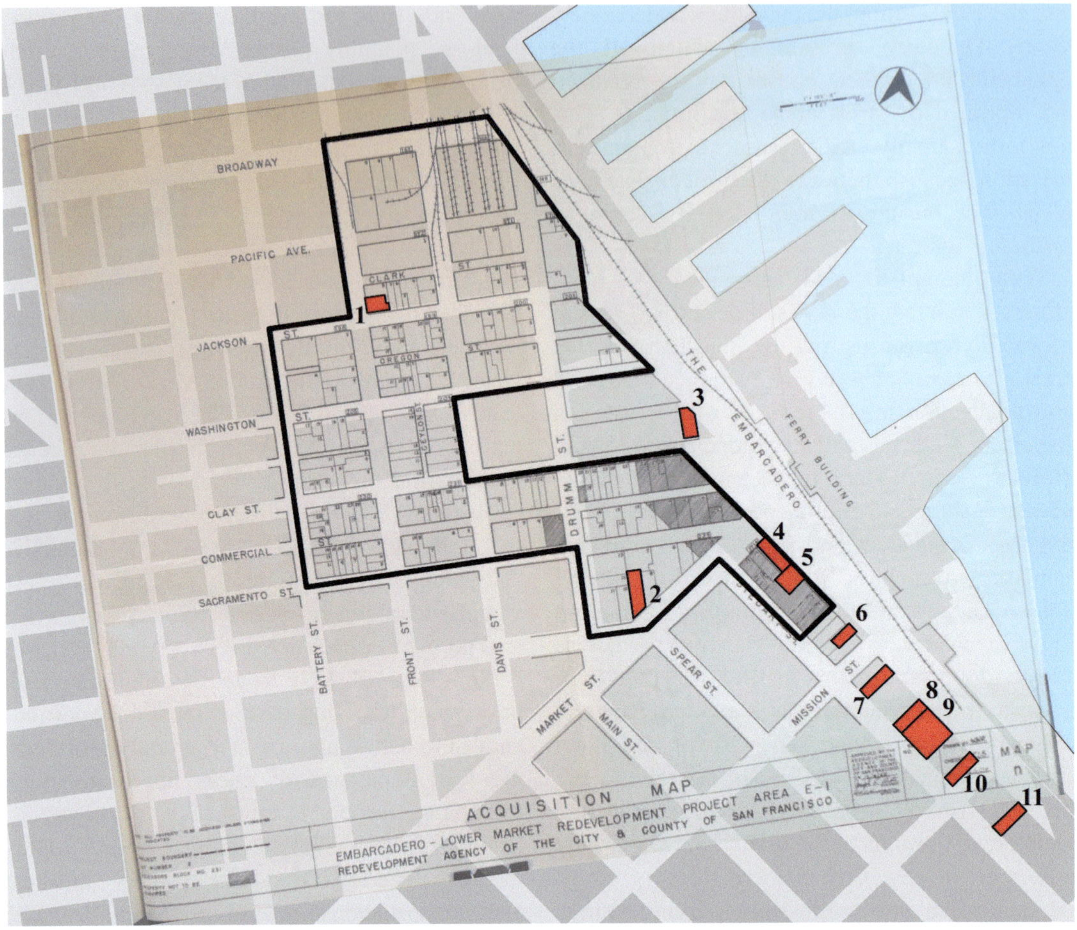

Fig. 5.3 Parcels with queer sites within and adjacent to urban renewal area project area E-1, 1950s and early 1960s. From left to right: (1) The Front; (2) Castaways, Talk of the Town and Last Resort at 90 Market; (3) Jack's Waterfront Hangout (first location); (4) Ensign Café and Broken Drum; (5) Lennie's; (6) Denny's Barrel House and Headhunters; (7) Sea Cow Café and Crossroads; (8) Chilli's 144 Club and Cove; (9) Army and Navy YMCA; (10) The Admiral and Sailor Boy tavern.; and (11) Edgewater Hotel, Jack's Waterfront (second location), and Original Don's. *Sources* Un-published Sites Database of the GLBT Historical Society and Acquisition Map (San Francisco Redevelopment Agency 1959)

detailed, synchronic map helps to more accurately and precisely register the mid-twentieth century queer topography of the waterfront (Fig. 5.3). As a complement to this new cartographic representation of the imbrication of planning and queer social history, the historical rise and fall of queer waterfront bars due to urban redevelopment comes into clearer focus by superimposing the timing of land acquisitions[3] on a time plot of queer sites (Fig. 5.4). Taken together, the map and the time-graph emplace queer life on the waterfront within the spatial history of the city's first urban renewal project (Hartman and Carnochan 2002; Rubin 2011; Howell 2015; Isenberg 2017).

The map highlights the concentration of queer sites along the Embarcadero clustered around the Army and Navy YMCA, a well-known destination for homosexual liaisons during the war

[3]See: Project Summaries-Acquisition: Western Addition A-1, Diamond Heights B-1, and Embarcadero Lower-Market Area E-1. Real Estate Division Files, San Francisco Redevelopment Agency Records. Office of Community Investment and Infrastructure, San Francisco.

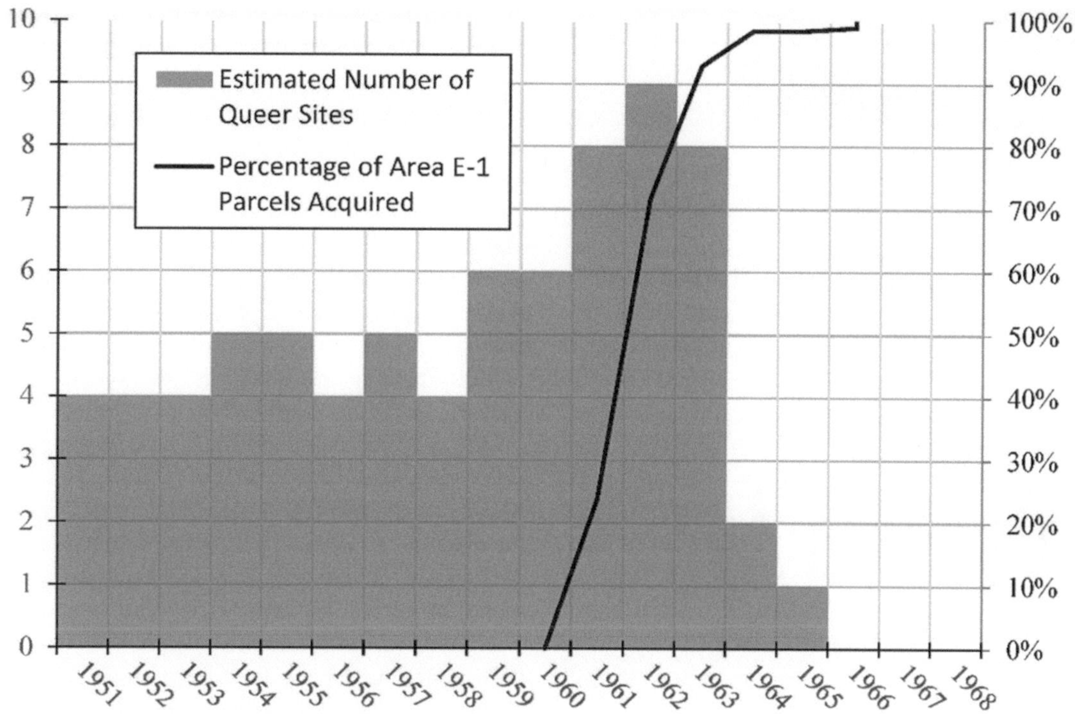

Fig. 5.4 Urban Renewal and the Growth and Decline of Queer Sites on the San Francisco Waterfront, 1951–1968. Sources: Unpublished Sites Database of the GLBT Historical Society and Unpublished Land Acquisition Documents in Urban Renewal Project Area E-1 from the San Francisco Redevelopment Agency Archives

years. Even before the war years, queer maritime workers, marines, and merchant mariners lived, worked, and socialized in the small waterfront hotels, restaurants, and bars in the stretch of the low-rise half-blocks known as "The Front" (Bérubé 2011, 2016). The graph—created by refining the estimated years of operation of particular queer sites from a wide array of sources—suggests three phases of queer life on the waterfront. Prior to the approval of the redevelopment plan in 1959, much of the 1950s was a period of relative stability with four or five sites in operation at a given time. The chart, however, masks the displacement, and subsequent relocation to new quarters elsewhere on The Front, of two sites as part of the planning and construction of the Embarcadero Freeway. This first phase (prior to 1959) was both an era in which city leaders struggled to secure public support for waterfront redevelopment and preexisting queer sites were disrupted and reconstituted.

In the second phase (1959–1962), the number of queer sites doubled while the San Francisco Redevelopment Agency (SFRA) acquired over 200 parcels of land within the urban renewal project area. Notably, this period also corresponds to growing social and political consciousness among gay bar operators, employees, and patron. For example, the League for Civil Education (L.C.E.) and its bi-weekly "publication of the day-to-day happenings in the 'community'" played an important role in expounding the need for collective social and political action, as a well as advertising commercial events like the Demolition Party and Grand Opening for "Jack's" when the bar moved from one end of the Embarcadero waterfront to the other (Fig. 5.5). The publication served as an important vehicle for voter mobilizations and fundraising drives for the "L.C.E. War Chest," a legal defense fund to defend liquor licenses during revocation proceedings. The increase in

Fig. 5.5 Advertisement in gay bar rag "The L.C.E. News" announcing the relocation of "Jack's Waterfront" prior to demolition of several city blocks to make way for a new set of on- and off-ramps to the four year old Embarcadero Freeway (Jack's Waterfront Demolition Party Advertisement 1962)

the number of queer waterfront sites might be an artifact of better, more systematic documentation of queer sites in the incipient gay press. Just as plausibly, the incipient gay press and the emergence of a socially and politically engaged "community" that was often detailed in pages of the L.C.E.'s Citizens News brought visibility and vitality to a growing queer commercial economy on the waterfront. In addition, the targeted "saturation" tactics of the San Francisco Police Department's "S-Squad" likely—and perhaps intentionally—concentrating and containing queer social life to the waterfront. Such a containment strategy would serve both to "clean-up" the Tenderloin and North Beach, while also driving property values down in and around the urban renewal area at precisely the time that the SFRA began negotiations with property owners over the fair market values of their holdings in the urban renewal project area E-1. In the third and final phase (after 1962), the number of queer sites on the waterfront dropped precipitously just as the SFRA finished taking possession of the few remaining parcels, evicting commercial and residential tenants, razing structures, and disposing of consolidated lots to private developers. The Ensign Club was located on one of the last parcels acquired by the city in the urban renewal project area, as well as the last queer site displaced from waterfront. According to the SFRA's property management records, the commercial tenant vacated his leased space in July 1965.

This new mapping of urban change suggests urban renewal—rather than the periodic policing of gay bars by law enforcement and liquor licensing (e.g. Agee 2014; Brown and Knopp 2016)—was overriding the cause of displacement of queer life from the San Francisco waterfront. In short and most significantly, these two renderings of the waterfront show that the closure of queer sites coincided with the completion of land acquisition activities in the urban renewal project area. It also suggests that after approving Area E as the city's top priority urban renewal project area queer social life expanded as crackdowns in the Tenderloin and North Beach shifted the center of queer nightlife from bars in those areas to new quarters at the foot of Market Street. The city's history of urban landscape change both complicates the task of drafting maps of the city's shifting sexual geography and implicates the broader transformation of the city's political economy in the respatialization to queer social life. Recuperating the social histories of marginalized groups prior to urban renewal disrupts that popular narrative by showing a complex mix of business, residential, and industrial uses within the urban renewal project area boundaries. This new mapping helps better understand what happened to mid-century queer sites on San Francisco's waterfront. These new renderings show that the physical transformation of San Francisco from an industrial port city into the urban core of an expanding metropolitan region disrupted established patterns of queer sociability.

5.5 Historical GIS and Social Histories of Urban Change

New spatial histories have been particularly powerful in recuperating the social and cultural histories of marginalized groups who have been left out of—or misrepresented in—popular narratives and academic treatments of the past. Urban planning historians are incorporating historical GIS, both as a research method and as a pedagogical tool to analyze the impacts and legacies of past planning interventions (Churchill and Hillier 2008; Hillier 2010; Hurley 2013). Urban scholars are deploying the analytical tools of GIScience to implicate racist planning policies and development practices in spatial and temporal patterns of racial and ethnic segregation (Gordon 2008; Gilliland et al. 2011; Gordon 2011; Beveridge 2014; Reece 2014; Shertzer et al. 2016; Sadler and Lafreniere 2017; Digital Scholarship Lab 2018). Labor historians, for example, have begun using historical-geographic methods to reconstruct the sites of working-class social formation and political organizing (Dunae et al. 2013; McFarland 2017). "Spatial" renderings of memoirs constructed by projecting in-text, locational references onto contemporaneous cartographic representations derived from extra-textual sources can recuperate the ways marginalized subjects in the past engaged in "claiming geographic ownership of significant parts of [the] City" (Lopez Santiago 2018, p. 152).

To date, geographers who foreground the interplay of sexuality, gender, and space in their research have largely taken a critical stance toward GIScience by rightly highlighting its epistemological pitfalls, even while gesturing toward its strategically essential merits in limited applications in which cartographic authority is deployed from below by queer subjects working collectively. Following a reflexive discussion of their role in a participatory, community-based GIS project to produce a public history brochure "Claiming Space: Seattle's Lesbian and Gay Historical Geography," Brown and Knopp stress the teleological dangers and heteronormative constraints of rendering the queer past with the explicit and categorical precision demanded by GIScience datasets and mapviews (Brown and Knopp 2008). Despite the "physical, visual, technical, and technological reifications" implicit in cartographically rendering the queer past, their "map not only relied on but produced—and continues to produce—new and different kinds of knowledge, by affirming identities, sparking imaginations, and inspiring activism (p. 55)." Their project—both in the collective spirit of cartographically "claiming" space and in its commitment to valuing communal attachments to place—is best understood—not as a critique of the epistemological and ontological limitations of GIScience, but a participatory mapping project very much in the vein of Bunge's work with the community of Fitzgerald of Detroit (Bunge 1971; Mountz 2011).

In the context of both contemporary sex-gender politics and rising urban land values, there is a growing urgency for new queer spatial histories that are informed by and sustain ongoing queer collective memory projects, LGBTQ historic preservation initiatives, and queer public history programs. For example, there is a very active effort both nationally and locally to document, landmark, interpret, and preserve sites of historical significance to LGBTQ + people as part of cultural heritage planning initiatives. As part of the National Park Services' "Telling All American's Stories," the federal agency sponsored an LGBTQ + initiative that produced a theme study to aid in identifying historically significant places, a guidebook for the public to get involved in local heritage and preservation projects, and a series of interpretive articles spotlighting national heritage sites. In another example, San Francisco's LGBTQ+ Cultural Heritage Strategy working group was conceived as a "living history" rather than a historical preservation initiative (LGBTQ+ Cultural Heritage Strategy Working Group 2018). The Strategy Working Group used a survey of nearly 1,500 residents to redefine "heritage" as not simply the conservation and interpretation of historic structures related to the queer past, but as a present and future-oriented imperative to "build a society that honors the legacy, ensures longevity, celebrates culture, nurtures well-being, and uplifts the most vulnerable of the LGBTQ+ community (p. 1)." The new historical framing of the city's past stresses active stewardship in continuing San Francisco's legacy as a "beacon for gender and sexual minorities" into the future. The group has endorsed the creation of several new cultural heritage districts, including the Compton's Transgender Cultural District, the Leather and LGBTQ Cultural District, and the Castro LGBTQ Cultural District as "place-based heritage conservation tools."

While the San Francisco Planning Department has welcomed and adopted a number of LGBTQ heritage projects—including the country's first sex/gender identity-themed historic context statement—much work remains to be done in reconstructing even the basic contours of the queer urban past in the city (Scott 2004).[4] The contemporary San Francisco Planning Department—in the rainbow-colored LGBTQ heritage data layer on its PIM platform—has perhaps missed the irony of misplacing the Ensign Club. The reason the Ensign "closed in the early 1960s" might have come into shaper relief if the footprint of the Ensign had been more accurately geo-referenced. The Ensign was one of the last queer sites on the waterfront (see Figs. 5.3 and 5.4) before the commercial building and city block in which it was located was razed to make way for a new park and plaza complex at the base of Market Street. Until very recently, the plaza bore the name of the Redevelopment Agency Director Justin Herman who took a strong hand in seeing the project through from its conception to its completion—making him a singularly significant figure in both the LGBTQ history and urban planning history of the San Francisco. Following a unanimous vote by the Board of Supervisors on a resolution introduced by Supervisor Aaron Peskin to rename the plaza, the city's Recreations and Parks Commission renamed the space "Embarcadero Plaza" in November 2017 citing Herman's role in displacing poor communities of color (Fracassa 2017). Over a year later, his name has not yet been excised from the underlying base map of the Planning Department's on-line PIM platform that is "powered by ESRI" (Fig. 5.1).

Much work is yet to be done to more carefully document and interpret the changing sexual geographies of the urban past. Just as Justin Herman's significance to the city queer past was hidden in plain sight on the map of lower Market Street, both the redevelopment archive and the LGBTQ public history projects need to be contextualized—or emplaced—within a carefully constructed, intertextual historiography of both queer- and urban planning history of the city. As this study illustrates, there are rich prospects of using GIScience as a tool for cultural and historical analysis by extracting and juxtaposing locational and temporal fixes from seemingly disparate, unrelated archives of personal experience, collective memory, administrative planning, and official recordkeeping. For the larger research project, the search for a serviceable base map to plot bygone queer sites led to the redevelopment agency archives where the author pored over a voluminous collection of moldering newspaper clippings, unprocessed boxes of planning studies and site reports, and land acquisition and property management records. As the redevelopment story came into greater focus, so did the queer history of the waterfront—principally from gay and non-gay press accounts and oral history transcripts. Street addresses and dates—rather than place names, surnames, or sexual nomenclature—linked these archives. The planning archive, authored by the dispassionate, disinterested voice of scientific officialdom, held scant, oblique references to sexual subjects in the Urban Renewal Project Area E-1. The land acquisition map and the summary report of real estate transactions in the Agency archive were, however, essential in more accurately plotting the physical location of the Ensign Club and situating its demise within the urban renewal history of the city. As a queer (strange) piece of metadata, these administrative records of the dispossession of real property by

[4]Prior to the 2016 "Citywide" historic context statement, the author prepared an earlier, more concise, targeted version for Friends of 1800, a historic preservation advocacy group, highlighting five different subareas and emphasizing the city as site of postwar "gay community" formation.

eminent domain could help the San Francisco Planning Department update its on-line GIS Property Inventory Management platform by more accurately identifying the former location of the Ensign Club and more precisely explaining in its rainbow-linked annotation why it "closed in the early 1960s."

As the author has highlighted in this chapter, there are productive possibilities of registering past queer places culled from textual sources onto urban planning maps generated to "renew" the city through large-scale demolition and rebuilding. By projecting the queer past onto official maps produced as part of a modernist planning project, this study's approach foregrounds the impact of urban landscape change on queerly and marginally positioned subjects on the San Francisco waterfront. While the author recognizes the epistemological and ontological critiques of GIScience in the production of social/spatial knowledge, his findings show that historical GIS can be useful as a tool for revising historical-geographic understandings of the changing literal and social positionality of sexual subjects. New spatial narratives can foreground the historically and geographically contingent processes that impinge upon, vitalize, and rework sexual subjectivities and demonstrate that contemporary sexual identities did not come out of nowhere, but were imbricated in the material transformation of urban space itself. Emplacing queer personal and social histories on the cartographic renderings and geo-spatial datasets generated by current and past urban planning projects has the potential to do more than "queer" the unequal power relations between state-sanctioned "big" dataset. Queer spatial histories can shift the focus from explicating the meaning and significance of the acts, behaviors, and identities of sexual subjects to interrogating the limits and liberties of the spaces and places in which those social relations were enacted in the past. In this case, the emergent queer spatial history of the San Francisco waterfront suggests that the state-directed management of urban land uses and the acquisition of real property, rather than liquor law enforcement, was the primary catalyst for mobilizing a socially articulated "gay community" in the city in the early 1960s.

References

Agee C (2014) The streets of San Francisco: policing and the creation of a modern liberal politics, 1950–1972. University of Chicago Press, Chicago, Historical Studies of Urban America

Ahmed S (2006) Queer phenomenology: orientations, objects, Others. Duke University Press, Durham

Anjali A, Geeta P (2016) Area impossible: notes toward an introduction. GLQ 22(2):151–171. https://doi.org/10.1215/10642684-3428687

Bérubé A (ed) (2011) My desire for history: essays in gay, community, and labor history. University of North Carolina Press, Chapel Hill

Bérubé A (2016) No red-baiting! no race-baiting! no queen-baiting!: the Marine Cooks and Stewards Union from the depression to the Cold War. outhistory. http://outhistory.org/exhibits/show/no-baiting/red-race-queen. Accessed 8 Mar 2019

Beveridge AA (2014) The development, persistence, and change of racial segregation in U.S. urban area, 1880–2010. In: Gregory IN, Geddes A (eds) Toward spatial humanities: historical GIS and spatial history. Indiana University Press, Bloomington, pp 35–61

Boschmann EE, Cubbon E (2014) Sketch maps and qualitative GIS: using cartographies of individual spatial narratives in geographic Research. Prof Geogr 66(2):236–248. https://doi.org/10.1080/00330124.2013.781490

Boyd NA (2003) Wide open town: a history of queer San Francisco to 1965. University of California Press, Berkeley

Brass K (2016) Is San Francisco's Castro district losing its gay identity? influx of wealthy tech workers in the city's famed heart has caused some friction. financial times. https://www.ft.com/content/20912e8e-a69e-11e6–8898-79a99e2a4de6. Accessed 8 Mar 2019

Brown M, Knopp L (2008) Queering the map: the productive tensions of colliding epistemologies. Ann Assoc Am Geogr 98(1):40–58. https://doi.org/10.1080/00045600701734042

Brown M, Knopp L (2016) Sex, drink, and state anxieties: governance through the gay bar. Soc Cult Geog 17(3):335–358. https://doi.org/10.1080/14649365.2015.1089588

Bunge W (1971) Fitzgerald: geography of a revolution. geographies of justice and social transformation. 2011 edn. University of Georgia Press, Athens

Castells M (1983) Cultural identity, sexual liberation and urban structure: the gay community in San Francisco. In: Castells M (ed) The city and the grassroots: a cross-cultural theory of urban social movements. University of California Press, Berkeley, pp 138–172

Churchill R, Hillier A (2008) Teaching with GIS. In: Knowles AK (ed) Placing history: how maps, spatial data, and gis are changing historical scholarship. ESRI Press, Redlands, CA, pp 61–94
Cockayne DG, Richardson L (2017) Queering code/space: the co-production of socio-sexual codes and digital technologies. Gend Place Cult 24 (11):1642–1658. https://doi.org/10.1080/0966369X.2017.1339672
Cosgrove D (1998) Social formation and symbolic landscape. University of Wisconsin Press, Madison
Cramer LJ (2014) Drag queens say farewell to mission insititution Esta Noche. SF Weekly. http://www.sfweekly.com/culture/drag-queens-say-farewell-to-mission-institution-esta-noche/. Accessed 8 Mar 2019
De Luca N, Ferber N, Atteneder H, Jekel T (2015) Feminist and queer approaches to education for spatial citizenship. GI Forum J Geog Info Sci 1:272–282. https://doi.org/10.1553/giscience2015s272
DeLaurentis IV VD (2018) Gay maps, queer reads: exposing violence in the spatial representation of Gay D.C. in Search of queer spatial potentials. ACME 17(1):109–143
Digital Scholarship Lab (2018) Renewing inequality: urban renewal, family displacements, and race 1955–1966. Robert K. Nelson and Edward L. Ayers (eds) American Panorama. https://dsl.richmond.edu/panorama/renewal/. Accessed 19 Nov 2018
Doan PL, Higgins H (2011) The demise of queer space? resurgent gentrification and the assimilation of LGBT neighborhoods. J Plan Ed Res 31(1):6–25. https://doi.org/10.1177/0739456x10391266
Domosh M (1997) With stout boots and a stout heart: historical methodology and feminist geography. In: Jones JP III, Nast HJ, Roberts SM (eds) Thresholds in feminist geography: difference, methodology, representation. Rowman Littlefield, New York, pp 225–240
Dunae PA, Lafreniere DJ, Gilliland JA, Lutz JS (2013) Dwelling places and social spaces: revealing the environments of urban workers in Victoria using historical GIS. Labour/Le Travail 72:37–73
Elwood S, Leszczynski A (2018) Feminist digital geographies. Gend Place Cult 25(5):629–644. https://doi.org/10.1080/0966369X.2018.1465396
Fracassa D (2017) SF park commission strips Justin Herman's name from Embarcadero plaza. San Francisco Chronicle. https://www.sfgate.com/politics/article/SF-parks-commission-strips-Justin-Herman-s-name-12363778.php. Accessed 8 Mar 2019
Frisch M (2002) Planning as a heterosexist project. J Plan Ed Res 21(3):254–266. https://doi.org/10.1177/0739456x0202100303
Ghaziani A (2015) There goes the gayborhood. Princeton Studies in Cultural Sociology. Princeton University Press, Princeton
Gieseking JJ (2017) Operating anew: queering GIS with good enough software. Can Geogr 62(1):55–66. https://doi.org/10.1111/cag.12397
Gieseking JJ (2018) Size matters to lesbians, too: queer feminist interventions into the scale of big data. Prof Geogr 70(1):150–156. https://doi.org/10.1080/00330124.2017.1326084
Gilliland JA, Olson SH, Gauvreau D (2011) Did segregation increase as the city expanded? the case of Montreal, 1881–1901. Soc Sci Hist 35(4):465–503. https://doi.org/10.1215/01455532-1381823
Godfrey B (1988) Neighborhoods in transition: the making of San Francisco's ethnic and nonconformist communities. University of California Press, Berkeley
Gordon C (2008) Mapping decline: St. Louis and the fate of the American city. Politics and culture in modern America. University of Pennsylvania Press, Pittsburgh
Gordon C (2011) Lost in space, or confessions of an accidental geographer. Int J Humanit Arts Comput 5 (1):1–22. https://doi.org/10.3366/ijhac.2011.0018
Graves DJ, Watson SE (2016) Citywide historic context statement for LGBTQ history in San Francisco. https://sfgov.org/sfplanningarchive/lgbt-historic-context-statement. Accessed 8 Mar 2019
Gustav-Wrathall JD (1998) Take the young stranger by the hand: same-sex relations and the YMCA. The Chicago series on sexuality, history, and society. University of Chicago Press, Chicago
Haag M (2017) Gilbert Baker, Gay activist who created the rainbow flag, dies at 65. New York Times, March 31, p B7
Hartman C, Carnochan S (2002) City for sale: the transformation of San Francisco, revised and updated edition. University of California Press, Berkeley
Hillier A (2010) Invitation to mapping: how GIS can facilitate new discoveries in urban and planning history. J Plan Hist 9(2):122–134. https://doi.org/10.1177/1538513210366964
Holmes K (2015) What's the tea: gossip and the production of black gay social history. Radic Hist Rev 2015(122):55–69. https://doi.org/10.1215/01636545-2849531
Houlbrook M (2001) Toward a historical geography of sexuality. J Urb Hist 27(4):497–504. https://doi.org/10.1177/009614420102700406
Howell O (2015) Making the mission: planning and ethnicity in San Francisco. University of Chicago Press, Chicago
Hurley JA (2013) Developing a digital map collection for research and teaching: the "Planning Atlanta: a new city in the making, 1930–1990s" collection. J Map Geog Libr 9(1/2):44–68. https://doi.org/10.1080/15420353.2012.726958
Isenberg A (2017) Designing San Francisco: Art, land, and urban renewal in the city by the bay. Princeton University Press, Princeton
Jack's Waterfront Demolition Party Advertisement (1962). L.C.E. [Citizen's] News, April 16, p 3
Knopp L (1992) Sexuality and the spatial dynamics of capitalism. Environ Plan D Soc Space 10(6):651–669. https://doi.org/10.1068/d100651
Legg S, Brown M (2013) Introduction to the special issue: moral regulation: historical geography and scale. J Hist Geog 42:134–139. https://doi.org/10.1016/j.jhg.2013.02.001

Leszczynski A (2009) Poststructuralism and GIS: is there a disconnect? Environ Plan D Soc Space 27(4):581–602. https://doi.org/10.1068/d1607

LGBTQ+ Cultural Heritage Strategy Working Group (2018) LGBTQ+ Cultural Heritage Strategy: Draft Executive Summary, Presented by the San Francisco Planning Department to the Historic Preservation Commission. https://sfgov.org/sfplanningarchive/LGBTQplusStrategy#hearings. Accessed 8 Mar 2019

Liquor Licenses of 2 S. F. Spots Are Threatened (1959) San Francisco Chronicle, December 4, p 36

Lopez Santiago AM (2018) The geography of Bernardo Vega's memoirs. Centro J 30(1):152–177

Lucchesi P (2015) Lexington Club announces a closure date in April; PlumpJack group moving in. SF Gate. https://insidescoopsf.sfgate.com/blog/2015/02/24/lexington-club-announces-a-closure-date-in-April-plumpjack-moving-in/#photo-599385. Accessed 8 Mar 2019

McFarland S (2017) The union hall was the center of the worker's life: spaces of class formation in the United Auto Workers, 1937–1970. J Hist Geog 55:17–29. https://doi.org/10.1016/j.jhg.2016.11.008

Meeker M (2006) Assembling a lavender baedeker. In: Contacts desired: Gay and lesbian communications and community, 1940–1970s. University of Chicago Press, Chicago, pp 201–224

Mountz A (2011) Classics in human geography revisited: commentary 3: reflections of Bill Bunge's Fitzgerald: geography of a revolution. Prog Human Geog 35(5):712–715

Oswin N (2015) World, city, queer. Antipode 47(3):557–565. https://doi.org/10.1111/anti.12142

Perhan C (2017) SF's Oldest gay bar will seriously become a 'Kung Fu' Laundromat: Kozy Kar's owner has applied to take over the gangway. Eater: San Francisco. https://sf.eater.com/2017/8/9/16120802/sf-gangway-oldest-gay-bar-kozy-kar-laundromat-owner. Accessed 19 Nov 2018

Podmore JA (2006) Gone 'underground'? Lesbian visibility and the consolidation of queer space in Montréal. Soc Cult Geog 7(4):595–625. https://doi.org/10.1080/14649360600825737

Podmore JA (2013) Lesbians as village queers: the transformation of Montréal's lesbian nightlife in the 1990s. ACME 12(2):220–249. https://www.acme-journal.org/index.php/acme/article/view/961. Accessed 8 Mar 2019

Podmore JA (2015) Queering discourses of urban decline: representing Montréal's post-World War II Lower Main. Hist Geog 43:57–83. https://ejournals.unm.edu/index.php/historicalgeography/article/view/3389. Accessed 8 Mar 2019

Podmore JA, Brown M (2015) Introduction to the special issue historical geographies of sexualities? Hist Geog 43:5–16. https://ejournals.unm.edu/index.php/historicalgeography/article/view/3480/Podmore-Brown. Accessed 8 Mar 2019

Rao R (2014) The locations of homophobia. Lon Rev Int Law 2(2):169. https://doi.org/10.1093/lril/lru010

Rao R (2015) Global homocapitalism. Rad Philo 194:38–49. https://www.radicalphilosophy.com/article/global-homocapitalism. Accessed 8 Mar 2019

Reece J (2014) Redlined: the history of race and real estate in Cleveland and its relationship to health equity today: an interim findings working session with the place matters team for Cuyahoga County, April 14th, 2014-Brooklyn, OH. Kirwan Institute for the study of race and ethnicity and the city and regional planning program at the Knowlton School of Architecture, The Ohio State University. http://kirwaninstitute.osu.edu/wp-content/uploads/2015/02/cleveland-place-matters.pdf. Accessed 8 Mar 2019

Roque Ramirez HN (2001) Communities of desire: queer latina/latino history and memory. San Francisco Bay Area, 1960–1990s. Dissertation, University of California, Berkeley

Rubin G (2000) Sites, settlements, and urban sex: archaeology and the study of gay leathermen in San Francisco, 1955-1995. In: Schmidt RA, Voss BL (eds) Archaeologies of sexualities. Routledge, New York, pp 62–88

Rubin J (2011) A negotiated landscape: the transformation of San Francisco's waterfront since 1950. Center for American Places at Columbia College, Chicago

Sadler RC, Lafreniere DJ (2017) Racist housing practices as a precursor to uneven neighborhood change in a post-industrial city. Hous Stud 32(2):186–208. https://doi.org/10.1080/02673037.2016.1181724

San Francisco Redevelopment Agency (1959) The Redevelopment Plan for the Golden Gateway: Embarcadero-Lower Market Approved Redevelopment Project Area E-1. Redevelopment Agency of the City and County of San Francisco, San Francisco

Sandercock L (ed) (1998) Making the invisible visible: a multicultural planning history. University of California Press, Berkeley

Scott D (2004) Sexing the city: The development of sexual identity based subcultures in San Francisco, 1933–1979, draft historic context statement prepared for the Friend of 1800 and presented to the San Francisco Landmarks Preservation Advisory Board, January 5, 2005. San Francisco Planning Department, San Francisco

Scott D (2008) The city aroused: sexual politics and the transformation of San Francisco's urban landscape, 1943–1964. Dissertation, University of Texas at Austin, Austin

Scott D (2015) Before the creative class: blight, gay movies, and family values in the Haight-Ashbury neighborhood, 1964. J Plan Hist 14(2):149–173. https://doi.org/10.1177/1538513214549430

Shertzer A, Walsh RP, Logan JR (2016) Segregation and neighborhood change in northern cities: new historical GIS data from 1900–1930. Hist Methods 49(4):187–197. https://doi.org/10.1080/01615440.2016.1151393

Smith N, DeFilippis J (1999) The reassertion of economics: 1990s gentrification in the Lower East Side. Int J Urb Reg Res 23(4):638–653. https://doi.org/10.1111/1468-2427.00220

Spring J (2010) Secret historian: the life and times of Samuel Steward, professor, tattoo artist, and sexual outlaw. Farrar, Straus and Giroux, New York

Strait G (1962a) The community. L.C.E. [Citizen's] News, September 17, p 1

Strait G (1962b) Roving report. L.C.E. [Citizen's] News, July 9, pp 7–8

Tolia-Kelly DP (2010) The geographies of cultural geography I: identities, bodies and race. Prog Human Geog 34(3):358–367. https://doi.org/10.1177/0309132509343378

Wilson MW (2015) New lines? enacting a social history of GIS. Can Geogr 59(1):29–34. https://doi.org/10.1111/cag.12118

Revisiting the Walking City: A Geospatial Examination of the Journey to Work

6

Don Lafreniere and Jason Gilliland

Abstract

The daily commute to work and its related social histories have long been of interest to historical geographers and urban historians. This article revisits the existing scholarship on the nineteenth-century journey to work and outlines a new methodological framework that uses a historical GIS to overcome many of the challenges identified in previous studies. These challenges include a reliance on small, atypical samples of workers, approximations of the spatial relationship between home and work, and unrealistic interpretations of journeys travelled by using only Euclidean paths. Combining city directories and decennial censuses through the use of probabilistic record linkage techniques uncovers the relationship between work and home for over 5,000 workers in London, Ontario in 1881. A GIS network-derived journey to work model re-creates more realistic journey that considers the many natural and built environment barriers that influenced the paths and distances workers travelled on a daily basis. Empirical results of the journey to work along the lines of occupational class, coincident home–work location, and gender are presented and contextualized to studies in other cities. The results highlight that the experiences of commuting differ widely along the lines of social class and gender.

Keywords

Journey to work · Space–time · Historical GIS · Census · Mobility

6.1 Introduction

The rapid industrialization and urbanization experienced in the late nineteenth century affected many aspects of daily life. For adults, one aspect that bridged all socio-demographic and economic groups was the journey to work. This unifying experience of daily life was a result of the growing spatial separation between home and the workplace due to changes in the nature of work, from primarily home-based artisanal work to wage labour. The journey to work also reflected larger evolving urban processes, such as suburbanization and segregation. In the second half of the nineteenth century, mid-sized Canadian cities were still largely dominated by commuters who travelled to work by foot, either on the streets, alongside the horse-drawn carts used for heavy or bulky loads, or on the rapidly emerging network of sidewalks. Commuters

D. Lafreniere (✉)
Michigan Technological University, Houghton, USA
e-mail: djlafren@mtu.edu

J. Gilliland
University of Western Ontario, London, Canada
e-mail: jgillila@uwo.ca

C. Travis et al. (eds.), *Historical Geography, GIScience and Textual Analysis*,
Historical Geography and Geosciences, https://doi.org/10.1007/978-3-030-37569-0_6

were closely connected to the physical and built environments they traversed, enjoying the pleasant fragrance of flowers wafting from the gardens of well-built homes or subjected to the foul odours that emanated from the breweries and cigar factories. They had to cope with the elements that the glass and steel of our modern automobiles shelter us from, such as July's intense sun and January's blowing snow and sleet. The daily commute provided individuals an intimate scale by which to experience the social environments present in the city. In the walking city, everyone else on the street was within earshot as well as within an easy visual distance, making the journey to work one of the periods with the highest potential for social interaction during an individual's daily time–space rhythm.

This chapter revisits the late nineteenth-century journey to work in the industrializing city, an area of research that has had little attention since the 1990s. Using the mid-sized Canadian city of London, Ontario as a case study, this paper extends the existing literature by exploring the spatial and temporal pattern of work at a scope and scale unprecedented in earlier studies. Through the use of a full count of the city's population in 1881 and 1891, which includes the workplace of nearly every male (and many female) workers, we can, for the first time, establish the relationship between home and work for an entire urban population. The dataset is interrogated using the spatial analytic capabilities of a GIS to re-create the daily time–space patterns of individuals from every socio-demographic and economic group in the city. This study of the journey to work was undertaken for an important methodological purpose—to re-create the spatial relationships between work and home. By having a sound methodology with which to re-create these relationships, other scholars can then re-examine the roles played by transportation, residential segregation, land use policies, and restrictive covenants on access to employment and their impact on demography, social mobility, and housing choice. This study looks to create a methodology that overcomes the many limitations noted in the literature of previous journey-to-work studies.

6.1.1 Related Work

The study of Canadian work history saw a resurgence in the late 1960s, thanks to British historian E.P. Thompson's call to recover the lives and experiences of working-class people 'from the enormous condescension of posterity' (Thompson 1980 [1963], p. 12). This call was answered by the likes of Kealey (1980) and Bradbury (1996), who both described the changes in working life during periods of industrialization in large Canadian cities (Toronto and Montreal, respectively). Several edited collections have been compiled, with chapters on the working lives of everyone from shantymen and railway workers to boys in mining camps and girls in textile shops (Craven 1995; Cross 1974; Opp and Walsh 2006). This scholarship provides a rich and invaluable narrative used by students of Canadian labour history across the country. However, the spatial relationships between work and home are less known. While there is little empirical evidence, especially for Canada, setting out details of the journey to work before 1900, the importance of the geography of the home–work relationship has been considered for several non-Canadian cities. The greatest attention has come from scholars who studied the dominant industrial cities of Britain, especially London (Green 1988; Lees 1969; Pooley and Turnbull 1997, 1999, 2000), as well as Birmingham (Vance 1967), Huddersfield (Dennis 1984), Liverpool (Pooley 1984), and Newcastle (Barke 1991).

In North America, work has been concentrated in large metropolitan areas such as Philadelphia; the journey to work was one of the areas of interest to the 12-year (1969–1981) Philadelphia Social History Project led by Theodore Hershberg (Greenberg 1980; Hershberg 1976; Hershberg et al. 1981). For Canada, Montreal has been the most productive laboratory, though we find the concerns of the relationship between work and home to be secondary to wages (Hoskins 1989), the process of industrial suburbanization (Gilliland 2004; Lewis 2000), segregation (Gilliland and Olson 2010; Gilliland et al. 2011; Olson and Thornton 2011), and the nature of the work itself (Bradbury 1996). An important exception to the wide-gap of empirical research on the journey to work in Canada prior to 1900 is Goheen's study of Toronto (1970). Goheen uses the daily commute to illustrate the extent of residential segregation among differing social classes in Victorian Toronto. There has been more interest in the Canadian experience of the journey to work after the turn of the twentieth century. Richard Harris and Victoria Bloomfield examined the topic while exploring the processes of decentralization of workplaces and the broader suburbanization of cities (Harris and Bloomfield 1997; Bloomfield and Harris 1997). Their landmark paper (Bloomfield and Harris 1997) outlined a methodology that has served as the inspiration for much of this paper.

6.1.2 Identifying Methodological Challenges

Despite what may appear at first blush to be a wealth of previous research, the study of the daily commute to work has not received significant attention. Geographer Richard Dennis suggests that one of the reasons may be that 'data for studying journeys to work are exceedingly scarce, particularly for quantitative research in which a single source records both place of residence and place of work' (Dennis 1984, p. 132). As a result, previous research on the nineteenth-century city has utilized sources (data) that either represent a small, non-random, possibly atypical sample of workers or only allow for the inference of home and work, rather than absolute spatial links (Dennis 1984, p. 132). For example, Green (1988) uses only a wage book for a single employer, as did Hoskins (1989) for his study of railworkers in Montreal. Barke (1991) used city directories to capture a set of nearly 2000 workers over four time slices. However, his sample only included middle-class business owners because he did not have the workplaces of non-owners, so it only represented 4% of the middle-class population of Newcastle. Dennis (1984), though recognizing the limitations of his methodology and calling for alternatives, elected to use the census returns for only one industry, silk manufacturing. Industry-specific studies have also been completed by Pred (1966) for five industries in New York City. The Philadelphia Social History Project also used five industries, derived from the Census of Manufacturing, though this sample was still robust with over 3700 workers represented. Bloomfield and Harris (1997) concerned themselves with a similarly small sample of employers.

Methodological decisions surrounding how to best re-create the needed home–work links have typically been constrained by limitations in the sources or lack of resources necessary for compiling and analyzing a more comprehensive database of employees and employers. These limitations have led authors of most previous studies to use methods of approximation. Vance (1967) saw the need to combine the census along with a comprehensive source of workplaces—the business directory. However, the 1851 British census that he utilized did not include workplaces for those enumerated, so he was unable to match individuals to their specific places of work. Vance instead created labour sheds, or areas where workers reside, and employment fields, areas of workplaces, and compared the distances between them. Dennis (1984), using the same sources as Vance, was likewise restricted to only inferring home–work links in his study of Huddersfield's silk industry. The Philadelphia Social History Project (Greenberg, 1980; Hershberg et al. 1981) also inferred the workplaces of its city's workers by creating arbitrary labour shed boundaries for each industry and then matching an individual's occupational title to the product manufactured within individual businesses. Pooley, in his study of the journey to work in late-nineteenth century Liverpool, used a 10% census sample and calculated distances from an individual's residence to the nearest potential workplace. He explains that 'for dock workers this is defined as the nearest operational dock; all office workers are assumed to work in the C.B.D. [central business district]' (Pooley 1984, p. 134).

The impact of a lack of quality sources has resulted in other difficulties for scholars of the nineteenth-century journey to work. Many scholars are interested in evaluating the differences in how the journey to work was experienced by members of various ethnic or racial groups. Mayer (1977), in his study of Milwaukee, used individuals' surnames, as recorded in the city directory, to infer their ethnic affiliation. Mayer acknowledges the difficulties of this approach, suggesting that even in the best of cases only 70% accuracy can be expected (Mayer 1977, p. 54). Deskins (1972) used directories in a similar manner to examine the racial differences in the journey to work in Detroit in the late nineteenth century. He derived race by a 'coloured' notation within the alphabetical directory rather than utilizing the more authoritative and complete decennial federal census. His decision not to use the census was not explained, nor did he compare the aggregate population totals by racial affiliation between the two sources to gauge the reliability of the 'coloured' notation in the directory. From these assumptions, significant historical generalizations of race-based behaviour were generated. These criticisms were noted by his colleagues and resulted in a heated debate within the pages of the *Professional Geographer* (Deskins 1975; Gale and Katzman 1975).

Deskins' work has other methodological limitations, the ones shared with nearly all previous journey to work studies—poor spatial accuracy in locating home and workplace as well as the methodology used to calculate the actual journey to work. Deskins' work, separate from his controversial derivation of race, suffered the greatest shortcomings. He was unable to link individuals to their workplaces, nor did he have the resources needed to map individuals to their specific residences. To compensate, he elected to approximate the location of work and home by using the mean centre of residences and the mean centre of workplaces while controlling for the extremes by calculating the standard distance around each mean. The resulting 'worktrip vector' is nothing more than a measure between the residential areas of Detroit and the non-residential areas. Other studies suffer from similar limitations, though not as severe. Green (1988) attempted to map each worker in Henry Poole's bespoke tailor shop to their exact residence, but found that addresses in London (UK) during the later decades of the nineteenth century were haphazard. His efforts were limited by a combination of address changes, street name changes, and sources that did not coincide temporally. His solution was to estimate each street's addressing by assigning civic address #1 at one terminus of the street and assigning numbers at equal intervals moving down the street (Green 1988, p. 186). Both Goheen and the Philadelphia Social History Project elected to use arbitrary grids to map residences and workplaces. The Philadelphia Social History Project created 660 × 775 foot grids and measured from centroid to centroid to calculate the length of the daily journey (Hershberg et al. 1981, p. 131). Additionally, several studies have assumed that workers were employed at the closest workplace for which they were qualified. This assumption skews the results to indicate shorter daily commutes than may have been realized.

Owing to the lack of computational power and robust datasets, all previous nineteenth-century journey to work studies elected to use Euclidean, or straight-line, distances in their calculations, rather than taking into account the actual distances that would have been travelled on the existing roadways. Even the most contemporary of studies, those that are beyond the scope of my current research, rarely utilize the existing geospatial technologies to re-create more realistic paths (Corcoran et al. 2009; Sultana and Weber 2007), and one of the only exceptions uses hypothetical journeys (Bousssauw et al. 2011). This chapter first overcomes the limitations outlined above that have influenced the results of previous journey to work studies. Secondly, it extends the existing journey-to-work methodologies by utilizing the spatial analytical capabilities of GIS and computer-based probabilistic record matching algorithms. Lastly, this chapter will outline some of the journey to work experiences along lines of socio-economic class, gender, and age.

6.2 Case Study: London, Canada

Previous journey to work studies have been primarily concerned with large urban centres, rarely concerning themselves with the more common medium-sized nineteenth-century towns. Debats (2009) outlines how studying large urban centres is not practical when the scale of analysis is the individual. We heed his advice in this study and apply the same 'small city research strategy' he extols. This case study is the mid-sized Canadian city of London, Ontario, Canada. London provides a classic example of a city that was changing from a commercial to an industrial town. London is located half-way between Toronto and Detroit, an advantageous position that was quickly recognized by many investors and industrialists. Several major companies began to employ Londoners in the last two decades of the nineteenth century. The Grand Trunk Railway built its train car factory in London East in 1887, which added over a hundred jobs to its already large operation in the city. Imperial Oil commenced refining in London East in 1880. Established companies such as the brewers Labatt and Carling, the confectioner McCormick, and McClary's foundry increased their workforce during this era as the demands for their goods grew faster than the efficiencies of mechanization. Despite these increases, London continued to be a diverse commercial city with most of its workers employed at small- or medium-sized firms in a wide range of industries. In 1881, there were an average of 2.5 workers per business; by 1891 it had risen to nearly 4 workers per firm.

The London of 1881 was a quintessential British industrial town on Canadian soil. Both the established citizenry and the waves of recent immigrants were predominantly from the British Isles, and Protestant faiths dominated the religious composition of the growing city. With a population of 19,746 at the time of the 1881 Census of Canada, London ranked 8th among Canadian cities and 86th in North America, making it comparable to cities such as St. John, NB, Lancaster, PA, and Des Moines, IA. For this study, we also included the three major suburbs of London: Petersville (aka London West), the Village of London East, and the urban section of Westminster Township (aka London South). Together these suburbs totalled 3,890 residents. It is critical to include these suburbs as they contributed significantly to the available employers and the diverse workforce found about the city. Additionally, suburban populations and industries are often overlooked in other studies, mostly because the census separates them from the urban core when disseminating their enumerations.

6.2.1 Overcoming Limitations: Larger Samples and Absolute Workplaces

To overcome the limitations of small sample sizes and the lack of representativeness found in previous studies, this paper utilizes the robust built and social environment stages model (Lafreniere and Gilliland 2015). Specifically, this study harnesses the geocoded decennial census that was captured through the creation of the social environment stage. As outlined by the tables in Fig. 6.1, the census was record-linked to the city directory, which facilitated the geocoding of nearly all individuals to their residential location. This dataset locates nearly 90% of the population in London and suburbs in 1881 and provides the needed variables to determine the home residence for each worker in the city, No other journey to work study has captured what is arguably the whole of the workforce in an urban location. Various methods of sampling have been utilized and rarely, with the exception to a certain extent of Deskins and Mayer, have researchers compared the distribution of their samples to the total enumerated population. It is generally accepted by historical geographers and historians alike that the federal census represents the gold standard record of a past population, though there are critics (Curtis 2001). Figure 6.1 (*Table 1*) compares the geocoded population used in this study to the full census enumeration collected in London for the year 1881. To illustrate how representative the 90% sample is, the population has been disaggregated using key demographic variables. It is observed that along the lines of religion, gender, and marital status, the variances are no more than 1%. For ethnic origin, we see an equally representative sample, with only a 2.55% overrepresentation of those of Irish decent and a 1.44% underrepresentation of Scottish in the geocoded sample versus the total population. Birthplace is also representative, with only a 1.92% underrepresentation of those born in Ontario. The representativeness of the sample population is without question. There are no concerns of an ecological fallacy (Sadler and Lafreniere 2017), as the sample is not only representative, but is also disaggregated to the level of the individual.

To overcome the limitation of previous studies that assumed workers were employed at the closest facility for which they were qualified, it was necessary to first establish the spatial link between each individual's home and their exact workplace. Workplaces varied widely in the industrializing nineteenth-century city, with many people still employed at home in semi-skilled trades and services, while others commuted daily to the factory or foundry. As London was also a commercial town, there was yet another significant cohort who travelled to work as clerks and support staff in the city's law offices, insurance companies, and financial institutions. Regrettably, an individual's specific workplace was not, and continues to not be, a question asked in the federal census. Occupation has been recorded since the first national census in 1871, but no question on workplace was asked until 1971. The modern questions on the journey to work are imprecise, with options such as 'worked at home', 'usual place of work', 'no fixed workplace address', or 'worked outside Canada'. Rather than a national enumeration, we instead turned to a local one, the city directory. City directories have been used, in some form, in many of the previous late nineteenth or early twentieth century journey to work studies, including Barke (1991), Bloomfield and Harris (1997), Deskins (1972), Goheen (1970), Harris and Bloomfield (1997), and Mayer (1977).

As with any source, they do have limitations. Harris (1986) reminds us that city directories were compiled for business purposes and likely underrepresented those who move often, including tenants and the poor. Bloomfield and Harris (1997) observe that working women are slightly underrepresented, a pattern also observed in the London directories. Regardless of these

Religion 1881			
	Geocoded Population	Census Population	Variance
	n = 21246	n =23636	
Protestant	83.46%	84.74%	-1.28%
Catholic	14.37%	13.33%	1.04%
Other	1.77%	1.45%	0.32%
Unknown	0.40%	0.48%	-0.08%

Gender 1881			
	Geocoded Population	Census Population	Variance
	n = 21246	n =23636	
Female	50.46%	50.45%	0.01%
Male	49.54%	49.55%	-0.01%

Marital Status 1881			
	Geocoded Population	Census Population	Variance
	n = 21246	n =23636	
Married	32.88%	33.49%	-0.61%
Single	62.68%	62.07%	0.61%
Widowed	4.44%	4.44%	0.00%

Origin 1881			
	Geocoded Population	Census Population	Variance
	n = 21246	n = 23636	
English	46.84%	46.61%	0.23%
Irish	28.81%	26.25%	2.55%
Scottish	17.69%	19.13%	-1.44%
German	2.02%	2.45%	-0.42%
French	0.96%	0.95%	0.01%
African	0.96%	0.94%	0.02%
Other European	0.70%	1.49%	-0.80%
Other	0.55%	0.49%	0.06%
Unknown	1.47%	1.69%	-0.22%

Birth Place 1881			
	Geocoded Population	Census Population	Variance
	n = 21246	n = 23636	
Ontario	57.60%	59.52%	-1.92%
England	18.78%	17.91%	0.87%
Ireland	8.81%	8.12%	0.69%
Scotland	5.52%	5.50%	0.02%
Canada	4.51%	4.16%	0.01%
United States	3.55%	3.49%	0.06%
Europe	0.77%	0.81%	-0.04%
Rest of World	0.29%	0.32%	-0.03%
Unknown	0.16%	0.18%	-0.02%

Fig. 6.1 *Table 1*. Geocoded census population compared to the total census population: London city and suburbs, 1881

limitations, Harris (1986) and Harris and Moffat (1986) support that city directories are arguably the most complete source of information on the specific employers of individuals. The study discussed in this chapter benefits from a collection of the London city directories that has been preserved in the Archives and Research Collections Centre at the University of Western Ontario and the London Room of the London Public Library. Four different publishers produced city directories between the years 1881 and 1891, with the London Publishing Company producing the directory of 1881 used in this study.

The London alphabetical directories include the name, residential address, tenure, occupation, and employer for each private citizen and business firm in the city. For widowed women, the name of their late husband is also noted. Those workers who did not work outside of their homes were also enumerated; the directory lists only a residential address or in some instances denotes this relationship between home and work with the word 'same'. These variables are captured from the social environment stage and used to establish the needed home–work link in this study. The geocoded business directory was utilized to determine the specific worksite of each employed person in the city. The business directory lists both major and minor employers, as well as sole proprietors, classified by their predominant industry or service (see Fig. 6.2 Upper). The name of the business and/or proprietor, along with a business address or the name of a significant commercial building is included in each listing. For those companies who paid an additional fee, a larger advertisement was included with some details of the services or wares for sale.

With the relationship between home and work established, it was necessary to record-link the combined city directory/census geodatabases with those of the business directory (Lafreniere and Gilliland 2018). To facilitate the record linkage across sources, we used *LinkageWiz*, a probabilistic record-matching software (*LinkageWiz* 2019). Records in both datasets were given a unique identifier, *WORK_ID* for the individual and *Business ID* for the business. Records were blocked using the municipal ward boundaries (captured through a spatial join with a temporally representative ward shapefile derived from a period map) and last names. Last names were blocked with various phonetic algorithms and string comparisons applied. A reference table was created due to the heavy use of short forms or nicknames used in the city directories for individuals' first names; examples include 'Wm' for William, 'Jn' for John, and 'Thos' for Thomas. Matching was then performed using match weights for last name, first name, gender (derived from a reference table of first names), occupation, and workplace (Fig. 6.2: Lower, *Table 2*). Last names had a variable weight determined by the number of instances within the dataset; this was used to overcome issues with commonality of

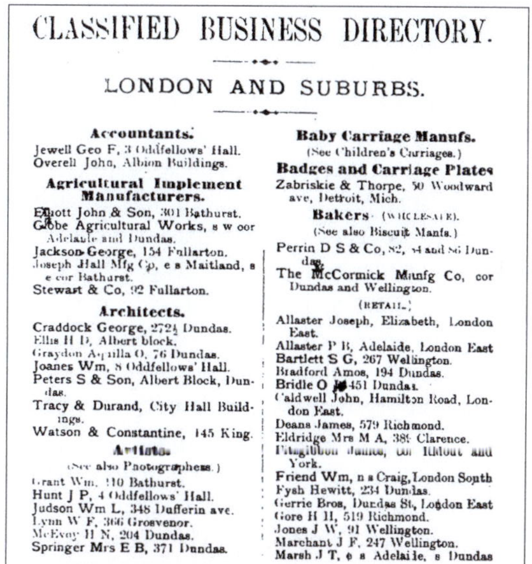

CLASSIFIED BUSINESS DIRECTORY.

LONDON AND SUBURBS.

Accountants.
Jewell Geo F, 3 Oddfellows' Hall.
Overell John, Albion Buildings.

Agricultural Implement Manufacturers.
Elliott John & Son, 301 Bathurst.
Globe Agricultural Works, s w cor Adelaide and Dundas.
Jackson George, 154 Fullarton.
Joseph Hall Mfg Co, e s Maitland, s e cor Bathurst.
Stewart & Co, 92 Fullarton.

Architects.
Craddock George, 272½ Dundas.
Ellis H D, Albert block.
Graydon Aquilla O, 76 Dundas.
Joanes Wm, 8 Oddfellows' Hall.
Peters S & Son, Albert Block, Dundas.
Tracy & Durand, City Hall Buildings.
Watson & Constantine, 145 King.

Artists.
(See also Photographers.)
Grant Wm, 110 Bathurst.
Hunt J P, 4 Oddfellows' Hall.
Judson Wm L, 348 Dufferin ave.
Lynn W F, 366 Grosvenor.
McEvoy H N, 204 Dundas.
Springer Mrs E B, 371 Dundas.

Baby Carriage Manufs.
(See Children's Carriages.)

Badges and Carriage Plates
Zabriskie & Thorpe, 50 Woodward ave, Detroit, Mich.

Bakers (WHOLESALE).
(See also Biscuit Manfs.)
Perrin D S & Co, 82, 84 and 86 Dundas.
The McCormick Manfg Co, cor Dundas and Wellington.

(RETAIL.)
Allaster Joseph, Elizabeth, London East.
Allaster P B, Adelaide, London East
Bartlett S G, 267 Wellington.
Bradford Amos, 194 Dundas.
Bridle O [illegible] 451 Dundas.
Caldwell John, Hamilton Road, London East.
Deans James, 579 Richmond.
Eldridge Mrs M A, 389 Clarence.
Fitzgibbon James, cor Ridout and York.
Friend Wm, n s Craig, London South
Fysh Hewitt, 234 Dundas.
Gerrie Bros, Dundas St, London East
Gore H H, 519 Richmond.
Jones J W, 91 Wellington.
Marchant J F, 247 Wellington.
Marsh J T, e s Adelaide, s Dundas

Variable	Match Weights
Last Name	Jaro-Winkler Variable
First Name	Exact Match: 5 Phonetic Match: 3 Nickname Match: 3
Gender	Exact Match: 1 Disagree: -1
Workplace	Exact Match: 8
Occupation	Exact Match: 5

Fig. 6.2 **Upper**: Excerpt of the London classified business directory, 1881 **Lower**: *Table 2.* Match weights used to record link census/city directory database to business directory

certain last names. For example, 'Smith' carries less weight than does 'Fowler' because there are 332 Smiths in the dataset versus 7 Fowlers. One-to-one matches were forced to ensure that a worker was linked with only one employer. A low minimum score of 10 for potential matches was maintained and only the most probable (over 16) were accepted as true matches. The potential matches were reviewed manually and adjustments to links made before they were included in the geodatabase. For workers who were employed by a large company with multiple worksites, such as the railway workers of the Great Western Railway, we used the occupation to estimate the most appropriate worksite. For example, engine builders were assumed to work in the engine house and wood and coal workers to work in the wood and coal yards. For complete unknowns, the main office was used, as it was assumed they would traverse there regularly for instructions and supplies or to retrieve wages. Each of these entries received a new record in the business directory geodatabase. For those who worked for an individual, such as gardeners or domestic servants, a new record was created in the business directory geodatabase for the home residence of the employer as listed in the city directory. To facilitate analysis, the field *WkPlaceID* was created to differentiate the worksites from the businesses.

The result is a robust sample of 5081 workers matched to their specific workplace out of the 2295 workplaces in the city. In previous studies, the largest sample of workers reported was 800, all from a single workplace (Harris and Bloomfield 1997). There were 14,707 adults between the ages of 17 and 63 in the city of London and its suburbs who were considered available for work. Age 63 was used as the upper limit as the number of individuals listed as 'gentlemen' was observed to increase significantly after this threshold. The reverse was observed for teenagers, with the rise in employment coming between their 16th and 17th year. It must be considered that 7,504 of the available workforce were women and this study was during a period where wage employment for women outside the home was still very low. It is estimated that 12% of working-aged women were engaged in wage labour or as proprietors in a home-based cottage industry.

Frances et al. (1996) found 11% of women were engaged in wage labour in Canada in 1891, while Costa (2000) found the number closer to 18% for the US in 1890. Bradbury (1984), on the other hand, estimates that 37% of Montreal women were engaged in wage labour during the same period. The numbers found for London appear to be in line, and any underrepresentation of women is likely an effect of the collection standards of the city directory enumerators. Finally, there were approximately 600 working-aged adults who lived in the city's asylum and old-age homes were included in the larger census but would be considered unfit for work because of various infirmities. The result is that the sample of 5081 workers used in this study represents 72% of the available workforce in the city of London and its suburbs in 1881, and although not as complete as the 90% sample of the total population, can still be considered the most robust ever compiled for a study of the home–work relationship in the late nineteenth century. Differences in the journey to work along ethnic and gender lines have been an area of considerable interest in the past, but little concern has been given to the representativeness of the samples which have been used to draw conclusions.

So, with the same concern as put forth previously for the representativeness of the geocoded 90% census sample, we must look to establish the reliability of the 72% workforce sample. These interrogations are critical, as subsampling inherently reduces the reliability of the inferences made about the larger population (Dixon and Leach 1977). This is particularly important in geodemographic studies where understanding the difference in behaviour or experiences across demographic variables and geographic space is the primary concern. Figure 6.3 (*Table 3*) provides a comparison between the working-aged population enumerated in the city directory and the workforce sample created in this study. Overall the sample is representative, with significant variances being found in a

Birthplace (% of Population)

	Geocoded Population	Directory With Workplace Link	Variance
Ontario	36.2	42.2	6.0
England	30.3	25.7	-4.6
Ireland	14.5	12.5	-2.0
Scotland	9.2	8.9	-0.3
Canada	4.2	4.8	0.6
United States	3.8	4.0	0.2
Europe	1.2	1.1	-0.1
Rest of World	.3	.4	0.1
Unknown	.2	.2	0.0

Occupational Class (% of Population)

	Geocoded Population	Directory With Workplace Link	Variance
1	11.3	14.8	3.5
2	6.2	7.5	1.3
3	10.3	14.7	4.4
4	42.7	44.0	1.3
5	6.4	5.6	-0.8
6	8.5	5.0	-3.5
7	4.5	6.9	2.4
8	.6	.1	-0.5
9	2.6	.9	-1.7
Unknown	7.9	.4	-7.5

Ethnicity (% of Population)

	Geocoded Population	Directory With Workplace Link	Variance
English	46.7	43.8	-2.9
Irish	29.1	30.9	1.8
Scottish	18.1	19.2	1.1
German	1.8	1.9	0.1
Unknown	1.3	1.2	-0.1
African	.9	.6	-0.3
French	.9	1.0	0.1
Other European	.7	.9	0.2
Other	.5	.6	0.1

Marital Status (% of Population)

	Geocoded Population	Directory With Workplace Link	Variance
Married	59.2	56.1	-3.1
Single	31.4	39.7	8.3
Widowed	9.4	4.2	-5.2

Gender (% of Population)

	Geocoded Population	Directory With Workplace Link	Variance
Male	81.2	85.8	4.6
Female	18.8	14.2	-4.6

Religion (% of Population)

	Geocoded Population	Directory With Workplace Link	Variance
Protestant	84.6	84.7	0.1
Catholic	13.1	13.2	0.1
Other	1.8	1.6	-0.2
Unknown	.5	.5	0.0

Fig. 6.3 *Table 3*. Comparison of geocoded city directory sample to journey to work sample, 1881

Occupational Class	Description
1	Merchant/Agent/Manufacturer
2	Professional
3	White Collar
4	Artisan
5	Semi-Skilled & Unskilled
6	Labourer
7	Servant
8	Farmer
9	No Occupation or Ambiguous *Note that this class includes students and gentleman

Fig. 6.4 *Table 4*. Modified Darroch and Orenstein occupational class scheme

select number of variables. For example, Ontario-born workers are overrepresented while those born in England are underrepresented. Single workers are overrepresented while those who are widowed are underrepresented. A brief explanation is in order for the occupational class variable outlined in Fig. 6.3 (*Table 3*). To facilitate the comparisons of the journey to work across the large number of occupations found in the sample, the author elected to use a modification of the occupational class scheme developed by Darroch and Ornstein (1980) for their study of the occupational structure in Canada in 1871. The scheme was modified by Dillon (2008) for use with the 1881 census and was provided for use in this study. The occupational class scheme is used as a proxy for socio-economic status since no wage information is available at the scale of the individual until the 1901 federal census (Fig. 6.4, *Table 4*). In addition to the lack of concern for representativeness in the sample datasets of previous studies is a complete lack of care to ensure adequate spatial coverage. No other study has considered whether the journey to work experiences they describe are representative of what individuals would have experienced across the urban fabric, from workers who lived in downtown residential hotels or above a store front, to those in the city's posh residential suburbs. Figure 6.5 (*Table 5*) outlines the spatial distribution of the workforce sample compared to those in the geocoded census with an occupation listed, by census geography, for each occupational class. Most of the variances are insignificant, with the major exception being that in Westminster

Geocoded Population vs. Total Population

By Census Geography

		Ward/Quartier No. 1			Ward/Quartier No. 2		
		Geocoded Population	Total Population	Variance	Geocoded Population	Total Population	Variance
occlass	1	11.19%	12.64%	-1.45%	15.99%	17.96%	-1.97%
	2	4.10%	3.97%	0.13%	8.15%	9.18%	-1.03%
	3	12.72%	13.36%	-0.63%	16.30%	14.04%	2.27%
	4	42.61%	43.32%	-0.71%	32.21%	33.13%	-0.92%
	5	9.05%	6.86%	2.19%	3.61%	4.33%	-0.73%
	6	9.56%	11.31%	-1.75%	4.70%	6.40%	-1.70%
	7	9.48%	7.22%	2.26%	15.52%	10.94%	4.58%
	8	0.09%	0.12%	-0.03%	0.55%	1.14%	-0.59%
	9	1.20%	1.20%	-0.01%	2.98%	2.89%	0.09%

		Ward/Quartier No. 3			Ward/Quartier No. 4		
		Geocoded Population	Total Population	Variance	Geocoded Population	Total Population	Variance
occlass	1	9.83%	11.20%	-1.37%	12.98%	17.11%	-4.14%
	2	4.62%	4.70%	-0.08%	12.98%	12.16%	0.81%
	3	9.56%	8.53%	1.03%	16.61%	15.05%	1.56%
	4	49.27%	51.88%	-2.61%	32.01%	33.40%	-1.40%
	5	8.42%	8.09%	0.33%	4.84%	4.33%	0.51%
	6	8.20%	7.59%	0.62%	2.77%	2.68%	0.09%
	7	7.71%	5.20%	2.51%	16.09%	13.81%	2.28%
	8	0.27%	0.72%	-0.45%	0.35%	0.62%	-0.27%
	9	2.12%	2.10%	0.02%	1.38%	0.82%	0.56%

		Ward/Quartier No. 5			Ward/Quartier No. 6		
		Geocoded Population	Total Population	Variance	Geocoded Population	Total Population	Variance
occlass	1	7.73%	8.16%	-0.43%	8.04%	10.01%	-1.97%
	2	4.84%	3.87%	0.98%	10.35%	11.28%	-0.93%
	3	10.39%	9.88%	0.51%	12.31%	12.92%	-0.61%
	4	48.82%	50.89%	-2.08%	42.03%	46.22%	-4.20%
	5	8.68%	7.73%	0.95%	6.22%	5.10%	1.13%
	6	11.69%	13.03%	-1.34%	6.85%	7.10%	-0.24%
	7	5.49%	4.58%	0.91%	10.56%	4.46%	6.10%
	8	0.59%	0.86%	-0.27%	0.28%	0.55%	-0.27%
	9	1.77%	1.00%	0.77%	3.36%	2.37%	0.99%

		Ward/Quartier No. 7			London (East/Est)		
		Geocoded Population	Total Population	Variance	Geocoded Population	Total Population	Variance
occlass	1	6.36%	6.14%	0.22%	7.62%	8.33%	-0.71%
	2	12.37%	13.05%	-0.68%	7.20%	2.78%	4.43%
	3	9.54%	9.40%	0.14%	5.05%	4.92%	0.13%
	4	38.52%	37.62%	0.90%	55.46%	61.41%	-5.95%
	5	7.60%	6.33%	1.26%	4.64%	3.85%	0.79%
	6	13.78%	16.31%	-2.53%	15.07%	15.13%	-0.06%
	7	8.48%	6.91%	1.57%	2.15%	2.06%	0.09%
	8	1.94%	3.07%	-1.13%	0.99%	0.72%	0.28%
	9	1.41%	1.15%	0.26%	1.57%	0.81%	0.77%

		Petersville (Village)			Westminster		
		Geocoded Population	Total Population	Variance	Geocoded Population	Total Population	Variance
occlass	1	7.11%	5.43%	1.68%	11.73%	5.34%	6.39%
	2	3.88%	3.04%	0.84%	7.62%	2.94%	4.67%
	3	10.56%	11.30%	-0.74%	9.79%	4.04%	5.75%
	4	53.45%	61.09%	-7.64%	36.03%	18.18%	17.86%
	5	9.05%	6.52%	2.53%	6.29%	2.67%	3.62%
	6	10.13%	8.70%	1.43%	11.61%	8.68%	2.93%
	7	2.37%	0.87%	1.50%	8.59%	4.24%	4.34%
	8	3.23%	3.04%	0.19%	3.14%	45.07%	-41.93%
	9	0.22%	0.00%	0.22%	5.20%	8.83%	-3.63%

Fig. 6.5 *Table 5*. Comparing spatial distribution, by census geography, of the workforce sample to the geocoded total population dataset distributed across occupational classes

Township artisans are overrepresented and farmers are underrepresented. This is due to the rural nature of the township and the methodology used in creating the social environment stage, which used fire insurance plans as the base dataset to capture civic addressing (Lafreniere and Gilliland 2015). This methodology suffers from a boundary effect between the edge of where coverage ceased for fire insurance plans and where the county atlases, which captured rural populations, began. Regardless, the sample used in this study is shown to be very representative of the overall population of London and its suburbs.

6.2.2 Calculating the Journey to Work

This chapter has thus far outlined a methodological framework for capturing a large sample of individuals at both their specific residential location and their specific daily worksite. Some studies have accomplished this feat, though not with the sample size and confidence we present here. What has limited all studies is the spatial accuracy of the calculated daily commute. Previous studies have used centroids of workplace and residential clusters, or arbitrary grids, and calculated the distances between them. Those studies that managed to map actual residences to workplaces used Euclidean distances rather than the more realistic street and pedestrian network paths. This straight-line approach removes the journey from the environment in which it took place and thus does not consider factors that may have influenced the commute. We know that the built form (building density, size, fencing), transportation networks (rail lines, roads, and pedestrian pathways), and natural barriers (rivers/lakes, elevation changes) influence the routes that individuals elect to take as they traverse the city (Ewing and Cervero 2010; Hanson 2004; Larsen et al. 2009, 2012). This study considers these environmental influences, providing a more reliable image of the daily commute to work in the late nineteenth century.

To begin, a built environment stage was harnessed to extract the 1881 road network for use in ArcGIS's network analyst. Possible pedestrian paths were added to the road network geodatabase by creating segments across obvious open space as denoted in the fire insurance plans. Two types of networks can be created for use in ArcGIS: a network dataset and a geometric network. It was elected to create a network dataset because it is optimized for undirected flow, whereas the geometric network is best used for directed systems such as rivers or utility lines. Default connectivity options were used and global turns were enabled as pedestrians are expected to be able to walk where they want and not have to follow restrictions typically imposed on vehicle traffic such as one-way streets (although there were no one-way streets in London in the nineteenth century). Length, expressed in meters, was the only cost calculated in the network; estimates of travel times were not added as they were too unpredictable to estimate for the 1880s because of poor roads and differences in personal behaviour. To calculate the path and distance to work, the shortest path route tools were utilized within the network analyst extension of ArcGIS 10. The first step was to use the calculate locations tool within network analyst to calculate and create a network location field for each business in the business directory shapefile extracted from the social environment stage. This is similar to snapping or linking the business to the road network, thus allowing the shortest path tool to more quickly and accurately locate the workplaces as stops. The distance from the centre of the businesses building polygon to the network was captured (an average of 24 m) to be added to the final calculated journey to work distance. The network dataset was used as the input analysis network. The search tolerance was set to 200 m in order to capture the few businesses that are far from the road network, such as the Great Western Railway and the Imperial Oil refining facility.

Numerous barriers to pedestrian travel were added before routes were calculated. These included dense multi-track rail yards and the two branches of the Thames River. Tolls existed on

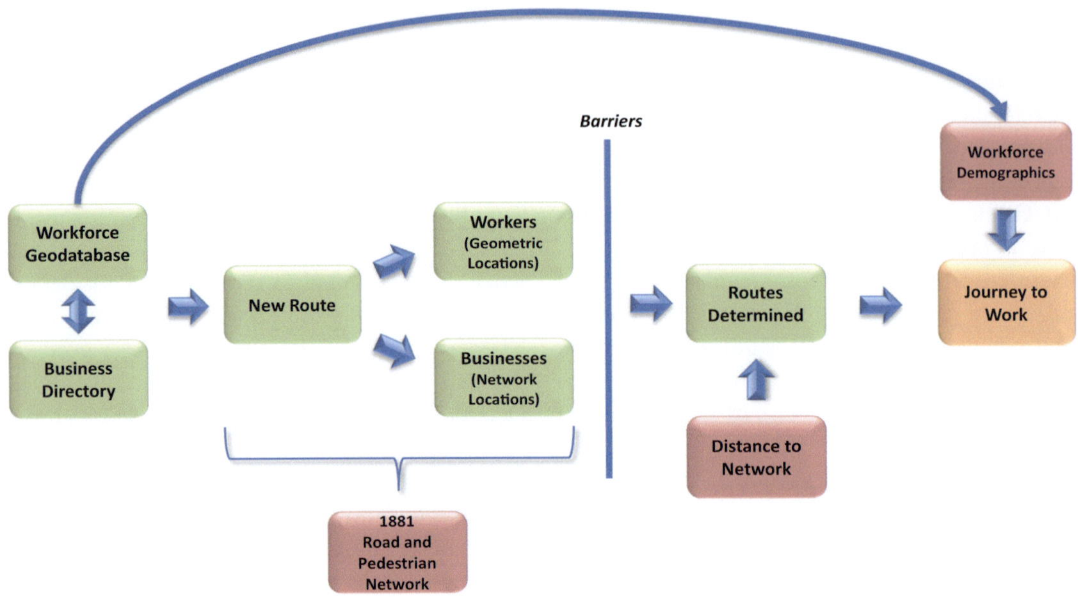

Fig. 6.6 Network-derived journey to work scheme

the many routes that emanated in and out from London's city limits, especially on bridges such as Clarke's Bridge on present-day Wellington Road. Diaries of the period suggest that these costs influenced travel paths for some of the lower income residents of the city and suburbs. While they may have had an impact on journeys, we elected not to include these tolls as barriers because of the difficulty in determining a toll's influence on a given individual. The workforce sample geodatabase and the business directory (with the newly created network location fields) were joined and a new route started in network analyst. The workers were loaded as stops using the default geometry search options. The businesses were then loaded as stops using the network location fields. The routes were solved, resulting in 4,945 routes calculated, and the distance to the network was added back to produce a final journey to work distance (Fig. 6.6). These routes represent the journeys of 97% of all workers. An examination of the failed routes showed that there were two workplaces, the insane asylum and Huron College, that were beyond the extent of the road network derived from the fire insurance plan and thus the routes to them were not able to be calculated. These workers were removed from the study. The final routes were then joined back to the original workforce geodatabase to allow for a socio-demographic relationship analysis.

The result is a reconstruction of the daily journey to work for some 5,000 workers from across the urban fabric of London and its suburbs. Figure 6.7 (Upper Left) outlines the distribution of commutes from the shortest to the longest. We see that the average commute for all workers is 1,071 m (the median is 954). 53% of workers travel more than 1 km but 88% travel less than 2 km. Figure 6.7 (Main) illustrates the completed journeys aggregated to each individual street segment. This provides a unique view of the daily commuting patterns in a nineteenth-century city. We can see the density of workers on street segments in the city's downtown core and the heavy use of arterial streets such as Richmond, Hamilton, King, and Queen. As we may expect, the pattern thins along the tentacles of the street network that reach out to the suburban fringe. In our largest contemporary cities, real-time traffic flow data is fed to our smartphones and vehicle GPS units or announced

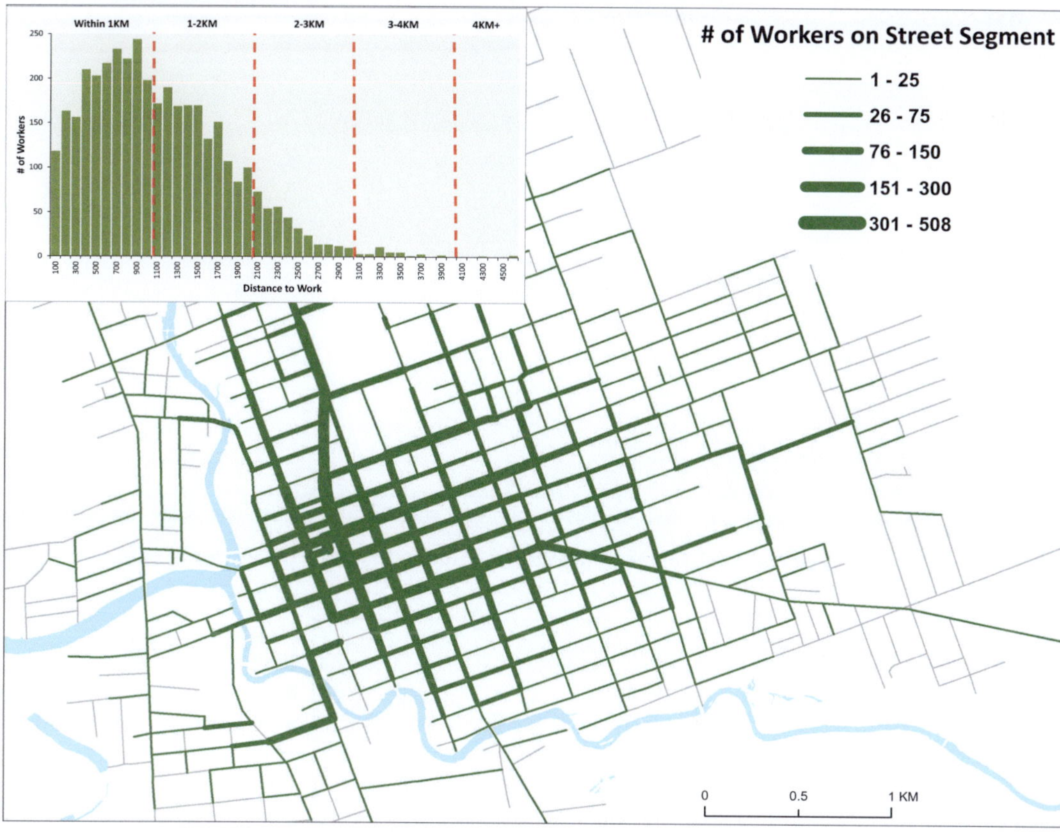

Fig. 6.7 **Upper left**: Distribution of distances travelled to work. **Main**: Daily traffic patterns in London and suburbs, 1881

to us over the radio. This view is the closest we have thus far to the same type of information for the pre-automobile city.

6.2.3 Comparing Euclidean and Network Methods

This study has depended upon the re-creation of daily journeys based on network distances rather than the Euclidean distances used in all previous studies. The use of network paths places the commutes in a more representative environment and allows for a deeper understanding of the role that natural and man-made barriers played on the routes individuals had available to them. But do the distances really change between the two methods of calculation? Figure 6.8 (Upper) illustrates that there is an overall average of a 20.8% underrepresentation when using the Euclidean method used in all previous studies versus the network method described here. We see that the Euclidean distance method used in previous studies has less effect on results as the journey lengthens; for the shortest journeys, distances are significantly underrepresented when the Euclidean method is used (Fig. 6.8, Lower).

The observed difference between the Euclidean and network-derived distances is a result of the aforementioned natural and man-made barriers. To illustrate, Fig. 6.9 (Left) outlines the difference between the Euclidean and network-based journeys of Ernest Smith, a bookkeeper who lived in the suburb of London South. His 1.5 km commute took Ernest across the iron York Street Bridge over the southern branch of the Thames River each day to his job at

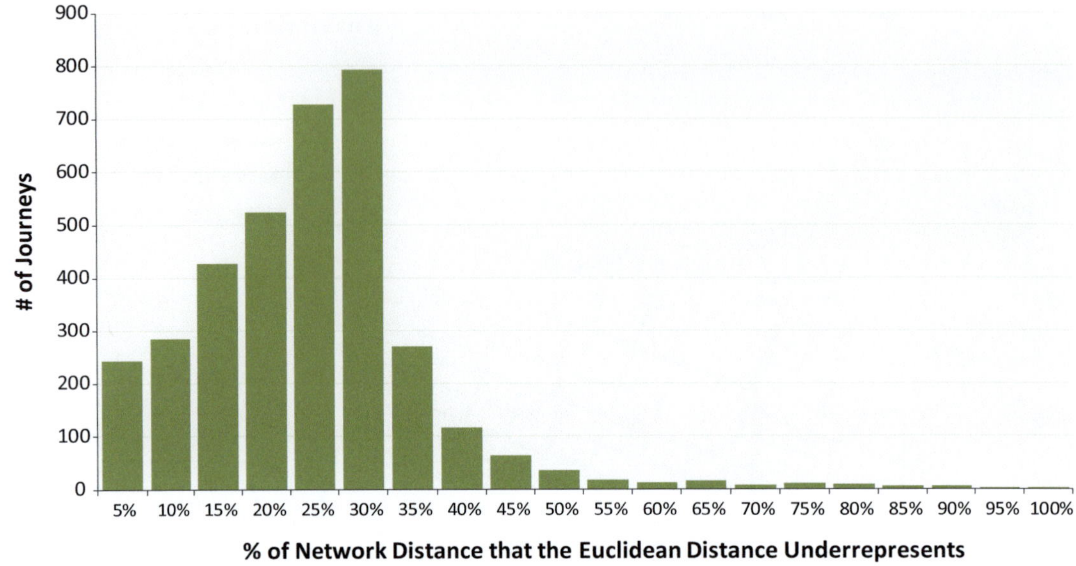

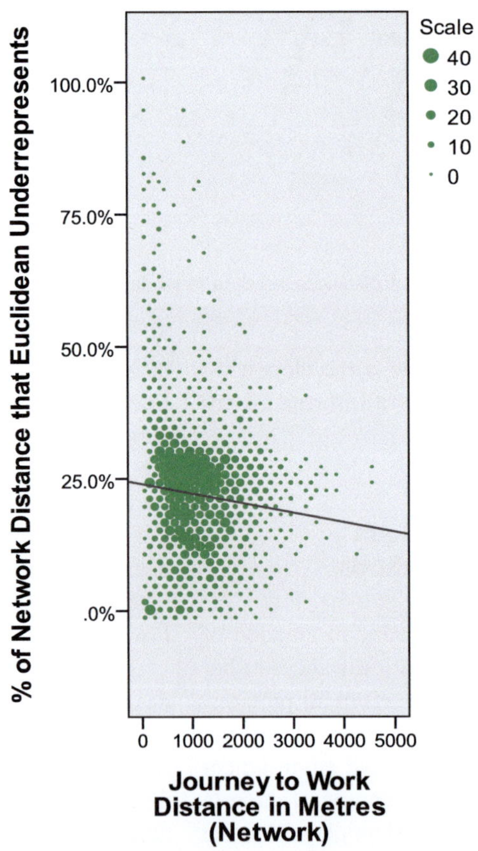

Fig. 6.8 **Upper**: Percentage of network distance that is underrepresented when the Euclidean method is used—by number of journeys * Euclidean mean is 848 m, network 1071 m. **Lower**: Percentage of network distance that is underrepresented when the Euclidean method is used as related to network-derived distance

the Bank of Montreal. Figure 6.9 (Left) illustrates that not only was a Euclidean journey literally not possible, but the difference between a Euclidean journey and a network one is substantial. What Fig. 6.9 does not show is the steep banks of the river that would have made it difficult to traverse any available shortcuts or alternative river crossings. If we look closer (Fig. 6.9 Right), we see the inability to continue with a more straight-line journey after crossing the river, as the density of the urban core of London restricted travel to the established roadways. These barriers resulted in a commute that was over 25% longer than the Euclidean distance, resulting in an extra 5–15 min' walk, depending on the weather and other factors.

6.2.4 Comparing to Assumed Workplaces Method

Many of the most influential studies of the nineteenth century journey to work utilized a method of approximation that assumed that workers were being employed by the closest employer for which they were deemed qualified to work (Dennis 1984; Deskins 1972; Greenberg 1980; Hershberg et al. 1981; Pooley 1984; Vance 1967). This paper suggests that this method of approximation leads to a significant misrepresentation of the true relationship between work and home, leading to an underestimation of the time spent travelling to work each day. A framework to overcome these assumptions has been outlined above, but the improvement has yet to be demonstrated.

To perform a comparative analysis between the two methods, it was necessary to find the closest workplace to each worker's home that would employ a worker with a related occupation, and then compare that distance to the actual distance the worker was travelling to work each day. To estimate what would be the closest workplace for which a given worker was qualified, it was necessary to link each worker with his/her industry of employment. However, due to the large number of industries (158) and workplaces (2295)—each of which would require a separate analysis—it was elected to use a sample. Two types of workplaces were chosen to represent two different industries, breweries and cigar factories. They were chosen because they represent workplaces that were not listed in multiple industry categories in the business directory, and thus provide a more direct relationship between home and a potential closest employer. Additionally, while they are both medium-sized industries in which workers commuted from a wide geographic area, they have differing levels of skilled labour and industrialization. Breweries employed 95 workers, a mix of skilled and semi-skilled labour, in a heavily mechanized environment. Maltsters, brewmasters, and coopers made up the skilled class, while semi-skilled workers included bottlers, packers, warehouse clerks, and teamsters. Some unskilled positions were also found in the city's breweries, such as those whose tasks were to shovel grain, load wagons, or clear rodents out of vats and barrels (Phillips 2000). Cigar manufacturing, a rapidly growing industry in 1881, was a modest operation with products made by hand by individual workers using a limited number of simple tools such as a rolling board, knife, and wrapping leafs. A total of 118 workers in our database were working in cigar factories. The 1889 Royal Commission on Labour and Capital includes testimonies from Londoners suggesting that many more individuals were employed in cigar making (Labour Commission 1889). They were primarily young women and children, recent immigrants who were 'secured by indenture' and paid the lowest of wages to complete the simple task of rolling tobacco leaves into cigars (Senecal 1889). We do not capture many of these indentured workers in this study. Those we do capture, however, represent a sample similar to that discussed in the Royal Commission reports, comprised of unskilled cigar makers, packers, and wood planers, including many who are women and teenagers.

To find the closest potential workplace on the road and pedestrian path network, we utilized the closest facility analysis tools in network analyst. Workplaces were loaded as facility locations using the network location fields created for the

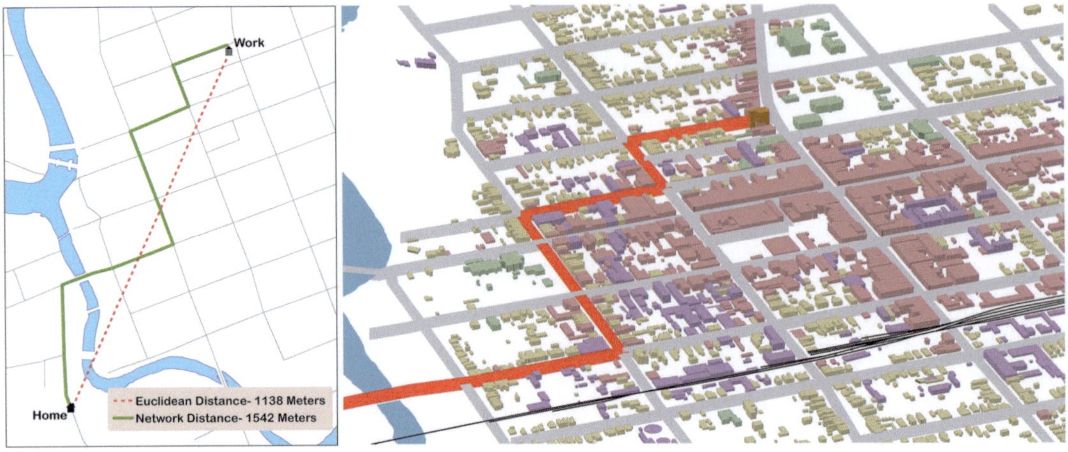

Fig. 6.9 **Right**: Ernest Smith's journey to work using Euclidean and network methods. **Left**: Closer look at Ernest Smith's journey to work in the dense urban core

journey to work analysis. The workers were loaded as incident locations, again using the default geometry location functions. Routes were calculated between each worker's residence and the closest workplace within their industry. The exported result has a name field that indicates both IDs from the facilities (workplaces) and incidents (workers) separated by a hyphen (e.g. 125-50). These IDs needed to be separated so the distance calculations to the closest facility could be joined to the dataset with the worker's actual distance to work. To facilitate, a new field was created and the following VBScript run in the field calculator: *Left([Name], ((InStr (1, [Name],"-",1) -2)))*. The difference between the actual and potential distances was calculated with a 0 difference denoting that the worker was working at the closest employer and any positive number reflecting the difference in metres. Figure 6.10 (*Table 6*) charts the results, while Fig. 6.11 (Left and Right) provides graphical representations of the differences between closest and actual employer for brewery workers and cigar makers, respectively.

It is observed that only 19 of 95, or 20%, brewery workers are employed at the closest brewery to their residence, resulting in a difference of nearly half a kilometre in their average journey to work. A closer look reveals that of the 19 workers who did work at the closest brewery, 10 were employed as maltsters, a job that requires careful attention to the germination of the cereals used to create the malt for beer production. Grains had to be crushed, weighed, and boiled into mash, then cooked, filtered and drained, and boiled again with hops before fermentation (Phillips 2000). This delicate chemical process required constant attention and maltsters were on call seven days a week, living near their workplace in case they were needed to suddenly stop a malting that might finish early. Of the remaining nine workers, two were watchmen and another two held positions of responsibility as 'engineer' and 'foreman'. For cigar factory workers, only 13 of 105, or 12%, were employed at the closest factory to their residence, resulting in a difference of a half a kilometre in their average journey to work, similar to the brewery workers. With the exception of one foreman, a John Gerlach of WT Rutherford & Co's, the availability of inexpensive, convenient accommodations appears to have influenced the housing decisions of those who worked at the closest employer. All of the other cigar workers who worked at the closest factory were boarding either in a residential hotel or in a dwelling with non-related individuals. More than half of the women were under the age of 20, and another two were 15-year-old boys who were neighbours and worked together at William Kelly and Sons

	Total Workers	Work at Closest	Actual Workplace	Mean Difference in Meters
Brewery	95	19	76	496 Meters
Cigar Factories	118	13	105	528 Meters

Fig. 6.10 *Table 6*. Working at closest workplace versus actual workplace: brewery and cigar factory workers

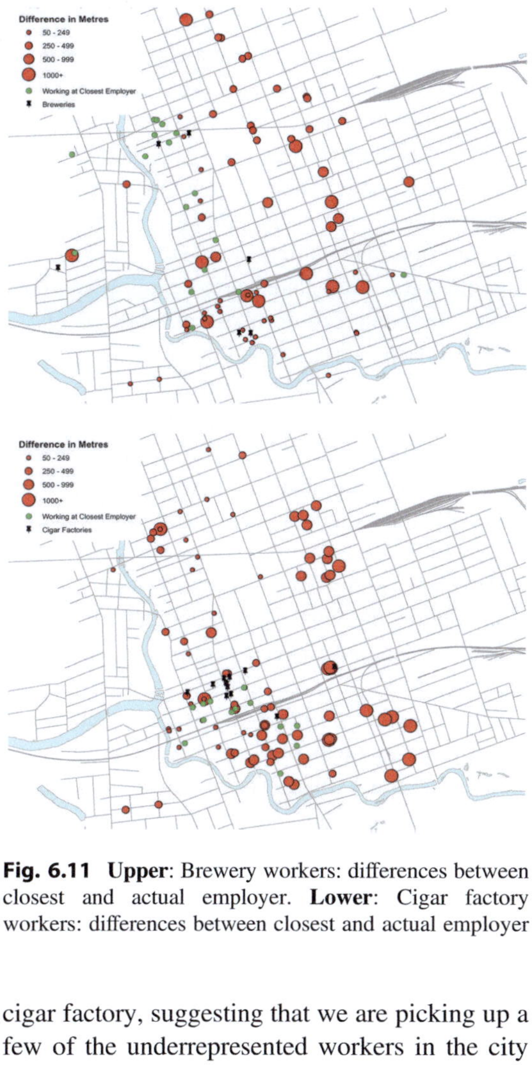

Fig. 6.11 **Upper**: Brewery workers: differences between closest and actual employer. **Lower**: Cigar factory workers: differences between closest and actual employer

cigar factory, suggesting that we are picking up a few of the underrepresented workers in the city directory. Most of the city's cigar factories were located in the downtown core among other medium-sized factories that used paper and wood products, such as furniture manufacturers, carriage factories, and paper bag makers. Rents were low in this district of the city with most residential offering being in hotels or above storefronts, thus allowing an opportunity for the lowest paid cigar factory workers to find accommodations close to work, reducing their commutes to help compensate for the long work shifts that were demanded of them. It is recognized that many factors go into residential and workplace choice; however, these two examples, each with their peculiarities in occupational demands or residential needs, help illustrate that there are significant errors and omissions if one assumes that people work for the closest employer for which they are qualified. By establishing true home–work links, we can not only better understand what was a more realistic daily journey to work but can also begin to provide some insight into residential choices made by workers in the nineteenth-century walking city.

6.3 The Walking City Observed

The robustness of the sample, the spatial accuracy of the geocoding, and the sophistication of a GIS-based approach to the historical journey to work allows researchers to interrogate the daily commute in ways other studies could not. The socio-demographic variables available in the social environment stage can be extracted and used with the home–work linked sample to interrogate the commuting patterns along lines of ethnicity, age, gender, occupational class, or any combination of these that may be of interest. This section illustrates the possibilities of interrogating journeys by occupational class, workplace location, and gender. Before completing the socio-demographic analysis, it was necessary to first separate out those workers who were working at home and those who had a commute. Those who lacked a commute to work were as important to the social geography of the industrial city as those who had a daily journey. Although many records in the city directory directly indicated that work was taking place in the home, there were numerous individuals for whom the information was either vague or missing. The expectation would be that the geocoding of home and work would then intersect; however, due to occasional imprecision in digitization or geocoding, some people who were indeed working at home were shown to have short commutes of anywhere from a fraction of a metre to 20 or 30 m. A decision was made to consider anyone working less than 50 m from home to be working at home. This threshold was selected as 50 m covers the maximum distance seen in the city lots from the front yard to a rear tenement, the residential experience for some of London's domestic servants. See Fig. 6.12 (Left).

6.3.1 Differing Commutes by Occupational Class

Interrogating how commutes vary across occupations, industries, or various occupational groups is a frequent and natural interest for researchers who examine the journey-to-work. Most of the studies outlined in the beginning of this chapter dissected the aggregate patterns of commuting along occupational lines. Goheen (1970), examining the journey to work in Toronto in 1860 and 1890, sampled select occupations to represent differing occupational sectors. For example, he uses a brewery and piano manufacturer to illustrate the industrial journey to work, and a retail outlet and life insurance company for the clerical journey. Lawton and Pooley (1976) followed a similar model using dock workers, shipbuilders, and office staff for their study of 1871 Liverpool. Aggregating individual occupations into three classifications is common in many studies, though the categories are not. Bater (1974) divided workers in St. Petersburg, Russia into managerial and professional, workshop owners and artisans, and factory workers. Barke (1991) divided his sample for Newcastle upon Tyne into the categories of professional, which included occupations such as accountants, lawyers and architects, producer-retailer, which included blacksmiths, watchmakers, and bakers, and finally retailers, which included pawnbrokers as well as those who sold fish, glass, or poultry. Taking a different approach, Hershberg et al. (1981) used industries to select their sample, starting first with ship building, sugar refining, iron rolling, banking, and morocco leather finishing. They were selected for their spatial coverage across Philadelphia, the ethnic composition of their workforces, and their respective longevity as industries in the city. To supplement their sample, they also used seven occupations to represent the swath of social classes, with physicians and lawyers representing the professional class, confectioners and bookbinders the small proprietors, and blacksmiths, cabinetmakers, carpenters, and the artisanal proprietors. Mayer (1977) categorized all occupations into one of eight

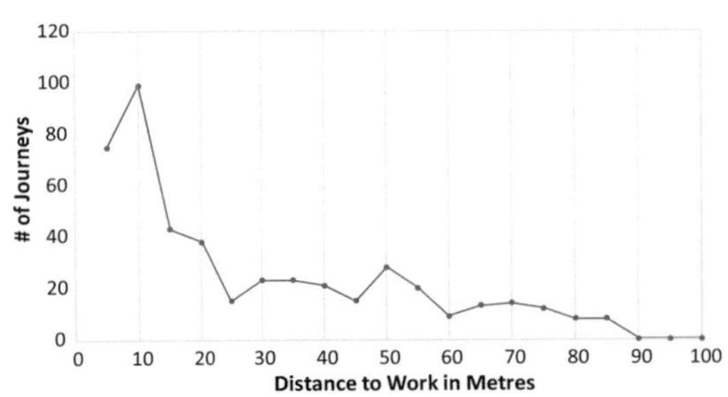

Occupational Class	% combining home & work	.2	.4	.6	.8	1.0	1.2	1.4	1.6	1.8	2.0	Median (km)
Merchants/ Agents/ Manufacturer	37.1	43.3	48.4	55.7	66.5	73.7	78.6	84.5	88.9	93.5	95.4	.86
Professional	26.2	31.8	36.5	44.1	51.7	61.7	68.0	75.9	80.3	85.0	89.8	1.04
White Collar	6.8	10.7	20.9	31.9	40.9	51.0	61.1	71.5	82.6	90.6	95.9	1.07
Artisan	14.6	19.7	28.0	36.3	47.2	58.1	66.8	73.9	80.5	85.9	90.0	.98
Semi-Skilled & Unskilled	22.6	31.7	39.8	49.3	58.8	65.3	70.9	77.0	82.9	88.0	90.8	.89
Labourer	13.0	14.6	23.0	38.3	46.0	57.1	67.4	74.3	79.3	84.7	88.9	.98
Servant	83.1	91.5	94.4	95.1	95.4	95.9	97.2	97.7	97.7	98.2	99.1	.19

Fig. 6.12 **Left**: Used to determine work at home versus work outside of the home status. **Right**: *Table 7*. Cumulative percentage at specified distances from workplace (in km)

occupational classifications using a scheme developed for the 1940 U.S. census by Edwards (1940) and later modified by Thernstrom (1973) and Conzen (1976).

To evaluate the differing journey-to-work across occupations in London, we follow Mayer's lead and utilize an occupational class scheme, though unlike Mayer, we employed one developed for the census that was used to build the dataset. This method allows us to easily examine all occupations across the entire spatial extent of the city rather than employing a sampling strategy, as done in the other studies. As all other studies have utilized Euclidean rather than network distances, it is impossible to accurately compare distances travelled across various cities. Additionally, the absolute distance travelled is largely influenced not only by residential and workplace choice, but by the size of the urban area, its density, the configuration of the street network, level of industrialization and commercialization, and development of transit systems, as well as barriers, both natural and man-made. Instead, patterns at the class level can be colloquially compared to previous studies by asking questions such as: are the distributions of the journey to work travelled different between occupational groups? Were labourers living near their factories and the professional classes near their offices? Do higher status workers travel greater distances than their lower class counterparts? *Table 7* in Fig. 6.12 outlines the journey to work experience for the seven urban occupational classes in the modified Darroch and Ornstein scheme. At first glance, with the exception of the very low median journey of servants, we see little difference in the median journey to work between the classes, with a difference of only 210 m between the longest and shortest median distances travelled by those who were not in service. As is observed in all other studies with a predominantly British population, most servants lived in their masters' domiciles, and those who did not lived very nearby, usually in boarding houses with other servants. However, if we examine the journeys closer by looking at the cumulative percentages at distance bands from 200 m to 2 km, we reveal a more complex pattern. Excluding servants for the time being, those in the highest occupational class of merchants, agents, and manufacturers, lived overwhelmingly closer to their workplaces than workers in any other group. Nearly 50% lived less than 400 m from work and three-quarters lived within a kilometre. They have the highest cumulative percentage of workers living close to work across every distance band but the very longest. Surprisingly, this pattern is not shared with the studies outlined above, where they found that factory owners and merchants, if they did not live at their businesses, lived a considerable distance away compared to other occupational groups. This difference may be a function of the very dense urban core of London that permitted a very short distance from the commercial and most of the industrial firms to the desirable neighbourhood on Talbot Street overlooking the Thames River and to the Woodfield neighbourhood adjacent to the city's only public park. The exception would be those industrialists involved in the growing rail and barrel works in the suburb of London East, as they lived in the same neighbourhood as their colleagues, but their businesses were significantly further away. When we turn our attention to the experiences of professionals and white collar workers, the patterns match many other studies. For example, Hershberg found that bank employees travelled further than their blue-collar counterparts. Barke found the same in Newcastle Upon Tyne, as did Layton and Pooley in Liverpool, Goheen in Toronto, and Mayer in Milwaukee. Interestingly, in London, one sees approximately a 10% increase in white-collar workers from one distance band to the next until 1.6 km, where the percentage catches up to the other occupational classes and the rate of increase slows. What is less pronounced in London is the number of labourers living close to their workplaces. This pattern was observed in Liverpool, Milwaukee, and somewhat in St. Petersburg. In London, however, labourers lag behind the other lower class group, semi-skilled and unskilled workers, until they meet at a 1.2 km journey. This may be explained by the high demand and limited inventory of housing in the core and nearby neighbourhoods. Instead, we

Fig. 6.13 Relationship between labourers and their workplaces

see labourers travelling from modest homes north-east of downtown as well as from the neighbourhood along the north shore of the south branch of the Thames to jobs along the rail lines in the centre of the city as well as the new industries in London East (Fig. 6.13).

6.4 Coincident Home–Work Locations

Studying the home–work relationship has primarily been the domain of feminist scholars who have been interested in the extent of unpaid domestic work by women and girls or the important role that piecework and cottage crafts played in supplementing the household income (Bradbury 1984, 1996; Bullen 1986). However, significant numbers of men still combined work and home. In the aggregate, we find that in 1881, 26% of London workers were combining home and work. 61% of the women worked at home whereas only 16% of men did. Women represent just over half (53%) of the work-at-home population in London in 1881. The methodology used to determine who worked at home versus outside the home was not clearly outlined in the few other studies that have directly investigated this relationship. The most comparable studies to this one are the work done by Hershberg et al. (1981) and Mayer (1977) who interrogated the frequency of coincident home–work locations in 1880 in Philadelphia and Milwaukee, respectively. Similar to this study, they relied heavily on the workplace information provided within the annual city directory; however, neither study outlines how they deemed a worker to be working at home. *Table 8* in Fig. 6.14 compares the frequency of workers who combined home and work, by occupation, for London to the case

Fig. 6.14 **Left**: *Table 8.* Percentage of workers with coincident home–work locations for select occupations: London, Philadelphia, and Milwaukee, 1880–1881

	London		Philadelphia		Milwaukee	
	N	% Combining Home & Work	N	% Combining Home & Work	N	% Combining Home & Work
Lawyers	14	0%	40	17.6%	57	10.5%
Physicians	38	84.2%	40	95%	44	65.9%
Carpenters	143	6.9%	115	53%	31	41.9%
Confectioners	28	14.2%	765	90.6%		
Bookbinders	16	18.7%	54	9.3%		
Blacksmiths	100	8.0%	112	23%		
Cabinetmakers	56	21.7%	129	54.3%		
Agents (unspecified)	40	5.0%			9	22.2%
Clerks (unspecified)	394	7.6%			19	26%
Tavern Proprietors	8	75.0%			206	89.3%
Barbers	24	20.8%			36	55.6%
Coopers	41	24.2%			34	64.7%
Foremen	59	22.0%			23	13%
Merchants	322	43.8%			54	29.6%
Butchers	54	46.3%				
Tinsmiths	35	8.6%				
Bakers	17	70.6%				
Servants*	511	87.9%				
Dressmakers*	59	55.9%				
Cooks*	40	72.5%				
Nurses*	44	97.5%				

* denotes occupations dominated by women

studies of Philadelphia and Milwaukee. All occupations listed in the comparable studies are included as well as select additional occupations for London. With the robustness of the London sample, we have the possibility to report every occupation, but have elected to only report some of the most notable here. Considering that each city has its own local industries that dominate the economy and the comparison cities have significantly larger populations than London (Philadelphia 850,000 and Milwaukee 115,000 in 1880), sample sizes for specific occupations are surprisingly similar across the cases.

Examining specific occupations, we see some immediate similarities between the three cases. No lawyers in London combined their home and work, and few did in Philadelphia or Milwaukee. Lawyers depended on centralized locations near the heart of the commercial district of the city, close to their clients in banking and insurance. However, they also had incomes that allowed for greater residential choice, with most electing to reside in some of the city's nicer neighbourhoods away from the downtown core (Fig. 6.15, Upper). The result is a long median journey to work of 1350 m, a pattern shared with the cohort in Philadelphia and Milwaukee. Unlike lawyers, in all three cities physicians rarely separated work from home. Instead, physicians operated out of their homes, in the transitional areas between the commercial core and the residential areas of the city. Two clusters are visible in London, one at the eastern edge of downtown bounded by Clarence and Colborne, largely along Dundas and Queen, and the other north of downtown between Talbot and Richmond. The five doctors who did have offices outside of their homes also located them in the same areas of the city (Fig. 6.15, Lower). One of the few offices in the commercial core was operated Dr. Oronhyatekha, a famous Mohawk doctor and businessman. His 1 km journey to work from his residence on Litchfield Street was the longest of all physicians in the city, where the median journey of physicians is a mere quarter of a kilometre. His election to operate his practice out of an office downtown on Richmond Street may be related to his appointment as the first Supreme

Fig. 6.15 **Upper**: Lawyer's residences and law offices in London, 1881. **Lower**: Doctor's residences and doctor's offices in London, 1881

Chief Ranger of the Independent Order of Foresters, whose offices in the Albion Buildings were a mere 150 m north of his office.

Skilled workers, such as carpenters, cabinetmakers, and blacksmiths, in London had a different home–work relationship than did their counterparts in Philadelphia. In London, these skilled workers rarely combined home and work. In Philadelphia, over half of carpenters and cabinetmakers, and nearly a quarter of blacksmiths, worked out of their homes. Hershberg et al. (1981, p. 138) suggest that Philadelphia saw an increase of small firms, employing 1–5 workers, working largely at home because the small profits associated with this artisanal production would have made operating a separate workshop cost-prohibitive. In London, carpenters and cabinetmakers had moved out of artisanal work by 1881 into the industrial setting, dominated by a couple of large factories. We find 45 (32%) of the carpenters were working at the Ontario Car Works in London East, building parlour train cars for the Great Western Railway. Another 26 (18%) were employed at Thomas Green's planing shop and sawmill. The remainder worked as carpenters in factories such as John Elliott's Phoenix foundry, where they made agricultural implements such as plows and shovels. Cabinetmakers follow much the same pattern, with over 80% being employed by three firms, the London Furniture Company, John Ferguson's coffin and cabinet factory, or the Bennett Brothers office furniture factory. London's blacksmiths were not found in a select few dominant factories, but rather spread across the industrial fabric of the city. The 100 blacksmiths in London were employed across 38 different factories, differing in form and function. Though not centralized, blacksmithing in London was still different from the more artisanal nature of blacksmithing reported for Philadelphia. The few blacksmiths who did combine home and work in London were all found in small homes with shops in the rear, located exclusively in largely industrialized areas of the city, presumably to reduce the possibility of irritating their neighbours.

Confectioners and coopers followed the same pattern, where few in London combined work and home compared to their counterparts in Philadelphia and Milwaukee. Most stark is the difference between London and Philadelphia for confectioners. A surprising 90.6% of confectioners in Philadelphia worked in their homes, versus a modest 14.2% in London. The only explanation provided for Philadelphia is that because they 'served a neighborhood market… they had little reason to separate their work and residence' (Hershberg et al. 1981, p. 137). In London, all but seven of the confectioners were employed by one of two firms, either McCormick's Manufacturing or Daniel Perrin's Biscuit Co. Both firms were substantial in size, employing over a hundred men, and had a large distribution of their confections across Canada. Coopers in London followed a similar pattern, with the majority working for either Seale and Child's Cooperage or that of William Hawkins. Both firms supplied Labatt and Carling the barrels needed for their beer and ale. In contrast, Milwaukee's coopers worked in a more artisanal setting, with one or two coopers working out of a home, utilizing the ready access to cheap lumber from Wisconsin's growing timber trade to supply barrels to one of Milwaukee's 40 or so breweries. Turning our attention from industrial labour to commercial work, we see an expected home–work relationship pattern in London. Tavern proprietors, like their counterparts in Milwaukee, primarily lived above or behind their pubs. Agents and clerks largely worked outside of the home, as they served the needs of either a specific organization or the public at large. Curiously, this does not correlate with the experience in Milwaukee, where a surprisingly high percentage of agents and clerks worked at their residences. Unfortunately, Mayer does not specifically address why this was observed, an omission that raises the curiosity of this researcher.

6.4.1 Women's Experiences

Women's relationship between work and home is not explicitly treated by Hershberg et al. or by Mayer, so comparisons are not possible. A glance at Fig. 6.16 provides some insight into the experiences of women in London. Women's paid labour was dominated by occupations that extended from their duties at home. Few married women had steady employment outside of the home and thus most did not report an occupation to a census enumerator. Bradbury (1984) outlines that women worked for wages occasionally, taking in washing, ironing, sewing, or performing other domestic duties for neighbours, relatives, or friends. Dressmaking and tailoring in particular were popular part-time occupations for women; most women were sewing at home for their families and the more entrepreneurial also made items for sale to supplement the family income. Few of these part-time dressmakers are captured in the city directories; however, we are

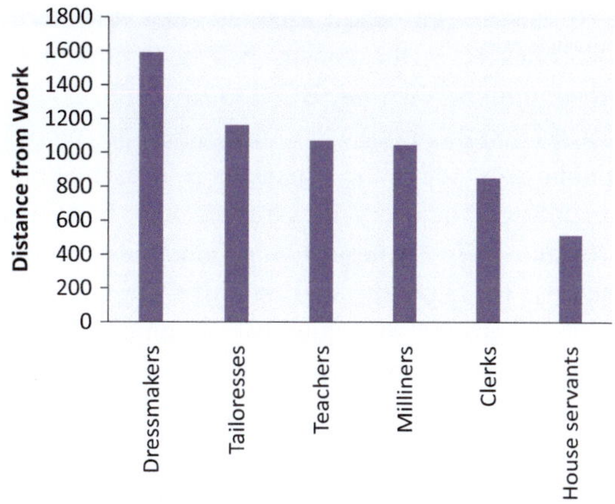

Fig. 6.16 Distance to work for women, select occupations, in metres

able to examine 59 dressmakers who appear to work on a more full-time basis producing dresses for the ladies of London.

Interestingly, only 56% of them work from home, with the remainder working primarily as one of many women at a dressmaking shop, such as those owned by Lydia Crone and by Mrs. Montague. These dressmakers had the longest commute of any women working outside the home, with a median journey of nearly a mile (1.6 km) (Fig. 6.16). The daily newspaper, the *London Advertiser*, had a steady increase in the number of notices in the female help wanted section for dressmakers and sewing girls through the 1870s. This increased demand for 'outsourced' dresses was partially a response to the School Act of 1871 which made schooling compulsory for all children, including the young girls who would have helped with the sewing (Ross 1896).

This loss of domestic labour, combined with the increases in real wages throughout the period, meant that women began to purchase dresses rather than make them themselves. The advertisements for dressmakers in the newspaper do not list a wage, though the frequency and voracity of the want ads suggest that the wages may have been higher than women would have made sewing at home, finding employment in the cigar factories, or working as cleaners in the one of the many hotels.

It should be acknowledged that the term 'dressmaker' in the census has been linked to prostitution in Victoria, BC (Dunae 2008, 2009). Census enumerators were instructed to record 'public women' in the census as a 'dressmaker'. Of course, not every dressmaker was in fact a prostitute. We harnessed a geodatabase of jail records for London and suburbs created for another project to quickly examine if the euphemism was possibly used in London, following the same methodology employed by Dunae. There were 54 arrests in 1881–1882 for the offence 'house of infamy', 'inmate of house of ilfame [sic]', or 'prostitute'. Six of these women had their occupation listed in the census as dressmaker, though none are found in our sample of workers. The enumeration goals of the city directory would have encouraged the directory enumerators to turn a blind eye to houses of ill repute. Although we don't have any 'prostitute dressmakers' captured in this study, the evidence suggests that it was possible that the environment found by Dunae in Victoria may also have existed in London.

The transition of girls to school and increasing household income resulted in more than just the need to purchase dresses from a dedicated dressmaker. It also helped support an increase in employment of domestic servants in Canadian middle-class homes. We found over 500 female domestic servants in our journey to work sample; this was the largest single occupation of women in the city. Nearly all (88%) lived in their masters' homes. The domestic servant was usually a single, young woman (median age in London was 23), and was primarily of English, Irish, or Scottish descent. Being 'in service', however, came with a stigma. Some scholars suggest that those women who had a choice preferred to work in factories, offices, or shops regardless of a difference in pay (Barber 1991, p. 8), while others argue that working-class parents preferred their daughters to work in private homes rather than being exposed to the dangers of the factory or receiving the low wages of the shops (Fahmi 1997, p. 73). We found the frequent advertisements for reliable servants to be increasingly dominant in the help wanted section of the newspaper from 1879 to 1891. The advertisements also served as a public notice that a household was capable of paying for domestic help, as the live-in servant was an important status symbol in Victorian Canada. Cooks and nurses represent another group of women who lived at work rather than working from home. London's cooks worked overwhelmingly in the city's hotels, working for both a wage and for room and board. They were older than domestic servants, with a median age over 30. The few that did not live at the hotel had a very short commute, averaging only a few hundred metres. All but a select few nurses lived at work, either caring for children in the Protestant Orphans' Home or for the infirm at the insane asylum. This section has illustrated the importance of interrogating the relationship between home and work, particularly for revealing the situation of a large portion of women. Women had long work days in the domestic sphere, and so did men in the factory. This was a period of industrialization when, in the aggregate, production shifted from being home-based and artisanal to factory-based. However, the decision to work outside or inside the home is still not always clear to historians and historical geographers. We need to continue to seek to understand the complexities of the geography of work in the nineteenth century.

6.5 Conclusion

This chapter has outlined a new methodological framework for re-creating the spatial relationships between work and home. The methodologies herein overcome many of the limitations of previous studies, helping to invigorate an area of research that has had little attention since the early 1990s. Although much of the data is the same as used by other researchers, by harnessing a complete digital record of the city's population and businesses, we can begin to understand the journey to work for not just a subset of the city's population, but for nearly every single individual, occupational group, workplace, or neighbourhood in the city. With a robust record linkage procedure, we have been able to make absolute links between workers and their workplaces and calculate their daily journeys on a model of the street and pedestrian paths of the city. This method overcomes the concerns that Dennis (1984) postulated about quantitative research on the journey to work due to a lack of data that directly links work and home. It must be acknowledged that the methodologies used herein are not perfect. Like any interrogation of the past, we are limited by the quality of the data recorded. In this case, we must remember that city directories were compiled for an explicit financial purpose, to sell advertising and subscriptions to businesses and citizens alike (H.H. B. 1913). It was very much like a paid version of the telephone book that was commonly distributed a few years ago before the creation of internet searches. The census was not perfect either, as Curtis (2001), Hamilton (2007), Steckel (1991), and many others have thoroughly outlined. Census enumerations were and continue to be wrought with imprecise methods and procedures resulting in marginalized individuals being undercounted, misrepresented, or omitted completely. For this study of the journey to work, the effects include a skew towards full-time wage-workers, rather than those who worked part-time, did piecework, or were chronically unemployed. Married women who did not work outside the home, as well as teenagers who worked after school and on weekends, are also missed. These issues affect this study and must be considered when findings are analyzed. The algorithms used may also skew results. The study utilizes shortest path techniques provided by the software company ESRI. Although the mathematical models used have been scrutinized and accepted by the larger GIS community, the assumption that individuals are most efficient in their commutes is still not confirmed. In the walking city, it would be expected that some individuals trip-chained or took differing routes to meet with a friend or relative or to patronize a preferred vendor or service provider.

In addition, this study provides a methodology to study the journey to work at one point in time. We know that populations are not static and residential moves were frequent; less so were changes in employer or the specific worksite in which one carried out one's daily tasks. We know that residential mobility happens for many reasons, among them the attempt to better not only the daily commute but also the size and quality of one's accommodation (Gilliland and Olson 1998). A natural next step is to attempt to follow workers longitudinally across time. Other journey to work studies has looked at how patterns change over time, but no other has examined the same individuals over time. By looking at how the daily commute changed for individuals, combined with how their residential environments changed, we can begin to untangle how individuals made choices to balance the daily commute and the desire to improve the environment they and their children lived in.

References

Barber M (1991) Immigrant domestic servants in Canada. Canadian Historical Association, Ottawa

Barke M (1991) The middle-class journey to work in Newcastle upon Tyne, 1850–1913. J Transp Hist 12 (2):107–134

Bater J (1974) The journey to work in St. Petersburg 1860–1912. J Transp Hist 2(4):214–233

Bloomfield AV, Harris R (1997) The journey to work; a historical methodology. Hist Methods 30(2):97–110

Bousssauw K, Neutens T, Witlox F (2011) Minimum commuting distance as a spatial characteristic in a non-monocentric urban system: the case of Flanders. Pap Reg Sci 90(1):47–65

Bradbury B (1984) Women and wage labour in a period of transition: Montreal, 1861–1881. Histoire sociale—Soc Hist 17(33):115–131

Bradbury B (1996) Working families: age, gender, and daily survival in industrializing Montreal. Oxford University Press, Toronto

Bullen J (1986) Hidden workers: child labour and the family economy in late nineteenth-century Urban Ontario. Labour/Le Travail 18:163–187

Conzen K (1976) Immigrant Milwaukee 1836–1860: accomodation and community in a frontier city. Harvard University Press, Cambridge

Corcoran J, Chhetri P, Stimson R (2009) Using circular statistics to explore the geography of the journey to work. Pap Reg Sci 88(1):119–132

Costa D (2000) From mill town to board room: the rise of women's paid labor. J Econ Perspect 14(4):101–122

Craven P (ed) (1995) Labouring lives: work & workers in nineteenth-century Ontario. University of Toronto Press, Toronto

Cross M (ed) (1974) The workingman in the nineteenth century. Oxford University Press, Toronto
Curtis B (2001) The politics of population: state formation, statistics, and the census of Canada, 1840–1875. University of Toronto Press, Toronto
Darroch G, Ornstein M (1980) Ethnicity and occupational structure in Canada in 1871: the vertical mosaic in historical perspective. Can Hist Rev 61(3):305–333
Debats D (2009) Using GIS and individual-level data for whole communities: a path toward the reconciliation of political and social history. Soc Sci Comput Rev 27 (3):313–330
Dennis R (1984) English industrial cities of the nineteenth century. Cambridge University Press, Cambridge
Deskins D (1972) Race, residence, and workplace in Detroit, 1880–1965. Econ Geogr 48(1):79–94
Deskins D (1975) A rejoinder to Gale and Katzman's comments on "Race, residence, and workplace in Detroit, 1880–1965". Prof Geogr 27(3):356–361
Dillon L (2008) 1881 Canadian census project, North Atlantic population project, and Minnesota population center. National sample of the 1881 census of Canada. Montreal, QC (Département de Démographie, Université de Montréal)
Dixon C, Leach B (1977) Concepts and techniques in modern geography. In: Sampling methods for geographical research, vol 17, Geo Books, Norwich
Dunae P (2008) Geographies of sexual commerce and the production of prostitutional space: Victoria, British Columbia, 1860–1914. J Can Hist Assoc 19(1):115–142
Dunae P (2009) Sex, charades, and census records: locating female sex trade workers in a Victorian city. Histoire sociale/Soc Hist 42(84):267–297
Edwards A (1940) Alphabetical index of occupations and industries. Government Printing Office, Washington D.C (Bureau of the Census, U.S Department of Commerce)
Ewing R, Cervero R (2010) Travel and the built environment. J Am Plan Assoc 76(3):265–294
Fahmi M (1997) 'Ruffled' mistresses and 'discontented' maids: respectability and the case of domestic service, 1880–1914. Labour/Le Travail 39:69–97
Frances R, Kealey L, Sangster J (1996) Women and wage labour in Australia and Canada, 1880–1980. Labour/Le Travail 38:54–89
Gale S, Katzman D (1975) Comments on race, residence, and workplace in Detroit, 1880–1965. Prof Geogr 27 (3):353–356
Gilliland J (2004) Muddy shore to modern port: redimensioning the Montreal waterfront time-space. Can Geogr 48(4):448–472
Gilliland J, Olson S (1998) Claims on housing space in nineteenth-century Montreal. Urban Hist Rev/Revue d'histoire urbaine 26(2):3–16
Gilliland J, Olson S (2010) Residential segregation in the industrializing city: a closer look. Urban Geogr 31 (1):29–58
Gilliland J, Olson S, Gauvreau D (2011) Did segregation increase as the city expanded? the case of Montreal, 1881–1901. Soc Sci Hist 35(4):465–503
Goheen P (1970) Victorian Toronto 1850–1900. Department of Geography, University of Chicago, Chicago
Green DR (1988) Distance to work in Victorian London: a case study of Henry Poole, bespoke tailors. Bus Hist 30(2):179–194
Greenberg S (1980) The relationship between work and residence in an industrialsing city. In: Cutler WW, Gillette H (eds) The divided metropolis: social and spatial dimensions of Philadelphia. Greenwood Press, Westport, Conn, pp 1800–1975
H.H.B. (1913) Making Toronto's Directory: A Prodigious Undertaking. The Toronto Star Weekly
Hamilton M (2007) "Anyone not on the list might as well be dead": 7 Aboriginal peoples and the censuses of Canada, 1851–1916. J Can Hist Assoc 18(1):57–79
Hanson S (2004) The context of urban travel: concepts and recent trends. In: Hanson S, Giuliano G (eds) The geography of urban transportation. Guilford Press, New York, pp 3–29
Harris R (1986) Housing in Canadian cities: an agenda and review of sources. Urban Hist Rev 14(3):258–266
Harris R, Bloomfield AV (1997) The impact of industrial decentralization on the gendered journey to work, 1900–1940. Econ Geogr 73(1):94–117
Harris R, Moffat B (1986) How reliable is the modern city directory. Can Geogr 30(2):154–158
Hershberg T (1976) The Philadelphia social history project: an introduction. Hist Methods Newsl 9(2–3):43–58
Hershberg T, Light D, Cox H, Greenfield R (1981) The "journey to work". an empirical investigation of work, residence and transportation, Philadelphia, 1850 and 1880. In: Hershberg T (ed) Philadelphia: work, space, family and group experience in the nineteenth century. Oxford University Press, New York
Hoskins R (1989) An analysis of the payrolls of the point St. Charles shops of the grand trunk railway. Cahiers de géographie du Québec 33(90):323–344
Kealey G (1980) Toronto workers respond to industrial capitalism 1867–1892. University of Toronto Press, Toronto
Labour Commission (1889) Royal Commission on the Relations of Labour and Capital in Canada, Report. Queen's Printer, Ottawa, ON
Lafreniere D, Gilliland J (2015) "All the World's a Stage": a GIS framework for recreating personal time-space from qualitative and quantitative sources. Trans GIS 19(2):225–246
Lafreniere D, Gilliland J (2018) Residential, occupational, and workplace mobilities, 1881–1891. In: Gregory I, Debats D, Lafreniere D (eds) The Routledge Companion to Spatial History, Routledge, London, pp 299–319
Larsen K, Gilliland J, Hess P (2012) Route based analysis to capture the environmental influences on a child's mode of travel between home and school. Ann Assoc Am Geogr 102(6):1348–1365
Larsen K, Gilliland J, Hess P, Tucker P, Irwin J, He M (2009) Identifying influences of physical environments and socio-demographic characteristics on a

child's mode of travel to and from school. Am J Public Health 99(3):520–526
Lawton R, Pooley C (1976) The social geography of Merseyside in the nineteenth century. Final Report to SSRC, Liverpool
Lees L (1969) Patterns of lower-class life: Irish slum communities in nineteenth-century London. In: City YC-C, Thernstrom S, Sennett R (eds) Nineteenth-century cities: essays in the new urban history. Yale University Press, New Haven, pp 359–385
Lewis R (2000) Manufacturing Montreal: the making of an industrial landscape. Johns Hopkins University Press, Baltimore
LinkageWiz (2019) LinkageWiz Data Matching Software. www.linkagewiz.net
Mayer J (1977) The journey to work, ethnicity and occupation in Milwaukee, 1860–1900. PhD Thesis, University of Michigan, Ann Arbor
Olson S, Thornton P (2011) Peopling the North American city: Montreal 1840–1900. McGill-Queens University Press, Montreal
Opp J, Walsh J (eds) (2006) Home, work, and play: situating Canadian social history, 1840–1980. Oxford University Press, Toronto
Phillips G (2000) The Odyssey of beer and brewing in Victorian London-Middlesex. Cheshire Cat Press, Sarnia
Pooley C (1984) Residential differentiation in Victorian cities: a reassessment. Trans Inst Br Geogr New Ser 9:131–144
Pooley CG, Turnbull J (1997) Changing home and workplace in Victorian London: the life of Henry Jaques, shirtmaker. Urban Hist 24(2):148–178
Pooley C, Turnbull J (1999) The journey to work: a century of change. Area 31(3):281–292
Pooley C, Turnbull J (2000) Modal choice and modal change: the journey to work in Britain since 1890. J Transp Geogr 8:11–24
Pred A (1966) The American Mercantile city: 1800–1840. Ann Assoc Am Geogr 56(2):307–338
Ross G (1896) The school system of Ontario Canada: its history and distinctive features. D. Appleton and Co, New York
Sadler RC, Lafreniere DJ (2017) You are where you live: methodological challenges to measuring children's exposure to hazards. J Child Poverty 23(2):189–198
Senecal A (1889) Report of the royal commission on the relations of labor and capital in Canada. Ottawa
Steckel R (1991) The quality of census data for historical inquiry: a research agenda. Soc Sci Hist 15(4):579–599
Sultana S, Weber J (2007) Journey to work patterns in the age of sprawl: evidence from two midsize southern metropolitan areas. Prof Geogr 59(2):193–208
Thernstrom S (1973) The other Bostonians: poverty and progress in the American metropolis. Harvard University Press, Cambridge
Thompson EP (1980 [1963]) The making of the English working class. Penguin Books, London
Vance J (1967) Housing the worker: determinative and contingent ties in nineteenth century Birmingham. Econ Geogr 43(2):95–127

Corruption and Development of Atlanta Streetcar Lines in the Nineteenth Century: A Historical GIS Perspective

7

S. Wright Kennedy

Abstract

This chapter uses historical geographic information system (HGIS) to trace the formative years of public transportation in Atlanta. Routes chosen for the early streetcar lines were examined to ascertain their effects on the mobility networks of city residents and residential patterns. This study finds evidence that Atlanta city leaders engaged in corrupt bargains to benefit themselves with little regard for the lasting effects on the city's transportation network and spatial development. Streetcar networks in postbellum Atlanta created demographic patterns that persist today. This chapter illustrates that HGIS can enable historians to distill new value from sources underutilized in urban history. Historians have used city directories and fire insurance maps as general reference material, but HGIS can operationalize such sources into an extensive archive of historical urban processes. This type of geospatial agency provides historians with new methods of investigation that can lead to new perspectives and understanding of the urban past.

Keywords

Historical geographical information system (HGIS) · Nineteenth-century atlanta · Streetcar line planning · Networks · Kleptocracy

7.1 Introduction

On December 21, 1835, the state government of Georgia passed an act to establish a centrally located rail junction at the foot of the Appalachian Mountains in northern Georgia. After years of preparation and construction, the Western and Atlantic Railroad reached the planned junction in 1842. Railroad employees and a few provisioners built shacks around the rail depot and referred to their town as Terminus, after the rail station (Barnwell 1867; Clarke 1881). Over the next decade, the Macon and Western Railroad and the Georgia Railroad reached the junction at the recently renamed Atlanta, solidifying the town's position as a transportation hub. By the 1870s, the rail lines had established Atlanta as a regional commercial and transportation center and as a major city in the southeastern United States. Railroads quite literally put Atlanta on the map and directed regional development around the city, yet the street railways built in the 1870s guided the evolution and expansion of the city itself. The early streetcar routes, and the people who chose these routes,

S. W. Kennedy (✉)
Columbia University, New York, USA
e-mail: w.kennedy@columbia.edu

C. Travis et al. (eds.), *Historical Geography, GIScience and Textual Analysis*,
Historical Geography and Geosciences, https://doi.org/10.1007/978-3-030-37569-0_7

played a consequential role in the development and the spatial arrangement of Atlanta. This chapter traces the formative years of public transportation in Atlanta. It uses geographic information systems (GIS) software to investigate the routes chosen for the early streetcar lines and to examine the resulting effects on the mobility networks of city residents and their residential patterns. This study finds evidence that city leaders engaged in corrupt bargains to benefit themselves with little regard for the lasting effects on the city's transportation network and spatial development. The streetcar network in postbellum Atlanta created demographic patterns that persist today (Klima 1982).

In Atlanta, as in many cities during the Gilded Age, corrupt politicians and unscrupulous people used city contracts to line their own pockets. Little incentive existed for participants to document most of these under-the-table deals, and in turn, their influences on long-term, path-dependent processes have been lost. Yet, traces of this corruption remain embedded in the spatial fabric of the city. By mapping the early history of the streetcar network in Atlanta, this chapter uncovers the perpetrators of this corruption and the spatial legacy within the city. The spatial history of streetcar routes is intrinsically linked to the politics and power struggles within the city. This study begins with the evidence embedded in the spatial fabric of the city (i.e., the streetcar routes) and works backwards to identify possible motivations for the decisions behind these routes. Compared to the geography of other nineteenth-century cities, Atlanta was less constrained by natural features and more by space and time. It was one of the only large U.S. cities at the turn of the twentieth century that was not located on a navigable waterway.[1] Because of this, the city was and remains a flexible urban system directed more by human-made features than geographic barriers. Public transportation compressed the space and the time-expense to travel through this space, in the city, which radically shaped the city. Streetcar lines changed the entire structure of space and time in Atlanta. Furthermore, this chapter argues that GIS and a geospatial framework enable historians to distill new value from sources underutilized in urban history. Historians have used city directories and fire insurance maps as general reference material, but historical GIS (HGIS) enable the use of these sources as an extensive archive of historical urban processes. Analyzing these processes in HGIS garners new perspectives and understandings of the past.

During the Gilded Age, corruption in city governments around the country was often connected with streetcar franchises. Rebecca Menes, an economic historian, states, "In city after city the books were (relatively) clean and city debt secure, but the street car franchises were almost universally tainted and insider trading on real-estate ran rampant" (2003, p. 2). Kleptocrats found allies in the local businessmen with city interests, often street railway and real estate development firms.[2] Not only could city officials financially benefit from corrupt deals with streetcar companies, but control over the streetcar companies gave the city officials more control over the city. They could direct development through the streetcar lines and control the flow of people through the city. City officials used their power to franchise companies as a means of control over streetcar companies. Officials could block franchises of companies, but they could also check the power of existing streetcar companies by "issuing new, competing franchises to new firms and forcing the old companies to buy out, or be bought out, to maintain the valuable monopoly rights" (Menes 2003, p. 21). Menes and other scholars have demonstrated that at the turn of the twentieth century, city officials in St. Louis, Philadelphia, and Los Angeles were involved in similar schemes and corrupt dealings connected with streetcar firms, yet, as this study will show, these patterns were present 30 years

[1]Denver and Indianapolis were the two other U.S. cities not on a navigable waterway during this period.

[2]Menes (2003, p. 11) defines a "kleptocrat" as a government official who is a thief or crook.

earlier in Atlanta (Glaeser and Goldin 2006; Steffens 1904). Previous studies have identified real estate speculation on the outskirts of town as the principal driver in the development of the early street railway network (Carson 1981, p. 2; Garrett 1969, p. 863; Klima 1982; Martin 1975). Klima (1982) linked the landholdings of Richard Peters and George Adair, who were wealthy civic leaders in postbellum Atlanta, with their role in the development of the first streetcar network in the city. Peters and Adair designed their streetcar network to provide direct service to the suburban residential developments in which they had invested. Residential developments of the late nineteenth century often succeeded or failed, Klima contended, based on their proximity to a streetcar line, and as such, demographic patterns created by streetcar lines had a lasting influence on spatial patterns in the city. While this accounts for the destination of a few streetcar lines, it does not explain seemingly arbitrary turns within early streetcar routes. This study uses GIS to examine these peculiarities and to argue that corruption was at the core of these decisions. This corruption benefitted a few people in power at the cost of an inefficient streetcar network that changed the residential landscape.

The HGIS constructed for this study combined a variety of textual and visual records of Atlanta to investigate the co-development of public transportation and residential areas. For historians exploring a new archive, it is important to understand the criteria used for archival curation; similarly, it is important to explain the decisions and methods used to construct the HGIS in this study. Broadly, this study used sources that detailed the historical streets, the changes in streetcar routes over time, the approximate daily movement of people within the city, and the locations of people and institutions who had the power to influence the streetcar routes. The infrastructure and built environment of Atlanta changed substantially during the twentieth century. In particular, the construction of interstate highways altered much of the historical street and address networks in the downtown core of the city (see Fig. 1). As such, this study used 1886 and 1899 *Sanborn Fire Insurance Maps* to reconstruct the historical infrastructure of the city.[3] After georeferencing these atlases, I digitized the historical road network of Atlanta from the Sanborn maps. I created a historical address locator from the historical address ranges labeled on the maps and listed in the city directories. I reconstructed the routes of the streetcar lines from historical maps and textual descriptions. I georeferenced the maps of streetcar lines found in Carson's (1981) *The Trolley Titans: A Mobile History of Atlanta*, and then I traced the routes to digitize them into the GIS. City directories (beginning in the 1872 directory) contained textual descriptions of the routes of streetcar lines, and I used these to reconstruct the routes as well. This study sampled city directories from 1871, 1881, 1891, and 1902 to build a database of the residents of Atlanta. The city directories contained information on each resident, including name, occupation, race, residential address, residential type (e.g., house, boarding house, rental), and place of employment. I used the business directories (located at the back of the city directories) to look up the business type and address and then I added this information to each resident's record. Linking home and work addresses proved laborious, so I created a subset to serve as a sample. I chose a street block with businesses that employed white-collar workers throughout the study period. White-collar workers could afford

[3]I georeferenced a variety of index maps, including an 1871 map from the city directory, ward maps from 1871 to 1883, and Sanborn index maps from 1886 to 1899. These proved useful for georeferencing the detailed Sanborn map sheets (14 map sheets from 1886 to 67 map sheets from 1899) in the numerous areas of Atlanta with substantial twentieth-century changes in the street network. Once georeferenced, I used the Sanborn maps as the primary cartographic source. Based out of New York City, the Sanborn Map Company began publishing fire insurance maps in 1867 and continued through the twentieth century. The Sanborn cartographers drew the maps at the scale of one inch to fifty feet (1–1.66 m), and this large scale allowed for the inclusion of very detailed features and notes. Many HGIS studies have used Sanborn maps to virtually reconstruct the historical built environment of cities (e.g., DeBats 2008; Gregory et al. 2018; Kennedy et al. 2015; Don Lafreniere and Gilliland 2014; Raymond 2011; Richardson 2018; Sauder 1980).

the higher streetcar fares in the early years and, in turn, they were more likely to use the streetcars for their daily commute. Additionally, the study block needed to have businesses that employed people of similar socioeconomic statuses. It was important to maintain a relatively consistent sample of residents/employees. Furthermore, I selected a principal commercial block near the streetcar hub in downtown Atlanta, since this area was well serviced by the early streetcar lines. I selected the first full block of Whitehall Street ("Whitehall Study Block") as the study block because it fit these criteria (see Fig. 1).[4]

I used the historical address locator to geocode residents' home and work addresses in HGIS. I used network analysis to calculate the shortest path along the street grid between these locations. This approximated the daily mobility network for residents who worked on the study block. Mobility networks estimate the "everyday" movement of people across a defined geographic space.[5] I used kernel density analysis to gauge the volume of daily traffic along each street, and to approximate the residential areas, I limited the overall geographic scope of analysis to an area that extended roughly 1.6 km (one mile) in every direction from the city center except for the area to the west (Fig. 7.1). Only a few people who worked on the study block resided in the western portion of the city. Additionally, blue-collar workers tended to live in the western residential area during the postbellum period, and apart from the West End neighborhood, the west side of the city did not get a streetcar line until the mid-1890s (Russell 1988).[6] The study area totaled approximately seven square kilometers and provided adequate coverage for the Atlantans who worked on the study block. Additionally, the study area provided good coverage for examining the creation and extension of the streetcar network in Atlanta since all the local streetcar lines in the nineteenth century used the downtown area as their main hub.[7]

In the second half of the nineteenth century, Atlanta transformed from an "overgrown country town" into a place of regional economic significance (Garrett 1969, p. 910). Atlanta's population grew from 9553 people in 1860 to 154,839 in 1910, an increase of more than 16-fold. In the early 1870s Henry W. Grady, who got his start as a newspaper editor for the *Atlanta Daily Herald*, began promoting the development of the industrial and commercial sectors of the Southern economy. He advocated for investment in new factories and mills, along with attracting new immigrants to the South to expand the workforce. Likewise, Grady promoted Atlanta as a commercial center of the South. He referred to this vision as "The New South", and Grady viewed Atlanta as the flagship city for the industrial and commercial development of the South.[8]

In the 1870s, the Atlanta business community attracted northern capital, enticed companies to relocate to Atlanta, and continued to encourage railroad companies to extend the rail lines connected to the city. The wholesale merchants and retail grocers of Atlanta used the rail network to grow their market area. The value of the Atlanta trade market continued to grow in the 1870s, despite the Panic of 1873, and in 1878 the *Atlanta Constitution*, a local newspaper, estimated the value of the market to be $40 million. By 1881, hundreds of traveling salespeople

[4]The historical addresses on the study block ranged from 24 to 64 Whitehall Street. In the twentieth century, the city renamed the street Peachtree Street.

[5]Many fields use mobility network modeling, including epidemiology, medical geography, urban planning, and disaster science (for a technical example, see Stephen Eubanks et al. (2004)). Reconstructing historical mobility networks is an emerging trend in HGIS research (Lafreniere 2014).

[6]For a short period of time the Taylor-Hill Line operated in this area (1873–1874). The Atlanta Street Railroad Company discontinued the route in 1874 because it was unprofitable.

[7]The transit station at Five Points is still considered "The Hub of Atlanta"; however, it has been moved underground.

[8]Garrett (1969, volume 1, pg. 909) identified a March 14, 1874 editorial in the *Atlanta Daily Herald* entitled "The New South" as one of the earliest instances in which Grady used the phrase. In this editorial, Grady reviewed the recommendations of the *Boston Daily Advertiser* editorial also entitled "The New South" (March 10, 1874). Grady would go on to promote the idea of the "New South" throughout the 1870s and 1880s.

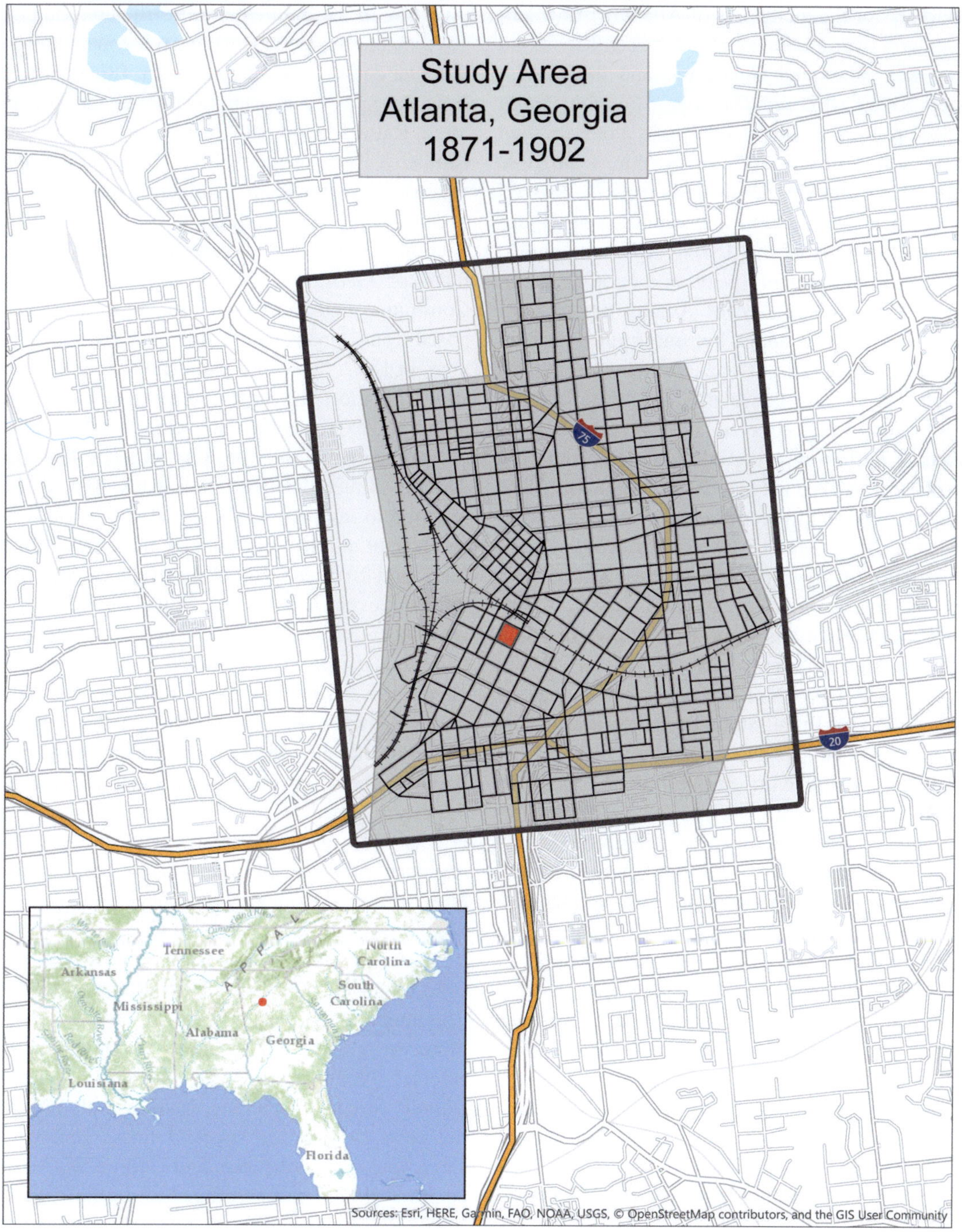

Fig. 7.1 The red rectangle shows the extent of the study area in Atlanta, Georgia. The study area extends approximately 1.6 km (one mile) in every direction from the city center except to the west. The map shows the 1899 street network constructed for this study overlaid on modern USGS base maps

working for the merchants of Atlanta "[travelled] regularly over all the country between Richmond and Key West and Charleston and the Mississippi River" (*Atlanta: The Gate City of the South* 1881, p. 6). The dry goods trade alone reached $10 million in 1881. This was more than triple the value of the dry goods trade in 1876 (Russell 1988, p. 125). Yet with the economic and population growths of the 1870s, Atlantans had to cope with the growing pains of a budding urban area. The infrastructure of Atlanta proved substandard in 1871, and it became further strained as the decade progressed. Municipal services, such as paved streets, water, gas, sewage, and fire protection, only covered the central business district during this period. One contemporary account described the streets of Atlanta as a "Red Sea of Mud, and the *Columbus Sun* referred to Atlanta as the "Mud City" (Garrett 1969, p. 863). Apart from a few principal avenues paved with granite macadam from Stone Mountain, Atlanta's street conditions were abominable. In 1880, Atlanta had 160 km of streets but only 5 km were paved. Paved streets were so rare in the city that a block of Broad Street became known as the "Granite Block" after local business paid to have the street paved with granite.[9] Pedestrians had trouble traversing the streets, and "carriages frequently bogged down on the principal business and residential streets" (Garrett 1969, p. 863). The street conditions were not only an inconvenience, but a hazard as well. In 1872 an ox died after "falling into a hole on Markham Street".[10]

Chartering private companies to develop streetcar lines seemed to be a shortcut to solve the city's transportation problems. Rather than investing in the municipal infrastructure, the city government decided to attach paving mandates to streetcar contracts. On February 23, 1866 the state legislature passed an act to incorporate the Atlanta Street Railroad Company, the city's first streetcar company. The act granted the City Council and the Mayor significant regulatory power over streetcar companies within the city (Act 207, Section II. Acts of the General Assembly of the State of Georgia 1866, p. 201). The Streets Committee and the Mayor reviewed each proposed route and granted approval by majority vote. In 1867, however, Alderman Richard Peters and others on the city council passed a prohibitively expensive paving mandate for street railway construction. For every streetcar line, the Atlanta Street Railroad Company would be required to lay flag stones on the entire width of each street and to "macadamize the width of the track, and for three feet [one meter] on each side of each and every line of track, as laid, that shall not have been previously macadamized or paved" (Atlanta City Ordinance Book 5 quoted in Martin 1975). The city justified the mandate by reasoning that the horses and mules pulling the streetcars would further dilapidate the muddy thoroughfares if they streets remained unpaved. Additionally, the city planned to collect an annual tax of five dollars ($5) for each streetcar in operation. In response to these strict mandates, the Atlanta Street Railroad Company postponed construction indefinitely. In 1869 the City Council repealed their earlier paving mandates and offered tax exemptions for fifty years (City Railroad 1869). The new mandate required paving only "in cases of necessity, to be judged of by the Committee on Streets" (Newman 1873, p. 130), which, with a friendly Streets Committee, would greatly reduce the costs of building streetcar lines. That very same year, Richard Peters and George W. Adair acquired the charter and franchise to the Atlanta Street Railroad Company, and in April of 1871 they took leading roles in the company. Peters and Adair oversaw the construction of Atlanta's first streetcar line, the West End Line. The line began operation on September 8, 1871 and was fully operational by the year's end (Fig. 7.2). The West End Line passed in front of both Peters' and

[9]William Hanleiter, publisher of the *Atlanta City Directory for 1871*, listed his office at the "Granite Block of Broad Street" (1871, Title Page).

[10]*Atlanta Daily Herald*, December 29, 1872, quoted by Russell (1988, p. 194). The loss of work animals due to poor street conditions was not unique to Atlanta. For example, in 1878 a mule drowned in a puddle on the streets of Memphis (Larsen 1985, pp. 116–122). Many Southern cities had extremely poor street conditions during this period (Goldfield 1982, p. 92). Newspaper editors in other cities used similar occurrences to illustrate the poor street and sanitary conditions and to advocate for infrastructure improvements.

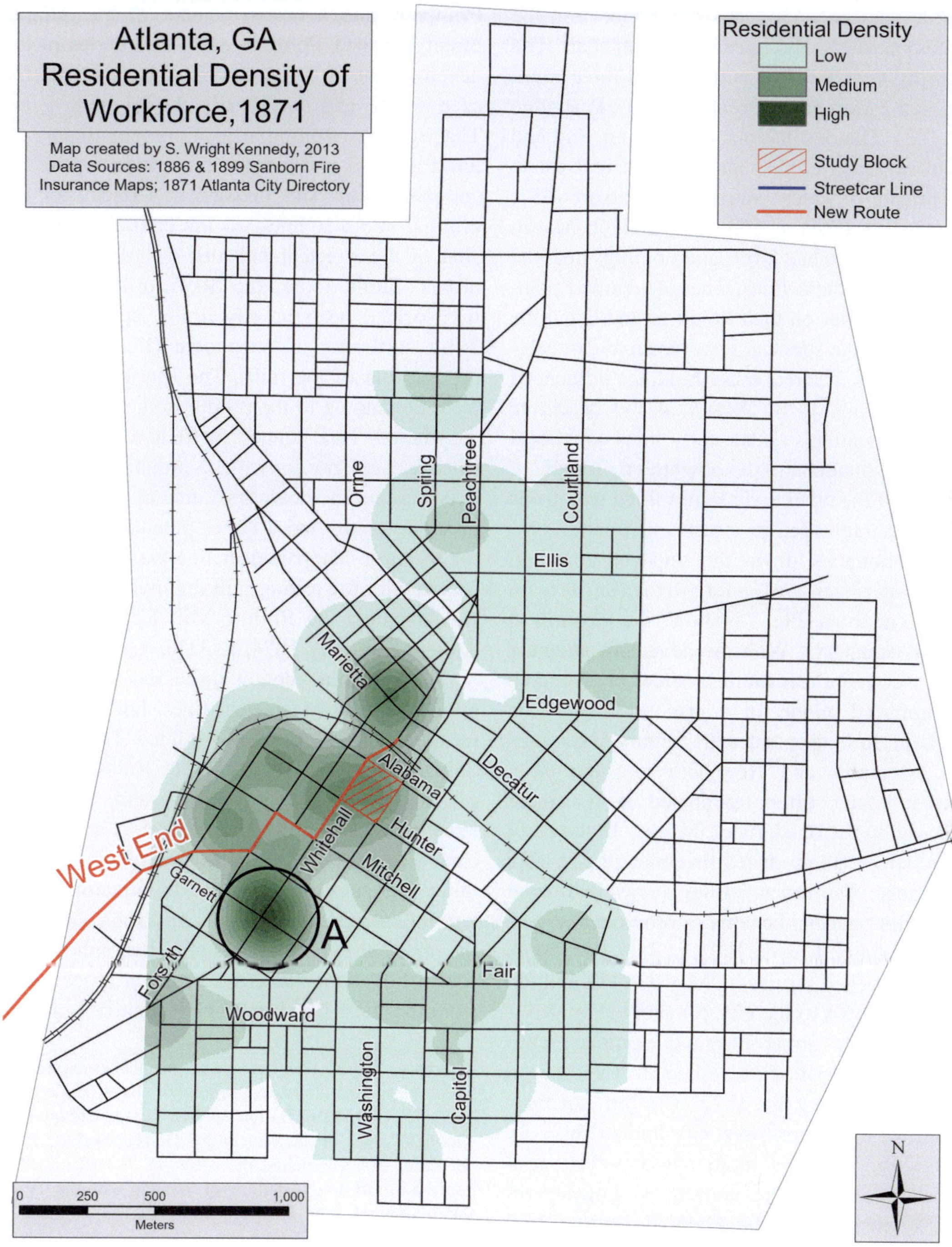

Fig. 7.2 In 1871, the Atlanta Street Railroad Company opened the West End Line, the first streetcar line in Atlanta. The map shows the Whitehall Study Block crosshatched in red and the residential density of the study block workforce in green. Many of the residents who worked on the study block lived nearby in 1871

Adair's residences, conveniently connecting the business partners who lived over 2 km apart. The following year the company opened three more lines on Marietta, Decatur, and Peachtree (Fig. 7.3). The destinations of the West End and Peachtree Lines corresponded to the real estate investments of Peters and Adair (Carson 1981, p. 2; Garrett 1969, p. 863; Klima 1982; Martin 1975). Both men had large land holdings near the terminuses of these lines. They developed planned communities on their holdings and listed the proximity of the streetcar lines as a main amenity (Klima 1982). The real estate holdings influenced the destinations of the West End and Peachtree Lines, and similarly, other early lines connected specific destinations to the downtown as well.

Some of the other early routes filled real needs within the city. Before streetcars, funeral processions bounced down the unpaved roads in wagons, carriages, and on foot to the cemetery on the edge of town (Fig. 7.6 shows the location of the city cemetery). With the advent of streetcar routes servicing cemeteries, streetcars became the preferred mode of transportation to the interment. In other nineteenth-century U.S. cities (e.g., Memphis and New Orleans), the early streetcar routes often terminated at the main cemetery on the outskirts of the city. In the early 1880s, streetcars produced revenues of $15–$25 per cortege. The Decatur Line was well situated to reap the financial benefits of providing service to the cemetery (Carson 1981, p. 3; Garrett 1969, p. 956). It began service on May 3, 1872 and ran from downtown to the city cemetery (Fig. 7.6).

The Atlanta Street Railroad Company constructed the Marietta Line in late 1871 and began service on January 14, 1872. The line ran from downtown to the northwest city limits (Fig. 7.3). By design, it served more working-class residential areas than the other lines. Lewis Scofield,[11] president of the Scofield Rolling Mill Company and a board member of the Atlanta Street Railroad Company, had an interest in the development of the streetcar network in Atlanta as a whole and specifically the Marietta Line. The Scofield Rolling Mill Company manufactured railroad and merchant bar iron. Scofield's company likely supplied the iron tracks for the original streetcar lines, so the continued expansion of the streetcar network meant more business.[12] Furthermore, the Scofield Rolling Mill was located near the city limits, approximately 2 km northwest of downtown (Fig. 7.6 shows the location of the mill). The Marietta Line ran from downtown to the rolling mill. Not only did the Marietta Line connect Scofield to his mill, but since a large portion of the Atlanta population lived around the downtown area at this time, the new line connected a larger potential workforce for the mill. The streetcar line was so interconnected with the rolling mill that by 1874 the line was renamed the Rolling Mill Line (Beasley's Atlanta Directory 1874, p. 317).

The previous section discussed the destinations of the early streetcar lines, but now let us turn to the actual routes of the lines. The board of directors for the Atlanta Street Railroad Company ensured they would benefit from the streetcar routes. In addition to Peters and Adair, early lines connected the homes of three of the other four men on the board of directors (Benjamin Crane, Lewis Scofield, and John James) to the streetcar network. Benjamin Broomhead, the fourth board member, lived only a block away from the West End Line. The Peachtree Line and

[11]Some secondary sources (Garrett 1969, volume 1, pg. 397, 427, 846, 863; Carson pg. 2) have spelled this board member's name as "Schofield"; however, the 1870 Directory, 1871 Directory, and 1872 Directory (pg. 160) list his name as "Scofield". Additionally, Scofield's company ran an advertisement in the 1871 Directory that spelled his name "Scofield". Garrett cited the 1872 Directory, pg. 160 as his source for the paragraph in which he spelled the name "Schofield". In other passages (pg. 315, 640, 819) Garrett spelled Lewis' surname as "Scofield". One explanation is that Garrett used the 1867 Directory that misspelled the name as "Schofield". Perhaps he mixed up the spelling of Scofield with the Union Major-General John M. Schofield, who participated in Sherman's Atlanta Campaign. Garrett spells General Schofield's name "Scofield" (pg. 604) at least once in the same volume.

[12]Since Scofield's company specialized in iron bars for railroad lines and he was on the board of directors for the Atlanta Street Railroad Company, it is likely that his company supplied the iron rails for the first streetcar lines; however, I have not found sources that document these transactions.

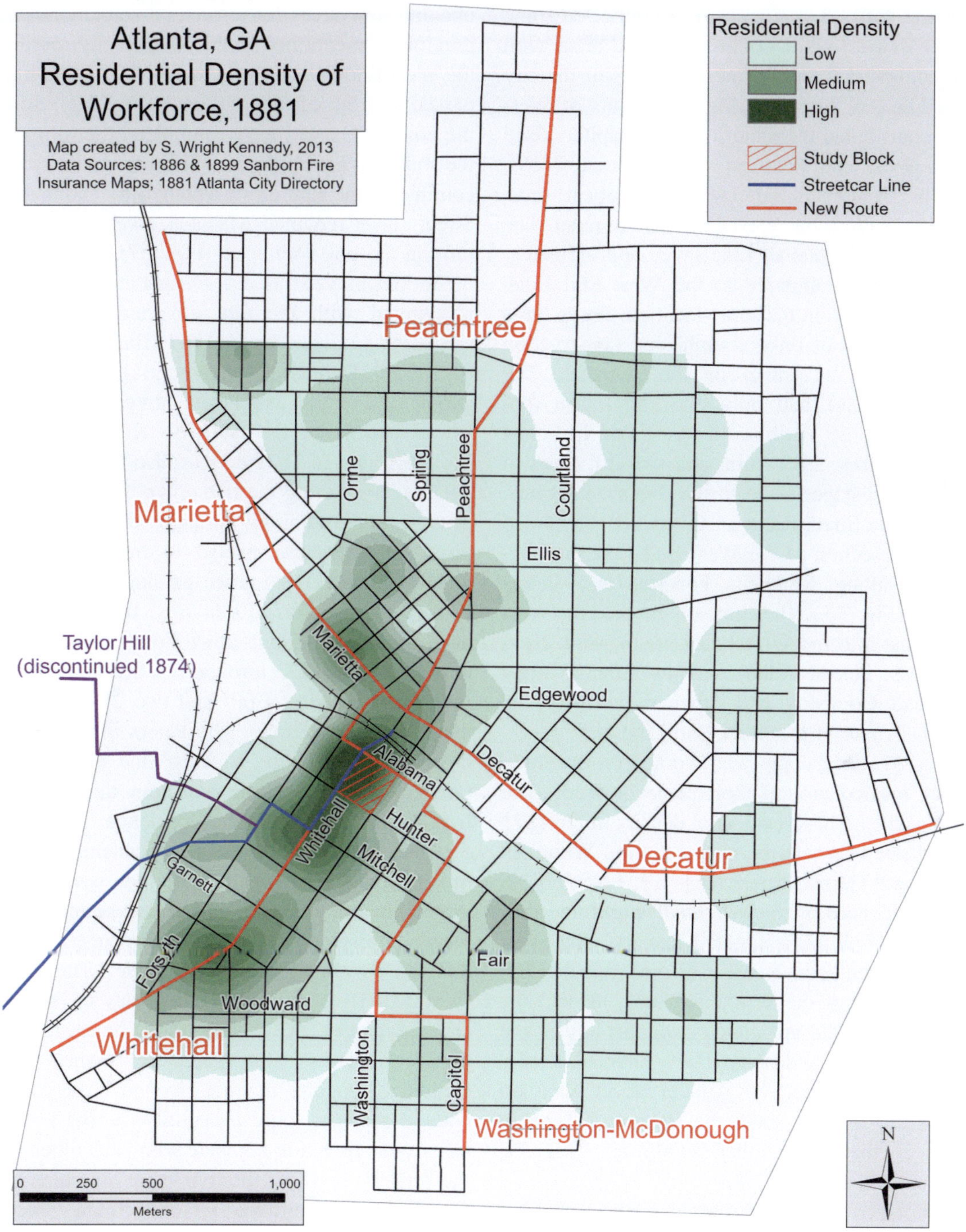

Fig. 7.3 After opening the West End Line (blue), the Atlanta Street Railroad Company constructed six additional streetcar lines (red) in the early 1870s

West End Line benefited the company owners and investors in their daily travels along with being profitable routes. In terms of financial feasibility, the Whitehall Line should have been the first route implemented in Atlanta, but the Atlanta Street Railroad Company did not open the line until 1874 (Fig. 7.3). It connected some of the densest white-collar residential areas with the commercial and business districts of the city (see Fig. 7.6, Location A).[13] Residents of West End contributed the majority of the capital raised for a West End streetcar (City Railroad 1869), and the Whitehall route was the shortest, most efficient path to the West End development. The route of the Whitehall Line saved approximately 100 m when compared to the West End Line, and both ended in the same location. Sharp turns in rail track not only complicated construction but slowed the travel of the streetcar. The Whitehall Line had only one slight turn (approximately 25°), while the West End Line had two 90° turns, a 60° turn, and two 20° turns in the same distance. Due to the duplicate service, the Atlanta Street Railroad Company abandoned the large section of the West End Line that zigzagged along Mitchell, Forsyth, and Peters Streets (Fig. 7.7). The company merged the West End Line into the Whitehall Line in 1882. Nevertheless, Peters, Adair, and the Atlanta Street Railroad Company had other reasons for the original route of the West End Line.

In 1871 the people who worked on the study block resided around their place of work. The area of the highest residential density in the 1871 workforce was near the intersection of Whitehall Street and Garnett Street (Fig. 7.2, Location A). The residences of the Whitehall workforce provide a sample of residential density. The West End Line ran between areas of moderate density, but apart from the Whitehall Study Block, the streetcar line did not directly connect any of the areas of moderate density. The residential density patterns show that the Atlanta Street Railroad Company did not route the West End Line to connect the white-collar workforce (Fig. 7.2 Location A and Fig. 7.7).

Streetcar routes constructed in the following decades show the stark difference between connecting the existing populations with the city and opening new areas for residential settlement. As previously mentioned in this chapter, by routing the West End Line down Mitchell Street, the line passed in front of Peters' house (Fig. 7.7). Still, the company could have minimized the turns in the route by avoiding Forsyth Street and instead continuing the line down Mitchell to Thompson Street (later renamed Madison Avenue) before turning the line southwest (Fig. 7.7).

Politics played a role. Richard Peters was well acquainted with the city government, having served as an alderman in 1867 (Barnwell 1867, p. 286).[14] Dennis F. Hammond lived on the corner of Forsyth and Peters Street. Hammond was a partner in the law firm Hammond and Wellborn, but in 1871 he was also the mayor of Atlanta. By routing the line down Forsyth Street instead of continuing straight to Thompson Street, the streetcar passed in front of Hammond's house. Even after passing in front of Peters' and Hammond's houses, the line could have continued straight down Forsyth Street and made just one 30° turn onto Whitehall Street rather than making a 60° and two 20° turns on the Peters Street route. The Peters Street route, however, brought the streetcar line in front of George Stewart's home and less than 200 m from Cornelius P. Cassin's residence (Fig. 7.7). Stewart was the city Superintendent of the Streets, and Cassin was an alderman who represented the First Ward on the city council. He served on five committees, but of interest was his service on the Streets Committee. The City Council's Streets Committee oversaw the streetcar companies in Atlanta. As Menes (2006, p. 77) explains, "The [streetcar] owner who sought a franchise did not have to bribe every council member, only a majority, so the politicians were forced to compete with each other for the bribes". In 1871, the four other aldermen on the Streets Committee were William Finch, Franklin P. Rice, Samuel W. Grubb, and Nicholas A. McLendon. Finch lived about 500 m from the city limits near the intersection of Ella

[13]In 1871, the downtown area had mixed land use, but the commercial and business properties centered on Five Points and the first blocks of Whitehall Street, including the Whitehall Study Block.

[14]Peters won a special election on February 19, 1867 after half of the aldermen resigned in protest of Radical Reconstruction.

Street and Foster Street (later renamed Edgewood; outside of the study area to the east). He was about the same distance from the end of the Decatur Line. Rice lived near the corner of Baker and Peachtree Street. The Peachtree Line ran past his house. After crossing Jones Street (later renamed Woodward Avenue), the route of the Washington-McDonough Line made some suspicious turns in relation to the residences of Samuel Grubb and Nicholas McLendon. The Washington-McDonough Line was the first line built (opened May 26, 1873) that had not been generally suggested in the 1866 city charter of the Atlanta Street Railroad Company. The second half of the Washington-McDonough Line route was similar to the West End Line in that it had a number of turns (Fig. 7.6). These were the only two routes in the 1870s that had any 90° turns.[15] This first half of the route connected City Hall and the group of four churches on Washington Street (Trinity Methodist Episcopal, Second Baptist, Central Presbyterian, and St. Phillip's Episcopal) to the downtown area. For Adair and Peters, it was an added bonus that their churches were located along the planned route.[16] At the corner of Washington and Jones Street, the same corner Alderman McLendon's house was on, the line turned east on Jones to McDonough Street (later renamed Capitol Avenue). At McDonough Street, the line went south to its terminus at Fulton Street. This was just far enough to reach Alderman Grubb's home on the corner of McDonough and Fulton Street (Fig. 7.6, Location B). These members of the city council benefitted in ways beyond just convenience.

Streetcar lines raised the values of land and houses in the surrounding area. The tracts of land on the edges of town owned by Peters and Adair greatly increased in value with the construction of the Peachtree and West End lines. The aldermen on the Streets Committee and Mayor Hammond also benefitted from the locations of the streetcar routes. When the streetcar lines began service, the aforementioned government officials witnessed their homes increase in value. Moreover, some aldermen, such as Hammond, boarded people in their houses and could justify higher rents with the added transportation amenity. The aldermen and mayor benefitted from the routes of the streetcar lines, and in return, they used their positions in the city government to help the Atlanta Street Railroad Company. A competing company, the West End & Atlanta Street Railroad Company (West End Company), was chartered on August 26, 1872. The West End Company, however, could not lay any track or begin operations until the city council granted the company a franchise. The city and the Atlanta Street Railroad Company successfully blocked the franchise of the West End Company throughout the 1870s. Following more than ten years of delay tactics by the city and the Atlanta Street Railroad Company, in 1883 Alderman Thomas Healey, the chair of the city council's street committee, "recommended the dismissal of litigation against the [West End] company" (Carson 1981, pp. 8–9). That same year the West End Company listed Benjamin Broomhead as an officer of the company. Broomhead got his start in the streetcar business as a founding director of the Atlanta Street Railroad Company in 1872. Now, just as the city voted to franchise the West End Company, Broomhead became an officer of the one-time rival company. Once franchised, the West End Company did not build lines in direct competition with the lines of the Atlanta Street Railroad Company.

Lincoln Steffens, a muckraker who detailed the corruption within city governments in a series of articles entitled *The Shame of the Cities*, described this process as "macing". "Macing", Steffens explained, "is a form of high blackmail. When [the ring members] have sold out all they have, the politicians form a competing company and compel the old concern to buy out or sell out" (1904, p. 222). After Alderman Healey ended Atlanta's lawsuit against the West End

[15]The Taylor Hill Line had a zigzag route as well, but it was only in existence for a year or less.

[16]Adair was a member of the Second Baptist Church, located on the corner of Washington and Mitchell streets. Peters was a vestryman of the St. Phillip's Episcopal Church, located on the corner of Washington and Hunter streets, one block north of Adair's church (from the 1867 to 1871 City Directories).

Company, he led a group of investors who bought the company. In 1887, the former alderman became president of the West End Company (Carson 1981). The city council and the Atlanta Street Railroad Company maced the West End Company and finally took it over in the 1880s. The Atlanta Street Railroad Company had what amounted to a protected monopoly on the streetcar transportation system in the city between 1871 and 1883. The company proved to be a financial success during this period, and in 1878 its capital stock was worth an estimated $300,000. Jean Martin, an historian of the public transit system in Atlanta, stated, "No doubt [the Atlanta Street Railroad Company's] monopoly in the field from 1871 to 1883 provided the impetus for its economic well-being" (1975, p. 10). The company's success was a direct result of "colluding with the government to manipulate markets" (Menes 2003, p. 9). Although the city officials helped to create the financial success of the company in the 1870s, city and state officials took it away in the next decade. In the 1880s, the state and city governments began dispensing numerous charters and franchises for new streetcar companies, and the Atlanta Street Railroad Company lost control of the market and fell victim to a macing itself. The city council chartered and franchised ten street railroad companies in the 1880s, and the new companies created a more competitive marketplace. Of the ten new companies, however, six functioned as shell corporations, never laying any track. In 1891, the Atlanta Street Railroad Company had 43 km (26.5 miles) of track, but the capital stock of the company had fallen to $40,000 (Martin 1975, pp. 10–11; Poor 1891, p. 1166). The same year, the newly formed Atlanta Consolidated Street Railway Company took over six streetcar companies, including the Atlanta Street Railroad Company (Carson 1981, p. 8). Although this marked the end of the Atlanta Street Railway Company, the corrupt dealings between the company officials and the city council members continued to influence the development in Atlanta. The streetcar lines of the 1870s were a step toward transportation infrastructure that would support a major metropolitan area. The streetcar network allowed Atlanta to continue growing without substantial government investment in street improvements. For the employees of the Whitehall Study Block, the streetcars quickly became the preferred way to travel to and from work. White-collar workers, such as those on the Whitehall Study Block, could now traverse the city in "decency", and the first routes of the streetcar lines primarily benefitted these "decent" citizens (Martin 1975, p. 2; Russell 1988, p. 191).

The increase in the workforce on the study block shows the major changes the streetcars brought to downtown Atlanta. Coupled with taller buildings, streetcars shifted the city toward single land-use blocks and increased the working-hours density of the central business district. In 1871, the city directory listed 121 people who worked on the study block (Hanleiter 1871). Ten years later, the workforce increased by 60% to 193 people (Sholes and Weatherbe 1881). By 1891, 326 people worked on the study block (69% increase), and in 1902 the workforce reached 642 people (97% increase) (Maloney and Godfrey 1902; Saunders 1891). The streetcars changed the patterns of business in the city, and by 1881 the streetcar lines had begun to influence residential patterns. Residents lived at higher densities along the streetcar lines than elsewhere in the city. The residential patterns show that between 1871 and 1881 the streetcar lines caused the workforce to shift towards the city's core (Figs. 7.2 and 7.3). While we tend to think of streetcar lines as creating sprawl (Warner 1962), the 1871 and 1881 residential density maps show that, initially, streetcars created density along the routes and caused shifts toward the urban core. As Figs. 7.2 and 7.3 show, in 1881 the workforce lived closer to both the streetcar lines and their places of work than in 1871. The first streetcar lines acted as an attractor for the inchoate downtown area. Over the next ten years, however, most of the workforce moved from the area around the Whitehall Study Block (Fig. 7.4) settling away from the downtown but along the new streetcar routes. By the 1890s, the downtown workforce regularly commuted by streetcar (Martin 1975).

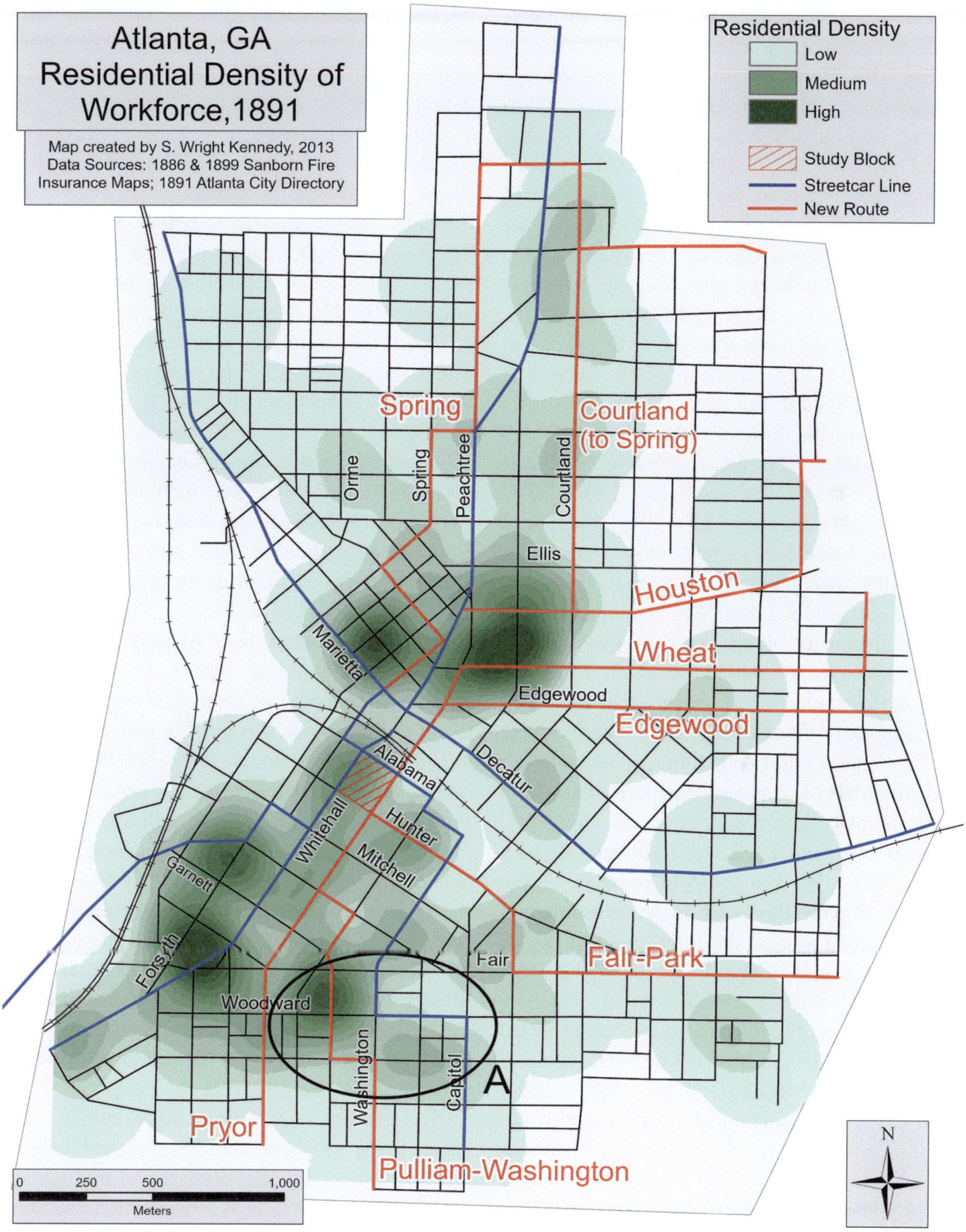

Fig. 7.4 In the 1880s, the city council chartered ten new streetcar companies. These companies opened lines that increased coverage in areas between the early lines. Notice how the Pulliam-Washington Line and the Fair-Park Line mirror the zigzag in the earlier Washington-McDonough Line (blue)

Although Lewis Scofield's rolling mill switched ownership and was re-chartered the Atlanta Rolling Mill Company in 1877, by 1881 the Marietta Line, which connected Scofield's mill to the city, had already begun to change the residential patterns in the northeastern part of Atlanta. Employees of the mill did not have to live in the immediate area, and people living near the rolling mill commuted on the Marietta Line to jobs in the downtown area. In 1902 two new streetcar lines, the Luckie and Orme Lines, further connected this area with the downtown, and consequently, a larger portion of the Whitehall workforce resided in this area. Figure 7.6 shows the location of the mill, and Fig. 7.5 Location A shows residents who lived near the mill and commuted to downtown in 1902. The Washington-McDonough Line did not foster residential areas for people who worked in white-collar occupations downtown until the 1890s. Between 1871 and 1881 the area had low residential density along much of the northern half of the Washington-McDonough Line. By 1891, more residents had moved near the intersection of Jones (Woodward) and Pulliam Streets (see Fig. 7.4 Location A) forming an area of high residential density centered on the Pulliam Line of the Metropolitan Street Railroad Company. Between 1891 and 1902, the residential density along the Washington-McDonough Line increased as well. Areas of moderate residential density formed around the intersection of Jones (Woodward) and Capitol Streets (Fig. 7.5). Residents of this area likely used the new Woodward Line, which provided a more direct route to the Whitehall Study Block. The route of the Washington-McDonough Line had a small direct effect on residential population density in this area; however, the line influenced the routes of other streetcars, which snowballed into a greater effect on the residential patterns. The zigzag route of the Washington-McDonough Line, devised to serve the residences of aldermen McLendon and Grubb in the early 1870s (Fig. 7.6, Location B), created a ripple effect in the subsequent routes of streetcar lines built in the area. While streetcars frequently ran on the same track, few crossed other tracks at 90° angles, especially the tracts of competing companies. For example, the development of a direct route from Mitchell through McDonough (Capitol) was essentially blocked by the Washington-McDonough Line. Likewise, shortly prior to consolidation, the Metropolitan Street Railroad Company opened the Pulliam-Washington Line, which mirrored the Washington-McDonough Line kink one block to the east (Fig. 7.4, Location A and Fig. 7.6, Location B). The Metropolitan Company's route serviced the southern portion of Washington Street, which had been neglected by the Washington-McDonough Line. It was not until the Georgia Railway & Electric Company consolidated all the streetcar companies of Atlanta in 1901 that many of the inefficiencies, often stemming from the routes laid out in the 1870s, could be remedied. Preston Arkwright, shortly after taking his position as the first president of the Georgia Railway & Electric Company, reorganized the streetcar lines to provide "the best and most economical service to the greatest number of people" (Carson 1981, p. 54). The Georgia Railway & Electric Company connected the disjointed streetcar lines on Washington Street, abandoned the Metropolitan Company's Pulliam-Washington Line on Pulliam and Clark Streets (between Pulliam and Washington), and abandoned the Washington-McDonough Line tracks on Woodward Avenue (formerly Jones Street) between Washington and Capitol Streets. The Georgia Railway & Electric Company rectified some of the consequences of corruption in the streetcar network, which began in 1873, although remnants within the residential patterns persisted. (Martin 1975, Klima 1982)

The Atlanta Street Railroad Company constructed the routes of the first streetcar lines in Atlanta, especially the West End Line and the Washington-McDonough Line, to benefit company and government officials personally. The circumstantial evidence of corruption is abundant in the spatial patterns of the city. The Atlanta Street Railroad Company built streetcar lines that directly served all but one of the city aldermen on the Streets Committee. These men benefited in both convenience and real estate appreciation

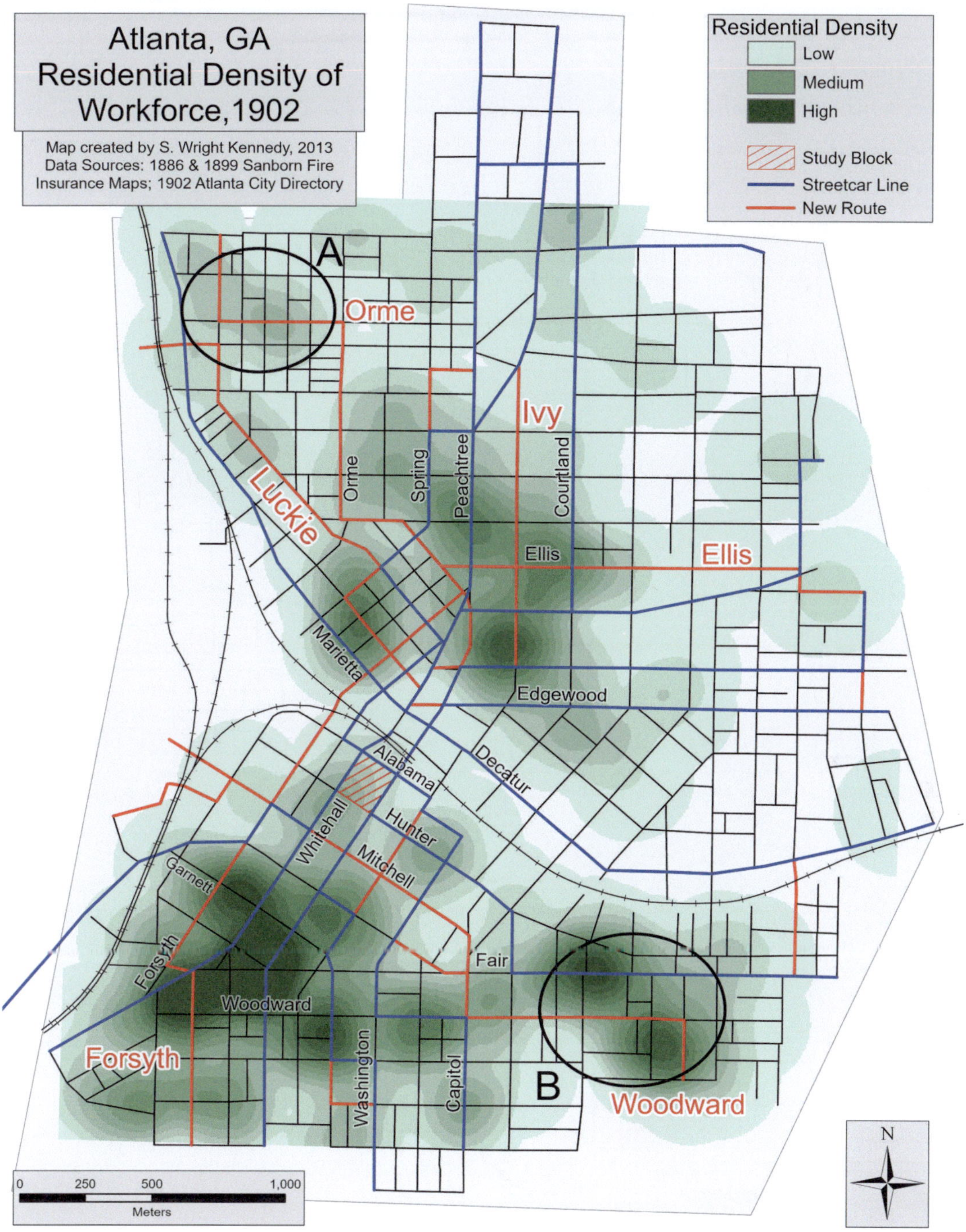

Fig. 7.5 In the 1890s, the inefficiencies of early routes resulted in worse service for city residents and difficulties for streetcar companies. The Woodward Line shifted residential areas eastward, mirroring the zigzag in the Washington-McDonough Line. In 1901, all streetcar companies were consolidated into the Georgia Railway & Electric Company, which reconfigured many of the lines to reduce the inefficiencies introduced by the early streetcar routes

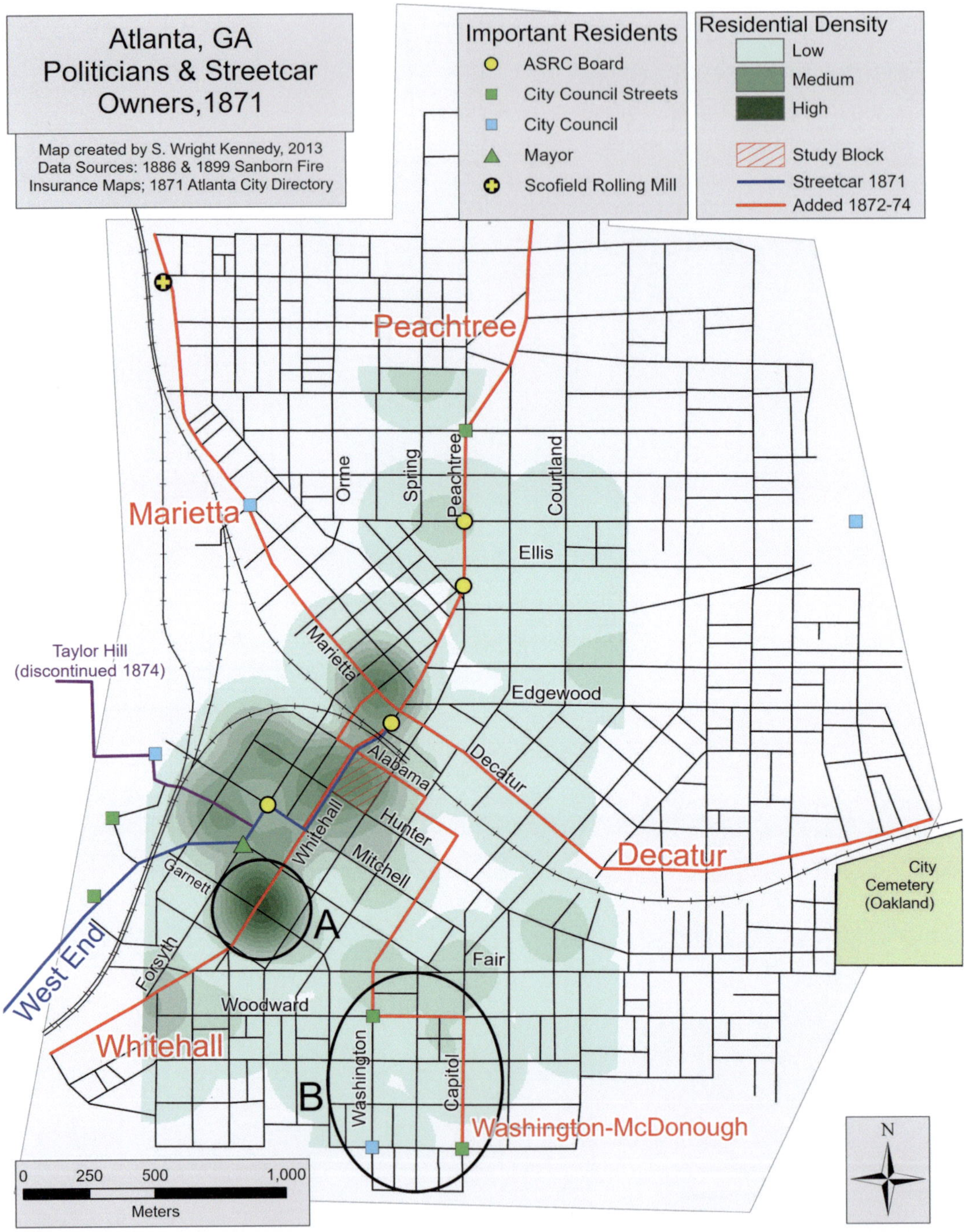

Fig. 7.6 The map shows the first seven streetcar lines in Atlanta. The Atlanta Street Railroad Company operated these lines for much of the 1870s. The Marietta Line connected the Schofield Rolling Mill to the downtown. The Decatur Line often transported corpses to the cemetery. The West End and Peachtree Lines connected the real estate developments of Adair and Peters to the downtown

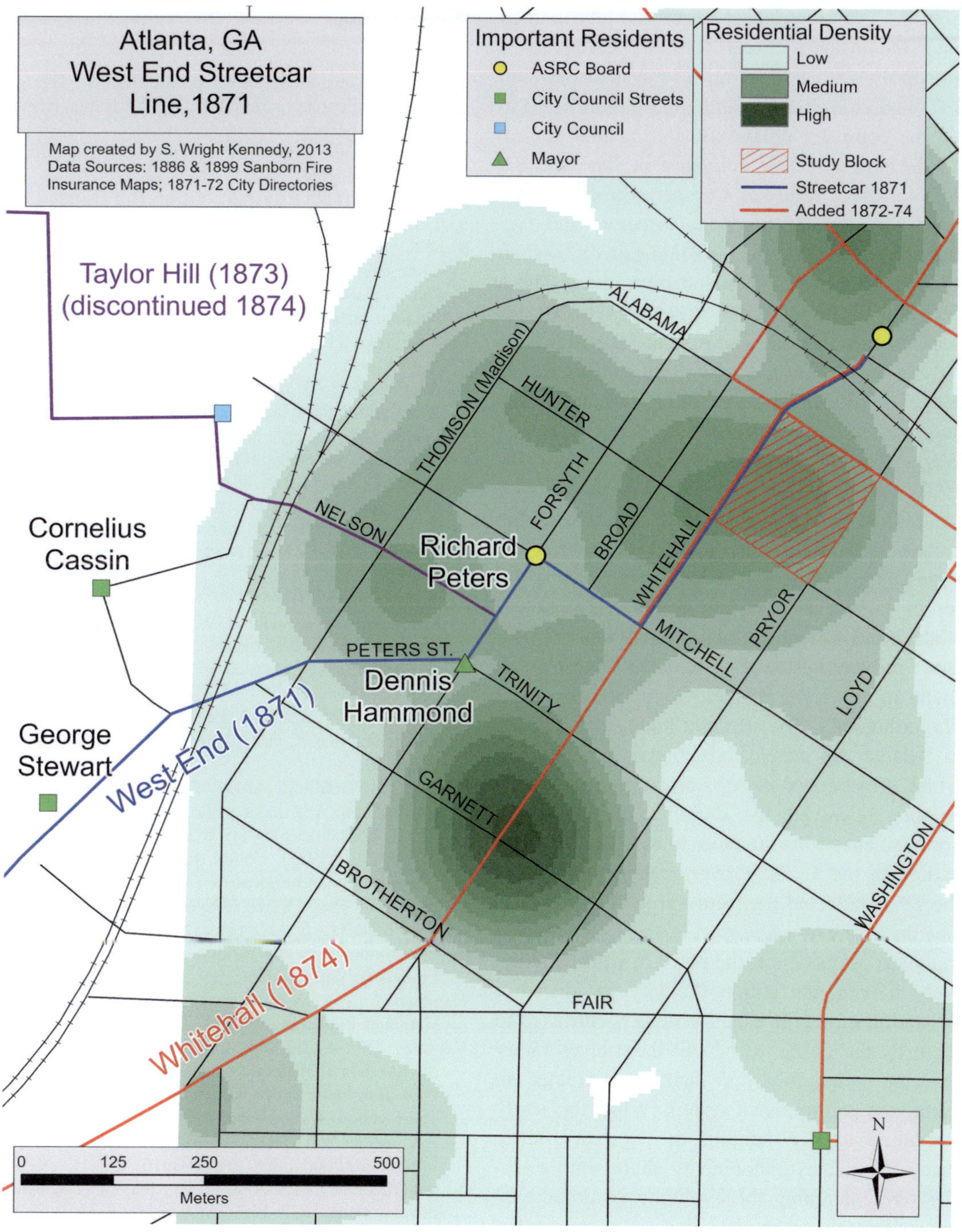

Fig. 7.7 This map shows the unusual routing of the West End Line. Notice how the route follows the residences of city and company officials rather than the residential areas. Peters, Hammond, Stewart, and Cassin benefitted from the original route of the West End Line. The Whitehall Line, opened three years later, offered better service to the same destination as the West End Line. Eventually, the Atlanta Street Railroad Company abandoned much of the West End Line

from the nearby streetcar lines. The city council gave the Atlanta Street Railroad Company a monopoly on the street railway system between 1871 and at least 1879, and city officials acted to aid the company unlawfully during this period. As the sole streetcar company in the city, the Atlanta Street Railroad Company profited from these actions. The company directors and city aldermen chose paths of personal gain rather than moral business practices and civic duty. These decisions influenced the streetcar network for decades to come and the residential patterns of Atlanta for an even longer period. The period between 1865 and 1902 was critical in the development of Atlanta as a city. Before the Civil War, Atlanta was a fledgling railroad town. By 1902, Atlanta's population had reached over 135,000 in the city and almost 150,000 in the greater metropolitan area. Not only was it the largest city in Georgia, Atlanta had become the fourth largest city in the South. Skyscrapers studded the downtown, and the city boasted of a street railway system that was the "finest to be found in any city" (Maloney and Godfrey 1902 "Introduction"). The streetcar lines created space in Atlanta for increasingly bigger populations, which drove economic and urban growth to new heights; however, the defining period in the arrangement of the streetcar lines came 30 years earlier, in the 1870s. Government corruption in the 1870s affected the future growth of the city. The decisions of a few men in the 1870s shaped the place of residence, the place of work, and the daily movements within the city for decades. A handful of people used the powers entrusted to them by the public to benefit themselves rather than the community, and their greed created an inefficient public transit system. While this study examined only a handful of streetcar routes, many other lines undoubtedly share similar stories. By examining the evolution of urban systems through a spatial lens, we can better understand the influences and under-the-table deals that created path-dependent processes and shaped the cities we live in today.

References

Acts of the General Assembly of the State of Georgia (January, February, March 1866) Boughton, Nisbet, Barnes & Moore, Milledgeville, GA

Atlanta: The Gate City of the South (1881) The Atlanta constitution, p 6

Barnwell VT (1867) Barnwell's Atlanta city directory, vol 1. Intelligencer Book and Job Office, Atlanta, GA

Beasley's Atlanta Directory for 1874 (1874) Beasley & Company Publishers, Atlanta, GA

Carson OE (1981) The trolley titans: a mobile history of Atlanta. Interurban Press, Glendale, Calif

City Railroad (1869, January 6) Daily Atlanta Intelligencer. p 1

Clarke EY (1881) Atlanta illustrated: Containing glances at its population, business, manufactures, industries, institutions, society, healthfulness, architecture, and advantages generally. Harrison & co, J. P

DeBats DA (2008) A tale of two cities: using tax records to develop GIS files for mapping and understanding nineteenth-century U.S. cities. Hist Methods: J Quant Interdiscip Hist 41(1):17–38. https://doi.org/10.3200/HMTS.41.1.17-38

Eubank S, Guclu H, Anil Kumar VS, Marathe MV, Srinivasan A, Toroczkai Z, Wang N (2004) Modelling disease outbreaks in realistic urban social networks. Nature 429(6988):180–184. https://doi.org/10.1038/nature02541

Garrett FM (1969) Atlanta and environs; a chronicle of its people and events, vol 1. University of Georgia Press, Athens

Glaeser EL, Goldin C (2006) Corruption and reform: lessons from America's economic history. https://doi.org/10.7208/chicago/9780226299594.001.0001

Goldfield DR (1982) Cotton fields and skyscrapers: southern city and region, 1607–1980. Louisiana State University Press, Baton Rouge

Gregory I, DeBats D, Lafreniere D (2018) The Routledge companion to spatial history. Routledge, Taylor and Francis Group, London

Hanleiter W (ed) (1871) Atlanta city directory for 1871. Hanleiter Publishing, Atlanta, GA

Kennedy SW, Curtis AJ, Curtis JW (2015) Historic disease data as epidemiological resource: searching for the origin and local basic reproduction number of the 1878 yellow fever epidemic in memphis, tennessee. Ann Assoc Am Geogr 105(5):987–1002. https://doi.org/10.1080/00045608.2015.1059167

Klima DL (1982) Breaking out: streetcars and suburban development, 1872–1900. Atlanta Hist J 26(2):67–82

Lafreniere D, Gilliland J (2014) "All the World's a Stage": a GIS framework for recreating personal time-space from qualitative and quantitative sources. Trans GIS. https://doi.org/10.1111/tgis.12089

Lafreniere D (2014) Reconstructing the spatial and temporal patterns of daily life in the 19th century city: a historical GIS approach. University of Western Ontario. Retrieved from Electronic Thesis and Dissertation Repository

Larsen LH (1985) The rise of the urban south. University Press of Kentucky, Lexington, Ky

Maloney TJ, Godfrey FM, Jr (eds) (1902) Atlanta city directory 1902, vol 26. Mutual Publishing Company, Atlanta, GA

Martin J (1975) Mule to MARTA, vol 1. Atlanta Historical Society Inc, Atlanta, GA

Menes R (2003) Corruption in cities: graft and politics in American cities at the turn of the twentieth century. Working Paper No. 9990. Retrieved from National Bureau of Economic Research website: http://www.nber.org/papers/w9990

Menes R (2006) Limiting the reach of the grabbing hand. graft and growth in American cities, 1880–1930. In: National Bureau of economic research. corruption and reform: lessons from America's economic history, pp 63–93. https://doi.org/10.7208/chicago/9780226299594.001.0001

Newman WT (ed) (1873) The code of the City of Atlanta: 1843-1873. W.A. Hemphill & Company, Atlanta, GA

Poor HV (ed) (1891) Poor's manual of the railroads of the United States, vol 24. Poor's Railroad Manual Company, New York

Raymond A (2011) Denny regrade, 1893–2008: a case study in historical GIS. Soc Sci Hist 35(4):571–597

Richardson G (2018) The vulnerability of New Orleans' black and foreign-born populations in the 1878 yellow fever outbreak: a reassessment. GeoJournal. https://doi.org/10.1007/s10708-018-9931-y

Russell JM (1988) Atlanta, 1847–1890: city building in the Old South and the new. Louisiana State University Press, Baton Rouge

Sauder RA (1980) The use of Sanborn maps in reconstructing "geographies of the past"; Boston's waterfront from 1867 to 1972. J Geogr 79(6):204–213. https://doi.org/10.1080/00221348008980649

Saunders HG (ed) (1891) Atlanta city directory for 1891, vol 15. R.L Polk and Company, Atlanta, GA

Sholes AE, Weatherbe CF (eds) (1881) Sholes directory of the city of Atlanta, GA for 1881, vol 5. Sholes and Company, Atlanta, GA

Steffens L (1904) The shame of the cities. McClure, Phillips & Company

Warner SB (1962) Streetcar suburbs: the process of growth in Boston, 1870-1900. Harvard University Press, Cambridge

"A Brother Orangeman the World Over": Migration and the Geography of the Orange Order in the United States

8

Cory Wells and Charles Travis

Abstract
Historical geographers' explorations of migrant networks have focused have focused on ethnic and religious organization and clustering in the host country as well the influence of the *sense of place* in the migrant's country of origin. This case study focuses on the Orange Order, a Protestant organization established in County Armagh in 1795, and its diasporic resettlement patterns from the Irish Province of Ulster to North America. Settlement patterns were typically anchored by Orange Lodges established as nodes on trans-national migration networks which stretch from Northern Ireland, the United Kingdom, North America, Australia, New Zealand and West Africa. As much work has been conducted on Orange Order Settlement in Canada, this case study specifically examines and maps early twentieth-century settlement patterns in the United States, particularly locations in New England and the Mid-West. This case study sources data from the *The International Bureau of 'Orange' Information*, a body established in 1903 to gather information on its diasporic communities. Historical geographical information systems (GIS) methods were employed in this case study to parse, map and explore demographics of Orange Order migrants' origins and destinations. Such methods have also been applied to explore how migrant identity is often tied to the sense of place carried from a migrant's native country such as Scotland and Nigeria. This study contributes to the literature and mapping of the Orange Order diaspora in north-east and mid-west regions of the United States.

Keywords
Orange Order · United States · Diaspora · Trans-national networks · Historical GIS

8.1 Introduction

Throughout the nineteenth century, tens of thousands of Irish Protestants who migrated to the United States joined the Orange Order in their newly adopted country. Formed in Ulster in the 1790s, the Loyal Orange Institution existed to maintain Protestant hegemony in Ireland. It quickly spread throughout the anglophone Atlantic, especially to Britain and Canada.

C. Wells (✉)
Independent Researcher, Dover, NH, USA
e-mail: cory.wells@mavs.uta.edu

C. Travis
Department of History, University of Texas, Arlington, TX, USA
e-mail: charles.travis@uta.edu

Trinity Center for Environmental Humanities, Trinity College, The University of Dublin, Dublin, Ireland

C. Travis et al. (eds.), *Historical Geography, GIScience and Textual Analysis*, Historical Geography and Geosciences, https://doi.org/10.1007/978-3-030-37569-0_8

Historical geographers have been at the forefront of the study of transnational Orangeism, especially its Canadian arm. Cecil Houston and William Smyth first described the geography of the Order in Canada, demonstrating it held widespread pan-Protestant appeal that transcended ethnic limitations to appeal to Canadians from non-Irish backgrounds. Houston and Smyth also compared the Canadian Order to its counterpart in America, noting that the southern version failed to compete with its northern cousin for obvious reasons, with the tension between loyalty to the crown and the republic featuring prominently.[1] When compared to Canada, the American Orange Order does appear to be a diminutive anomaly. However, when placed in American context, the question quickly becomes: how did it exist at all? As Houston and Smyth found with Canadian Orangemen, geography and demographics reveal much about the migration processes for members of the American Orange Order.[2] Indeed, instead of placing the two in contrast, a more useful framework may be to examine them as part of the same phenomenon.

National borders, despite existing as political constructs that often ignore the realities of topography, population distribution, and culture, certainly exert a tangible effect on groups whose identity is not predicated upon adherence to a particular nation-state. The Orange Order was no different, and the American version tempered the original organization's loyalty to the crown and supplanted it with loyalty to the Republic. But the Loyal Orange Institution's maintenance was dependent on steady streams of immigration from both Ulster and Canada, and it exhibited a surprising durability that was partly facilitated by its ability to transcend national boundaries through fostering migration networks and a transnational identity. As in Orange movements throughout the Anglophone world, the Orange Order in North America was primarily populated by first-generation Irish migrants and their immediate descendants, and first-generation representation was especially pronounced in lodges in the United States. By locating individual members of these lodges in census and immigration records, we can see strong indications of familial migration and personal networks that likely influenced individual decisions to migrate. The contours of these networks come into sharp contrast when approached with historical-geographic framework and the application of GIS methodology. GIS has exploded within the humanities in recent decades, finding its way into fields as diverse as historical geography, history, international relations, area studies, and higher education, to name a few.[3] A small number of humanities and social science scholars began utilizing GIS methods and technologies in the late 1990s to better represent spatial data and address epistemological problems presented by traditional political maps.[4] Within only a decade, historical geographers

[1]Cecil J. Houston and William J. Smyth, "Transferred Loyalties: Orangeism in the United States and Ontario," *American Review of Canadian Studies* 14, no. 2 (August 1, 1984): 193–211.

[2]Cecil J. Houston and William J. Smyth, *The Sash Canada Wore: A Historical Geography of the Orange Order in Canada* (Toronto: University of Toronto Press, 1980).

[3]Georg Christ, "Mapping Change: A Collaborative GIS-Based Cue Card System for the Humanities," in *Scientific Computing and Cultural Heritage*, ed. Hans Georg Bock, Willi Jäger, and Michael J. Winckler, Contributions in Mathematical and Computational Sciences (Berlin: Springer, 2013), 109–17; Nicholas Karanikolas, "Understanding Place: GIS and Mapping across the Curriculum," *Cartographic Perspectives*, no. 63 (2009): 71–73.

[4]Janette Lee, "Redistributing the Population: GIS Adds Value to Historical Demography," *History and Computing 8*, no. 2 (1996): 90–104; I. N. Gregory and R. G. Healey, "Historical GIS: Structuring, Mapping and Analysing Geographies of the Past," *Progress in Human Geography 31*, no. 5 (2007): 638–53; Anne Kelly Knowles and Amy Hillier, eds., *Placing History: How Maps, Spatial Data, and GIS Are Changing Historical Scholarship* (Redlands, CA: ESRI Press, 2008); Eric Gidal and Michael Gavin, "Introduction: Spatial Humanities and Scottish Studies," *Studies in Scottish Literature 42*, no. 2 (2016): 143–150; Jordan Branch, "Geographic Information Systems (GIS) in International Relations," *International Organization 70*, no. 4 (2016): 845–69; Diana Stuart Sinton, "Roles for GIS within Higher Education," *Journal of Geography in Higher Education 33*, no. sup1 (2009): 7–16; Charles Travis, *Abstract Machine: Humanities GIS* (Redlands, CA: ESRI Press, 2015).

accelerated the use of GIS in their research, bringing about a new field dubbed "historical GIS."[5] The possibilities for going beyond simply "new varieties of narrative histories" and instead reframing research with entirely new questions soon became apparent to many historians who acknowledged that virtually all documents contain spatial information. The interpretation of spatial data could be married with narrative explanation to bring new contours to historical research.[6] Proponents of historical GIS have found such methods especially useful in examining how identity is often tied to a *sense of place*, like Ireland, Scotland, or Nigeria, and thus fertile ground for spatial humanities.[7]

8.1.1 The International Bureau of 'Orange' Information

When the Loyal Orange Institution's international body, the Imperial Grand Orange Council, met in New York in 1900, it attempted to mitigate the long-standing problem of allowing Orange men and women to be recognized when visiting lodges outside their own. They created an "International Bureau of Information," under the office of the Imperial Grand Secretary, James Rice. By 1903, Rice was still hard at work on the project, and in his address to the convention he called for a streamlining of the process based on increased uniformity of Orange procedures and admonished primary lodge secretaries for their tardiness in responding to inquiries. This criticism was especially directed at Irish secretaries, who received the most requests because of high rates of emigration. Rice lamented that, "The brethren in the United States have had much to contend with in the difficulty surrounding the securing of certificates for members from the old country who would like to join in the United States but find unexpected barriers." Rice's desire for reform also included the adoption of a universal password and ritual so "an Orangeman should be a brother Orangeman the world over," a potential procedural shift for which the body had created a committee in New York in 1900.[8]

The issue was raised again in 1906, and by this time Rice was thoroughly frustrated with his fellow Imperial officers, who, like himself, traveled a great distance to be part of the Council. He admonished them, "Such conduct I simply describe as childish. What is the use of coming three thousand miles to do business and then going back three thousand miles to repudiate what you have done? I like straightforward conduct on the part of a man, much more so on the part of a Brother Orangeman."[9] Rice was clearly upset that his efforts to strengthen the international Orange community had been thwarted by the perceived laziness of those whom he considered part of a special community. Rice was not unique in his views; staying connected was a perennial challenge for Orangemen around the world, and much of their effort was dedicated to fostering and strengthening transnational ties. Migration out of Ireland was so high after the Famine that it became a permanent feature of life that affected both Catholics and Protestants. Within only a few short years of the Orange Order's formation in the 1790s, Orange men and women had migrated throughout the English-speaking world, maintaining ties to the "old country" everywhere they

[5]Ian N Gregory and Paul S Ell, *Historical GIS Technologies, Methodologies and Scholarship* (Cambridge: Cambridge Univ. Press, 2012); Gregory and Healey, "Historical GIS"; Alistair Geddes and Ian N Gregory, *Toward Spatial Humanities Historical GIS and Spatial History* (Bloomington: Indiana University Press, 2014).

[6]Knowles, *Placing History*.

[7]Julius Komolafe, "Nigerian Migration to Ireland: Movements, Motivations and Experiences," *Irish Geography 41*, no. 2 (2008): 225–41; Gidal and Gavin, "Introduction: Spatial Humanities and Scottish Studies"; Samuel M. Otterstrom and Brian E. Bunker, "Genealogical Geography and the Generational Migration of Europeans to America," in *The Routledge Companion to Spatial History*, ed. Ian Gregory, Don DeBats, Don Lafreniere (New York: Routledge, 2018).

[8]Loyal Orange Association of the World, Imperial Grand Orange Council, *Report of Proceedings of the Triennial Session of the Imperial Grand Orange Council of the Loyal Orange Association of the World* (Glasgow, Scotland: Published at the Office of Imperial Grand Secretary, 1906), 12, 16, 37.

[9]Imperial Grand Orange Council, *Report* (1906), 16.

went. As an inherently transnational organization, the Loyal Orange Institution employed a range of methods to maintain an "Orange Diaspora."[10]

8.2 The Orange Order as Diaspora and Transnational Network

The Loyal Orange Institution evolved out of agrarian vigilante societies formed to insure the implementation of laws that gave Protestants (especially Anglicans) social, political, and economic control of Ireland. The Order formally coalesced by 1795, and by 1797, it was the largest political organization in Ireland. Within only a few short years, Orange men and women had migrated throughout the English-speaking world, and they frequently clashed with Catholics throughout the Atlantic. The Orange Order's displays of anti-Catholicism even found a receptive audience in the late-nineteenth-century United States, where their July 12 parades, which celebrated William of Orange's defeat of James II in 1690, frequently devolved into the kind of rioting we associate with modern sectarian parades in Northern Ireland. To ease the migration process for fellow Orange men and women, the Order developed a loosely uniform membership system based upon transfer certificates that extended to all lodges throughout the world. When an Orangeman moved to a new location where a lodge existed, he simply asked the new lodge to write to his old one asking for a certificate, for which the lodge usually asked a small fee, with the "amount being forwarded to the old country for [the] same."[11]

The new lodge would duly admit the applicant at the first meeting following the receipt of the certificate. While we cannot tell from meeting minutes if migrants chose to move to locations because there was already an Orange presence there (and possibly an acquaintance or relative), we can be certain that transfer certificates helped ease migrants into their new homes and provided them with an almost instant social circle. In both English and American lodges, the members' origins varied by location, but virtually all were majority foreign-born well into the twentieth century. This steady stream of new blood from abroad was vital to sustaining the movement, which frequently struggled to retain membership in both countries.[12]

The case studies discussed below demonstrate the intersections between geography, occupation, and migration among American Orangemen. From 1883 to 1974, 106 Orangemen submitted transfer certificates to Lily of the Valley Loyal Orange Lodge (LOL) no. 167, located in the northern suburbs of Philadelphia (Fig. 8.1). Of the total number, 43 transferred directly from Ireland. Another 56 transferred from other Pennsylvania lodges, although they may have been born outside of Ireland and these may not have been their first lodges.[13] Those who transferred from local lodges likely knew some members of the new one from previous interactions between lodges, as local lodges frequently collaborated on events, and individual members sometimes visited other lodges in their area. Membership applications to Lily of the Valley Lodge survive from as far back as the first decade of the twentieth century to well into the 1970s. Thankfully for the historian, these applications collected a great deal of demographic information, such as place of birth, occupation, residence, denomination, and, sometimes, previous

[10]Donald M. MacRaild has used the term in "The Orange Atlantic," in *The Irish in the Atlantic World* (Columbia, SC: University of South Carolina Press, 2012), 307–26, and "Wherever Orange Is Worn: Orangeism and Irish Migration in the 19th and Early 20th Centuries," *The Canadian Journal of Irish Studies* 28/29 (2002): 98–117.

[11]Donald M. MacRaild, *Faith, Fraternity and Fighting: The Orange Order and Irish Migrants in Northern England, C. 1850–1920* (Liverpool University Press, 2005), 77; Purple Star Loyal Orange Lodge no. 40, Meeting minutes, Nov. 12, 1899, June 12, 1900, Aug. 25, 1900, Collection D0462 M98-14, box 4, Historical Society of Pennsylvania, Philadelphia, Pennsylvania (Hereafter cited as, Purple Star minutes).

[12]For England, see MacRaild, *Faith, Fraternity and Fighting*, 4.

[13]Loyal Orange Institution of the United States of America, Records, 1883–1974, Collection MSS 103, Historical Society of Pennsylvania, Philadelphia, Pennsylvania (Hereafter cited as, LOI USA).

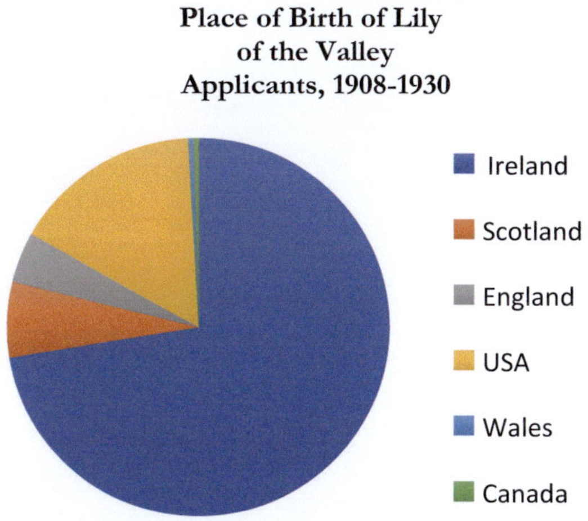

Fig. 8.1 Place of birth of Lily of the Valley applicants, 1908–1930

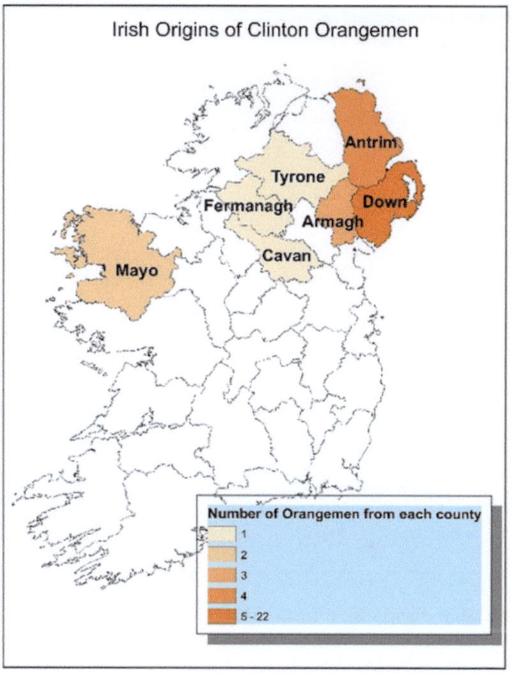

Fig. 8.2 Irish origins of Clinton Orangemen

Orange Lodge affiliations. A look at the records of Maiden City Loyal Orange Lodge no. 40 in Clinton, Massachusetts, is also revealing, especially regarding origin and occupation (Fig. 8.2). Most importantly, members of both lodges were overwhelmingly Irish. While American Orange Lodges fielded members of multiple nationalities, the majority of members in both Philadelphia and Clinton were born in Ireland. From 1908 to 1974, 225 of 378 applicants to Lily of the Valley (virtually all of whom were accepted) were of Irish birth, representing 60% of the total. Another 30 (8%) were born in in Scotland, and just 12 (3%) in England, for a total of 71% of foreign birth (Fig. 8.3). In the period before 1930, the numbers lean even more heavily toward foreign birth, at 78% of the applicants. While 95 of the Irish-born applicants listed only "Ireland" as their place of birth, we can reasonably assume from the rest that these came almost exclusively from the northern province of Ulster. County Tyrone, with twenty-seven applicants, led the way, with Antrim (including Belfast) contributing twenty-one, and Derry (twenty-one) and Donegal (sixteen) close behind. Only two listed County Down as their place of origin.[14]

The Irish origin of the members of Maiden City were more skewed than those of Lily of the Valley. Of the 57 members who appeared in Maiden City's meeting minutes, 41 (72%) were born in Ireland. Surprisingly, given its relative proximity, Canada only provided five members. The US also accounted for only five members, and England just one, although this member bore the distinctly Scottish name of James Campbell. Thirty-one of the members' birthplaces can be narrowed down to a specific place within Ireland, and of those for whom this information is available, twenty-two (71%) were born in County Down. Even more striking, fifteen (26% of members found from 1897 to 1915) of the Down natives were from the village of Rathfriland or nearby. This evidence is strengthened when we look at the employment of the Maiden City members. Most of those whose occupations are recorded in census records were employed in some way or another by the mills of Clinton, most of them at Lancaster Mills. These mills were known for both producing cotton gingham and employing large

[14]Membership applications, Lily of the Valley Loyal Orange Lodge No. 167, folders 2, 4–9, 12–18, MSS 103, HSP.

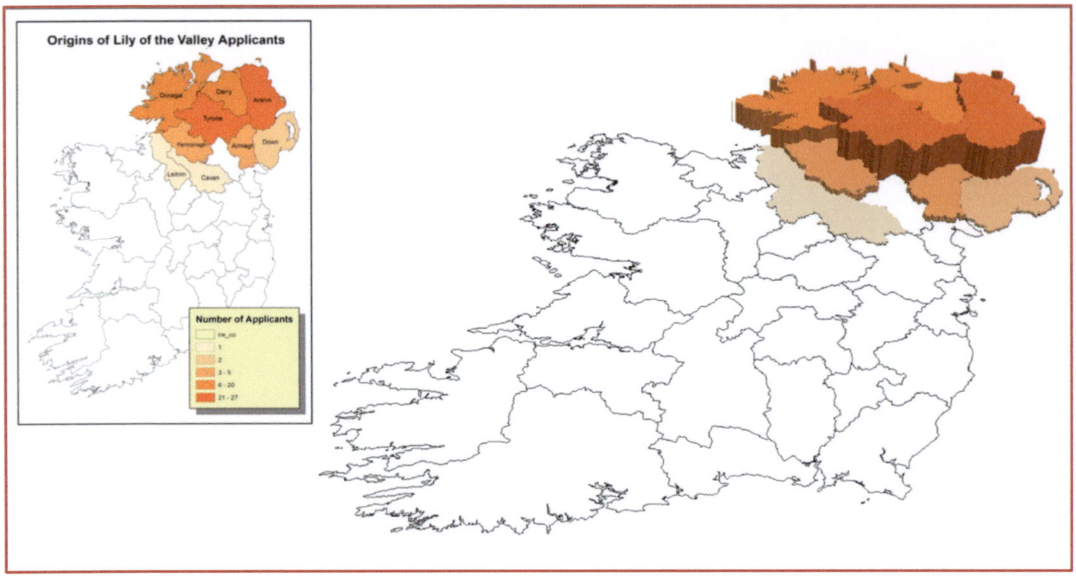

Fig. 8.3 Proportion from Irish Counties, applicants to Lily of the Valley lodge. Although this data is limited, it strongly suggests a process of migration from specific villages based on information sent back to Ulster from friends and family who had already made the journey and secured employment ("US Federal Census Collection, 1790–1940," digital images at *Ancestry.com;* Purple Star minutes, HSP. Bruce S. Elliott examines the phenomenon of network migration among Protestants from Tipperary to Canada in *Irish Migrants in the Canadas: A New Approach* (Kingston, ON: McGill-Queen's University Press, 2004), chapter 5.)

numbers of Irish migrants.[15] Of the 54 members for whom we have occupational data, 37 (61%) either held occupations in the textile industry or were explicitly listed as employees of the mills in other capacities, such as watchman, driver, or machinist. The single most common occupation among the Orangemen of Maiden City, with 11 among their number, was weaver. Loom fixers, teamsters, firemen (those who stoked the fires of steam-driven machinery), machinists, and dryers also filled the ranks of the lodge. American Orangemen strongly encouraged their recently arrived brethren to become citizens, and petitions for naturalization of the members of Maiden City also suggest the existence of migration networks among Ulstermen who left for America. These petitions could only be filed once an individual had been in the country for 5 years and had made a declaration of intent three or more years prior. While I have located only 14 such records, a few trends emerge regarding the origins of both the petitioners and the individuals who played key roles in their networks. Several of these men originated from a relatively small geographic area, implying they had some sort of connection before migrating. As more individuals used these channels, the network itself migrated and became stronger in the process.[16] As with the larger group for whom we can identify a place of birth, this smaller sample is also overwhelmingly represented by members born in County Down. The county of birth is unknown for three individuals, but like the lodge members as a whole, a majority of the men for whom we do have records (six) hailed from the small village of Rathfriland, with three others from the villages of Kilkeel and Saintfield, and the town of Banbridge. The remaining two

[15]Andrew Elmer Ford, *History of the Origin of the Town of Clinton, Massachusetts, 1653–1865* (Clinton, MA: W. J. Coulter, Courant Office, 1896), 222–225.

[16]Charles Tilly, "Transplanted Networks," in *Immigration Reconsidered*, ed. Virginia Yans-McLaughlin (Oxford: Oxford University Press, 1991), 84.

originated in Tyrone and the border county of Cavan.[17]

These petitions also give us some clues as to Orangemen's direct personal connections with each other. The application required two US citizens to sign and attest to the applicant's suitability for citizenship. For most of the men in our sample group, the signers were fellow Irishmen, although the connection to a particular place in Ireland was not quite as apparent. Only four of the signers can reliably be traced to County Down, although three were from Rathfriland. While most of these signers appear only once, a few signed for multiple petitioners. Henry Thompson, from Rathfriland, signed for four other men, while Benjamin Dixon, of Cavan, signed for three, as did Joseph Crothers. Crothers' birthplace in Ireland is unknown, but it is likely he is related to two other Crothers, Samuel and James, who themselves signed for petitioners and whose names appear frequently in Maiden City's records. These Crothers likely came from Castlereagh, Down.[18] Not surprisingly, many of the signers for these Orangemen were Orangemen themselves. Just under half (eight out of twenty) of the signers can be found in the membership records of Maiden City, although it is probable that others were also fellow lodge members. The Orange Lodge must have provided a place for Protestant Irishmen to socialize within a familiar institution from the homeland and to make connections that would benefit them in their new homes. Although it is unclear whether these men joined Orange Lodges because they already knew men who were members or if they got to know these other men after they joined, the lodge undoubtedly provided a place for men to strengthen such connections. Clearly, these Irishmen frequently looked to their Orange brethren when they needed someone to vouch for their personal character.[19]

For some Orangemen, the fictive kinship between brethren became literal, as they might be related by marriage. Of the 15 Maiden City Orangemen from Rathfriland, 8 married verifiably Irish-born spouses, and many of these marriages exhibited Rathfriland ties. George Stewart, who would later serve as Supreme Grand Master, married Marguerite Dougan, whose brother Robert was a Maiden City member. Although Stewart and Dougan were both from the Rathfriland area, they married after settling in Clinton. Sisters Bessie and Mary Ann Hamilton also married men from their hometown of Rathfriland—John D. Johnston and John Wright, respectively—after migrating, making their husbands brothers-in-law of their Orange "brother" Matthew Hamilton. Esther Hutton's marriage to John Martin probably connected him through her maternal grandparents to brothers (in both senses) David and William Gilmore, as well as to father and son James and Robert Crothers. Familial connections extended the Rathfriland contingent, too, as when Rathfriland-born George McCullough wed Elizabeth Mayberry, whose family came from Westport, Mayo, and provided Maiden City with at least two members: Elizabeth's brothers George and Charles.[20] Orange and occupational ties are difficult to untangle because many Orangemen worked alongside their fellow lodge members. While these individuals could draw from among their friends, neighbors, family members, and lodge brethren to find witnesses for naturalization applications, they usually had to look no further than the shop floor where they worked. But again, it is unclear which came first: the personal connection or the job. It does seem clear that many of these migrants worked together because they came from proximate locales throughout Ireland, where they all received similar information from the same friends and relatives about

[17] "US Naturalization Records, 1840–1957," digital images at Ancestry.com; Purple Star meeting minutes.

[18] Ibid.

[19] Ibid.

[20] Ibid; "1911 Ireland Census," digital images at Ancestry.com; "Massachusetts, Marriage Records, 1840–1915," digital images at Ancestry.com; "Records of the Immigration and Naturalization Service, 1787–2004," digital images at Ancestry.com; "US Federal Census Collection, 1790–1940," digital images at Ancestry.com.

the availability of jobs in Massachusetts, Pennsylvania, and other spots throughout the US and Canada. Once they arrived, they certainly would have sought out friends and relatives, or at least fellow countrymen, who had already made the journey. This search could have led them in many directions, of course, but two of these paths would have ended (or converged) at the workplace and the Orange Lodge. Partly because of this phenomenon, the typical Orangeman in Philadelphia was likely to be a gardener from Tyrone, while in Clinton, Massachusetts, he was apt to be a weaver from County Down. By looking at the examples of Lily of the Valley Lodge and Maiden City Lodge, we see definite signs of network migration. In both cases, Irish Protestants migrated to areas with long-established routes forged by their ancestors and countrymen. These streams of migration ebbed and flowed in response to many historical contingencies but saw an increase in the late nineteenth century, leading to pronounced occupational trends that intersected with migration streams. The textile industry had long been a major feature of the Ulster economy, first with wool, and later linen, which was a cottage industry.[21] Of course, many Irish Catholics worked in the textile mills of the northeast (most famously Lowell, Massachusetts, just a few miles away), and many Protestant migrants were also likely compelled to seek these same jobs, especially when guided by friends and family who had already migrated.

8.3 Canadian Migrants and the Orange Order in the United States

An Irish-born majority was not always to be found in American Orange Lodges, but the organization nevertheless typically drew its membership from immigrants—including many from Canada. Historians Bruno Ramirez and Randy William Widdis have shown that some generalizations can be made about Anglo-Canadian migrants to the United States, a group that subsumes those of Irish Protestant background within state records. The larger context of Anglo-Canadian migration can therefore help us make sense of many of the occupational and migratory characteristics found among Canadian-born American Orangemen. The late nineteenth century was a dynamic time for transatlantic migration, and Canada received the third-largest number of European migrants between 1870 and 1914, after the United States and Argentina.[22] By 1900, Canadians were the third-largest nationality in the U.S., after the Irish and Germans, with 2.5 million having arrived since 1860. The British element of the imagined Canadian culture was assumed to be dominant, with indigenous and French-Canadians being peripheral. This assumption, along with the Loyalist origins of early Anglo-Canadians, played into the loyalist bent that made it possible for the Orange Order to thrive in Canada. British migration to Canada accelerated throughout the nineteenth century, and most of these migrants came from the peripheries of the British Isles: Scotland, Ireland, and Northern England. By 1842, the Irish were the largest immigrant group in Canada, but this included both Protestants and Catholics. During the Famine, about 30% of those who left Ireland landed in Canada, with Protestants generally gravitating toward rural areas. But many British migrants were craft workers, attracted by semi-urbanized, diverse labor markets.[23] Many of the Canadian Orangemen who migrated to America and formed Orange Lodges there were the children of the British and Irish migrants who came to Canada a generation before, and all of them were shaped by the sense of a Protestant identity. It was

[21] James G. Leyburn, *The Scotch-Irish*, (Chapel Hill: UNC Press, 1962), 159; Kerby A. Miller, *Emigrants and Exiles: Ireland and the Irish Exodus to North America* (Oxford: Oxford University Press, 1988), 40.

[22] Walter T. K. Nugent, *Crossings: The Great Transatlantic Migrations, 1870–1914* (Bloomington: Indiana University Press, 1995).

[23] Randy William Widdis, *With Scarcely a Ripple: Anglo-Canadian Migration into the United States and Western Canada, 1880–1920* (Kingston, ON: McGill-Queen's University Press, 1998), xx, 3, 6, 9, 12, 62.

during the decades surrounding the turn of the twentieth century that the US-Canadian border began to truly exist as an obstacle to migrants, but Canadians still enjoyed an easier time crossing than most. They were exempted from a head tax upon entry, and the required documentation generated made it easier for them to acquire naturalization in the future than most other groups. Only about one in ninety Canadians was denied entry, and they were exempt altogether from the quota restrictions of 1921 and 1924, even though Canadian migration had increased during the First World War while migration from Europe had slowed.[24] Geographic constraints typically shaped migration from Canada to the US. Borderlands scholars have divided the vast international border between the US and Canada into several trans-border geo-economic regions, including the "Greater New England" region and Great Lakes region. The Greater New England region, characterized by its economy of "wood, wind, and sail," includes New England in the US, and extends all the way to Labrador on the northeast coast of Canada. Shipbuilding, lumbering, and fishing were the most important industries, and these geo-economic ties fostered cultural affinities, too, as the Atlantic Provinces were somewhat isolated from the rest of English Canada, being separated by French-dominated Quebec.[25] The New England states were also the most common destination for migrants leaving the Maritimes for the US. Between 1871 and 1901, close to 250,000 individuals (10.5% of the Maritime population) made this journey, with half choosing Massachusetts as their destination.[26] Boston was the primary destination for Maritime migrants, and there they enjoyed socio-economic rankings comparable to native-born Americans. Only 4% were unskilled laborers, according to a 1903 study, and Canadian men dominated the carpentry and shipbuilding trades.[27] Women, who outnumbered male migrants, mostly worked as domestic servants, but there were also many nurses, saleswomen, and office workers.[28] The membership records from Bostonia Lodge no. 505, in Boston, provide an example of how Anglo-Canadian migration affected American Orange lodges. Whereas Canada did not provide an especially large contingent within Maiden City or Lily of the Valley, just over half of the members of the Boston lodge were Canadian-born. The members of Bostonia 505 prove to be more elusive in census records, partly because of the greater likelihood that several residents of the larger city shared the same name, but it is also possible that these men were mobile at the moment the records captured them and they did not linger in Boston permanently. Of the 94 members listed in the roll book up to 1928, only forty-seven can be confidently tied to a place of birth. Of these 47, however, 24, or 51%, were Canadian-born, and these men were overwhelmingly from the Maritimes and Newfoundland. Only one originated in Ontario. After Canada, the US was the most represented country of origin, with fifteen members. Surprisingly, Ireland accounted for only two of those whose birthplace could be found, tying with Scotland and even ranking

[24]Bruno Ramirez, *Crossing the 49th Parallel: Migration from Canada to the United States, 1900–1930* (Ithaca: Cornell University Press, 2001), 45, 51.

[25]Graeme Wynn, "New England's Outpost in the Nineteenth Century," in Stephen Hornsby, Victor Konrad, and James Herlan, eds., *The Northeastern Borderlands: Four Centuries of Interaction* (Orono and Fredericton: Canadian-American Center, University of Maine and Acadiensis Press, 1989), 64–90; Patricia A. Thornton, "The Problem of Out-Migration from Atlantic Canada, 1871–1921: A New Look," *Acadiensis* 15, no. 1 (October 10, 1985): 3–34.

[26]Alan Brooks, "The Exodus: Migration from the Maritime Provinces to Boston During the Second Half of the 19th Century." (Ph.D. diss., University of New Brunswick, 1978), 72–75.

[27]Frederick A. Bushee, *Ethnic Factors in the Population of Boston* (New York: American Economic Association, 1903), cited in Ramirez, *Crossing the 49th Parallel*, 25–26.

[28]Betsy Beattie, "'Going Up to Lynn': Single, Maritime-Born Women in Lynn, Massachusetts, 1879–1930," *Acadiensis* 22, no. 1 (1992): 68.

behind of England, which provided three members.[29] What is most surprising about the presence of so many Canadian-born men is not necessarily that they were so numerous in Boston, but that there were not more of them elsewhere. Wynn's analysis of the New England Borderlands region is based upon a core and periphery model, which posits Boston as the core metropole which exploited its peripheral hinterlands of northern New England and the Maritimes of Canada for both labor and resources. Established even before 1812, this exchange of people and goods, especially lumber and coal, persisted even after Confederation and the push for westward expansion on both sides of the border.[30] Clearly, Boston was an attractive urban destination well into the twentieth century for those migrating from eastern Canada. We can find a similar trend in the leadership of Maine lodges in this period. A directory of British associations in North America in 1914 provides the names of the Master and Secretary for 173 Orange lodges, and census records show that twenty-seven out of the nineteen Maine members whose birthplaces can be located were Canadian-born. As in Bostonia Lodge, nearly as many were born in the US, with only two having been born in Scotland and none in Ireland. Not surprisingly, most of the Maine lodges listed were clustered in the eastern half of the state, near the border with New Brunswick. In the records of the State Grand Lodge meetings in the first 3 years of the 1920s, the Canadian presence is even more pronounced. Of the members present whose birthplaces are known, two-thirds were Canadian, with the remaining members being American-born. While there were no known members of Irish birth, the Orange Order in Maine was still very much an organization populated by immigrants, albeit of Canadian birth. As in Bostonia Lodge, the Canadian-born Orangemen were primarily from New Brunswick, Nova Scotia, and Newfoundland. While the number of known occupations (only fifteen) is too small to make any sweeping generalizations, there are still hints at some migratory links facilitated by occupation. For example, the three representatives from the town of Mexico and the two from Millinocket all worked in paper mills. Two of them, George and Wesley Leckey, were electricians at the mills and likely related. Seth Myers and John Spinney were both from the easternmost town in the US, Eastport, and were co-workers at the sardine cannery, a major business in the town (Fig. 8.4).[31]

A second trans-border region developed around the Great Lakes, with what would become Ontario on one side, and Michigan and upstate New York on the other.[32] Canadians began migrating into Michigan in waves in the 1830s, spurred by numerous factors that affected demand for labor, including changes in reciprocity laws between the US and Canada, a shift toward dairy production, and mechanization. By the 1930s, 76% of Anglo-Canadian migrants in Michigan were from Ontario, as were 66% of those in New York. Detroit was the primary destination for Ontarians, and by 1880 they represented 23.6% of its foreign-born population. For Ontarians, this was a relatively short-distance migration, as many had only to cross the Detroit or St. Clair Rivers.[33] Canadian-born Orangemen were also, if we can gauge by the leadership in 1914, the majority in Michigan lodges. Of the 37 listed as lodge masters and secretaries, the birthplaces of all but four

[29]Bostonia Loyal Orange Lodge no. 505, Meeting minutes, 1924–1929, Collection D0462 1993-102, box 1. Historical Society of Pennsylvania, Philadelphia, Pennsylvania (Hereafter cited as Bostonia meeting minutes).

[30]Wynn, "New England's Outpost.".

[31]"US Federal Census Collection, 1790–1940," digital images at Ancestry.com; Bostonia meeting minutes.

[32]For recent scholarship on the Great Lakes borderlands region, see Janet Dorothy Larkin, *Overcoming Niagara: Canals, Commerce, and Tourism in the Niagara-Great Lakes Borderland Region, 1792–1837* (Albany: State University of New York Press, 2018); Holly M. Karibo, *Sin City North: Sex, Drugs, and Citizenship in the Detroit-Windsor Borderland* (Chapel Hill: University of North Carolina Press, 2015), and Karl S. Hele, ed. *Lines Drawn upon the Water: First Nations and the Great Lakes Borders and Borderlands* (Waterloo: Wilfrid Laurier University Press, 2008).

[33]Ramirez, *Crossing the 49th Parallel*, 10, 27, 31–32, 106, 109.

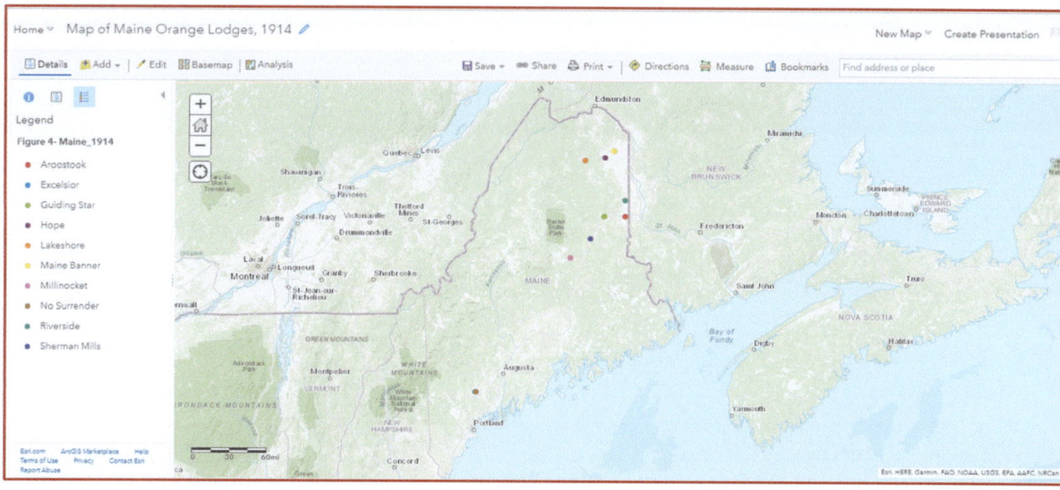

Fig. 8.4 Map of Maine Orange lodges, 1914

are available in census records. Of these 33, 18 (55%) were born in Canada, and of the 11 whose province of birth we know, all were from Ontario.[34] The connection with Ontario is to be expected because of its proximity, with little but the St. Clair River separating it from Michigan. Unfortunately, such a small sample size does not allow for much generalization, but a few elements beg for future investigation. Three of the four Orangemen in Sault Ste. Marie originated in Grey County, Ontario, which included townships whose Irish Protestant population exceeded 40% in the 1870s. The area was also home to a healthy number of Orange Lodges from the 1830s into the late twentieth century.[35] This over-representation in leadership could simply be a fluke of that year's elections within the two represented lodges, but two of the men were brothers who migrated to the US at different times, suggesting the likelihood that when John Syer relocated in 1885, he was acting upon information gathered from his older brother Thomas over the previous 3 years (Fig. 8.5).[36]

8.4 Conclusion

Several arguments can be made about the membership of the Orange Order in North America, broadly, and the United States, specifically, when we explore the records discussed above. First, there is clear evidence that occupation and migration were often connected and intersected in ways that are difficult to untangle but originated with ties formed in migrants' countries of birth (Ireland or Canada). The jobs in which Orangemen worked could be both specific to their location and their points of origin. When migrants from Ulster migrated to New England, where textiles were the cornerstone of the economy, they often found positions in the mills of Massachusetts, where they could use their skills from home. But those who went to the suburbs of Philadelphia may have been more likely to take the jobs that were available working for that city's upper classes, such as drivers and gardeners. Likewise, both of these examples indicate that Protestants from Ireland took many of

[34]D. MacDougall, *The American Year Book-Directory of Scottish Societies & British Associations in the United States, Canada & British Possessions* (New York: Caledonian Publishing Company, 1914); "US Federal Census Collection, 1790–1940," digital images at Ancestry.com; "Canadian Census Collection, 1851–1916," digital images at Ancestry.com.

[35]Houston and Smyth, *The Sash Canada Wore*, 33, 39, 41.

[36]"US Federal Census Collection, 1790–1940," digital images at Ancestry.com.

Fig. 8.5 Michigan Orange lodges, 1914

the same migratory paths as their Catholic countrymen. Irish Catholics famously worked the mills of New England and as domestics in virtually all the cities of the eastern seaboard, including Philadelphia. Other scholars have also documented instances of workspaces shared between Irish Catholics and Protestants, from the anthracite fields of Pennsylvania to the foundries of Troy, New York.[37] It is also clear from these examples that migration networks were at work, helping individuals make choices regarding resettlement and making their transitions successful. In each locale studied, the majority of Orangemen were from a distinct region—usually a county, if not a specific town or village. It is very likely that a "charter" group (or individual) made its (or his) way to the area of settlement and sent back information to friends and family about job prospects, the cost of living, and other important details. Those who received this information in Ireland were likely optimistic about the opportunities and their ability to transfer skills already learned at home in the new environment to their advantage. Others may have followed long-established patterns of migration from Ulster, based on less specific, more common information, such as those who went to Philadelphia.[38] Lastly, the maintenance of the Orange Order in the United States was dependent upon continued immigration, both from Ulster and Canada. At the peak of the Order's size in the first decades of the twentieth century, its membership was overwhelmingly foreign-born. In Philadelphia and Clinton, members came almost exclusively from Ulster. While it is unclear how connected to Ireland they remained, the presence of network migration suggests that their ties were at least somewhat significant. In Maine, Michigan, and Boston, the proximity to Canada and the region's role as a transnational borderland rendered these areas viable destinations for Canadian-born Orange men and women, most of whom were the children of Irish Protestants. Together, all of these individuals, whether they were Canadian carpenters in Boston or Ulster weavers in Clinton, brought with them attitudes and practices that were distinctly

[37]Kevin Kenny, *Making Sense of the Molly Maguires* (New York: Oxford University Press, 1998); Miller, *Emigrants and Exiles*, 548.

[38]Lesger, Lucassen, and Schrover have argued for the importance of non-network migrants who follow labor demands. While many travel for work because they have specific information based on their network connections, some migrate with only general knowledge of the likelihood of opportunities in a particular location. Cle Lesger, Leo Lucassen, and Marlou Schrover, "Is There Life Outside the Migrant Network? German Immigrants in XIXth Century Netherlands and the Need for a More Balanced Migration Typology," *Annales de Démographie Historique*, no. 2 (January 1, 2002): 29.

shaped by Irish ethnic politics, and which they would adapt to unique American circumstances. If in good standing with the Order, a member was immediately welcomed as a Brother or Sister anywhere in the world. Because the Order was an institution that was essentially Irish in nature, migration to the rest of the world was a fact of life for which practical considerations routinely had to be made. The Order dealt with this by attempting to create a unified transnational system based on the premise that the institution itself constituted a community that traversed national boundaries. Orange men and women in America clearly saw themselves as part of a larger diaspora. Long distances and the vagaries of human nature certainly made this system imperfect, but the dedication of its participants ensured the maintenance of a shared identity, where "an Orangeman should be a brother Orangeman the world over."

References

1911 Ireland Census. Ancestry.com

Beattie B (1992) 'Going up to Lynn': single, maritime-born women in Lynn, Massachusetts, 1879–1930. Acadiensis 22(1):68

Bostonia Loyal Orange Lodge no. 505, Meeting minutes, 1924–1929, Collection D0462 1993-102, box 1. Historical Society of Pennsylvania, Philadelphia, Pennsylvania

Branch J (2016) Geographic information systems (GIS) in international relations. Int Organ 70(4):845–869

Brooks A (1975) The Exodus: migration from the maritime provinces to Boston during the second half of the 19th century. PhD dissertation, University of New Brunswick (1978), pp 72–75

Bushee FA (1903) Ethnic factors in the population of Boston. American Economic Association, New York

Canadian Census Collection (1851–1916) Ancestry.com

Christ G (2013) Mapping change: a collaborative GIS-based cue card system for the humanities. In: Bock HG, Jäger W, Winckler MJ (eds) Scientific computing and cultural heritage. Contributions in mathematical and computational sciences. Springer, Berlin, pp 109–117

Elliott BS (2004) 2004 Irish migrants in the Canadas: a new approach. McGill-Queen's University Press, Kingston, ON

Ford AE (1896) History of the origin of the town of Clinton, Massachusetts, 1653–1865. W.J. Coulter, Courant Office, Clinton, MA, pp 222–225

Geddes A, Gregory IN (2014) Toward spatial humanities: historical GIS and spatial history. Indiana University Press, Bloomington

Gidal E, Gavin M (2016) Introduction: spatial humanities and Scottish studies. Stud Scott Lit 42(2):143–150

Gregory IN, Healey RG (2007) Historical GIS: structuring, mapping and analysing geographies of the past. Prog Hum Geogr 31(5):638–653

Gregory IN, Ell PS (2012) Historical GIS technologies, methodologies and scholarship. Cambridge University Press, Cambridge

Houston CJ, Smyth WJ (1984) Transferred loyalties: Orangeism in the United States and Ontario. Am Rev Can Stud 14(2):193–211

Houston CJ, Smyth WJ (1984) The Sash Canada Wore: a historical geography of the Orange Order in Canada. University of Toronto Press, Toronto (1980)

Hele KS (ed) (2008) Lines drawn upon the water: first nations and the Great Lakes Borders and Borderlands. Wilfrid Laurier University Press, Waterloo

Karibo HM (2015) Sin City North: sex, drugs, and citizenship in the Detroit-Windsor Borderland. University of North Carolina Press, Chapel Hill

Karanikolas N (2009) Understanding place: GIS and mapping across the curriculum. Cartogr Perspect (63):71–73

Kenny K (1998) Making sense of the Molly Maguires. Miller, emigrants and exiles, vol 548. Oxford University Press, New York

Knowles AK, Hillier A (eds) (2008) Placing history: how maps, spatial data, and GIS are changing historical scholarship. Esri Press, Redlands, CA

Komolafe J (2008) Nigerian migration to Ireland: movements, motivations and experiences. Irish Geogr 41 (2):225–241

Larkin JD (2018) Overcoming Niagara: canals, commerce, and tourism in the Niagara-Great Lakes Borderland Region, 1792–1837. State University of New York Press, Albany

Lee J (1996) Redistributing the population: GIS adds value to historical demography. Hist Comput 8(2): 90–104

Lesger C, Lucassen L, Schrover M (2002) Is there life outside the migrant network? German immigrants in XIXth century Netherlands and the need for a more balanced migration typology. Annales de Démographie Historique (2):29

Leyburn JG (1962) The Scotch-Irish. UNC Press, Chapel Hill, p 159

Loyal Orange Institution of the United States of America, Records (1883–1974) Collection MSS 103, Historical Society of Pennsylvania, Philadelphia, Pennsylvania (Hereafter cited as, LOI USA)

Loyal Orange Association of the World, Imperial Grand Orange Council (1906) Report of proceedings of the triennial session of the Imperial Grand Orange Council of the Loyal Orange Association of the World. Published at the Office of Imperial Grand Secretary, Glasgow, Scotland, 12, 16, 37. 16

MacDougall D (1914) The American year book-directory of Scottish societies & British associations in the United States, Canada & British possessions. Caledonian Publishing Company, New York

MacRaild DM (2002) Wherever Orange is worn: Orangeism and Irish migration in the 19th and early 20th centuries. Can J Ir Stud 28(29):98–117

MacRaild DM (2005) Faith, fraternity and fighting: The Orange Order and Irish migrants in Northern England, C. 1850–1920. Liverpool University Press, p 77

MacRaild DM (2012) The Orange Atlantic. The Irish in the Atlantic world. University of South Carolina Press, Columbia, SC, pp 307–326

Massachusetts, Marriage Records (1840–1915) Digital images at Ancestry.com

Membership applications, Lily of the Valley Loyal Orange Lodge No. 167, folders 2, 4–9, 12–18, MSS 103, HSP

Miller KA (1988) Emigrants and exiles: Ireland and the Irish exodus to North America. Oxford University Press, Oxford, p 40

Nugent WTK (1995) Crossings: the great transatlantic migrations, 1870–1914. Indiana University Press, Bloomington

Otterstrom SM, Bunker BE (2018) Genealogical geography and the generational migration of Europeans to America. In: Gregory I, DeBats D, Lafreniere D (eds) The Routledge companion to spatial history. Routledge, New York

Purple Star Loyal Orange Lodge no. 40, Meeting minutes, Nov. 12, 1899, June 12, 1900, Aug. 25, 1900, Collection D0462 M98-14, box 4, Historical Society of Pennsylvania, Philadelphia, Pennsylvania (Hereafter cited as, Purple Star minutes)

Ramirez B (2001) Crossing the 49th parallel: migration from Canada to the United States, 1900–1930, vol 45. Cornell University Press, Ithaca, p 51

Records of the Immigration and Naturalization Service (1787–2004) Ancestry.com

Sinton DS (2009) Roles for GIS within higher education. J Geogr High Educ 33(sup1):7–16

Thornton, PA (1985) The problem of out-migration from Atlantic Canada, 1871–1921: a new look. Acadiensis 15(1):3–34

Tilly C (1991) Transplanted networks. In: Yans-McLaughlin V (ed) Immigration reconsidered. Oxford University Press, Oxford, p 84

Travis C (2015) Abstract machine: humanities GIS. Esri Press, Redlands, CA

US Federal Census Collection (1790–1940) Digital images at Ancestry.com. Purple Star minutes, HSP

US Naturalization Records (1840–1957) Digital images at Ancestry.com. Purple Star meeting minutes

Widdis RW (1998) With scarcely a ripple: Anglo-Canadian migration into the United States and Western Canada, 1880–1920. McGill-Queen's University Press, Kingston, ON, xx, 3, 6, 9, 12, 62

Wynn G (1989) New England's outpost in the nineteenth century. In: Hornsby S, Konrad V, Herlan J (eds) The Northeastern borderlands: four centuries of interaction. Canadian-American Center, University of Maine and Acadiensis Press, Orono and Fredericton, pp 64–90

Part III

Climate, Weather, Environment

Mining Weather and Climate Data from the Diary of a Forty-Niner

9

Jase Bernhardt

Abstract

Primary sources such as personal diaries can provide insight into weather and climate conditions in times and places where quantitative instrumental observations are unavailable. The diary of Gideon Nichols provides an especially compelling case study of how such sources can be used to determine spatiotemporal patterns in meteorological conditions. Nichols, a farmer in Long Island, New York, elected to venture across the United States in 1849 to partake in the California Gold Rush, remaining there for 2 years before returning home via both oceanic and overland routes. Using content analysis, this chapter undertakes an investigation of his detailed records of weather conditions throughout his travels, as well as his firsthand account of major events such as the Sacramento flood of 1850. His daily recordings are supplemented by regular letters to relatives back home on Long Island, which contain ample details and emotional descriptions of his surroundings and how he experienced them. The result is a unique snapshot of the mid nineteenth century climate of numerous physical geographic regions across North America, along with a novel record of weather conditions during the early stages of the California Gold Rush. Moreover, Gideon's meticulous attention to detail, especially geographic location, permits the spatial analysis of these patterns, using both a physical geographical approach (e.g., his comparisons of different climate regions) and chronological approach (e.g., tracking extreme weather events over the course of a year). Thus, this ongoing research complements past work by introducing a spatiotemporal component into the human interpretation of weather conditions, and can be replicated using the diaries of other pioneers who regularly observed environmental conditions.

Keywords

Historical climatology · Climate history · U.S. Gold Rush · Historical GIS

9.1 Introduction

Historical climatology involves the utilization of unconventional, often qualitative records such as diaries, logs, and newspapers to gain insight into past climate (White et al. 2018). Methodologies such as content analysis are employed to evaluate the qualitative data in a more quantitative fashion (Moodie and Catchpole 1976). Such an approach is necessary to fill spatial and temporal gaps in

J. Bernhardt (✉)
Department of Geology, Environment and Sustainability, Hofstra University, Hempstead, NY, USA
e-mail: Jase.E.Bernhardt@hofstra.edu

C. Travis et al. (eds.), *Historical Geography, GIScience and Textual Analysis*, Historical Geography and Geosciences, https://doi.org/10.1007/978-3-030-37569-0_9

climate data and is often undertaken for places or times where instrumental weather observations are unavailable. Applications of historical climatology vary widely, including the examination of newspaper articles to aid in estimates of hurricane track and intensity (e.g., Businger et al. 2018), the analysis of microclimates from two farmer's diaries (e.g., Bernhardt 2015), and the reconstruction of climate change and societal collapse using long-term written accounts of the weather (e.g., Butzer and Endfield 2012). The utility of work involving diaries is readily apparent. They are frequently the products of long-term, dedicated observers who provide keen insight into local weather and climate patterns (Adamson 2015). Close reading of diaries can allow researchers to develop a familiarity with the author's terminology and biases, while the human element to the observations can allow a certain depth to the understanding. For example, vivid and descriptive language is often used to reflect more severe or uncommon occurrences, such as an ocean vessel encountering a tropical cyclone (e.g., Chenoweth and Mock 2013). Some authors place historical climatology at the intersection of climate science and environmental history (e.g., Pfister 2006), as both can be studied through the lens of human culture. Others affirm that work in this realm falls under the umbrella of historical geography, as taking a spatial approach to creating historical narratives can reveal important contextual details and result in a richer understanding (e.g., Raymond 2011). Visualizations and spatiotemporal analysis through the use of Geographic Information Systems (GIS) and other advanced tools is often a keystone of such research and necessary to fully interpret the many social dynamics at play (e.g., Gregory and Healey 2007; Fyfe et al. 2009). The diary of Gideon Smyth Nichols (1810–1894), a farmer based in Hempstead, New York, on Long Island, is ideally suited for work in the area of historical climatology with a spatiotemporal focus. In particular, Nichols describes a 3-year journey, from 1849 to 1852, in which he travelled west to join the California Gold Rush, spending nearly two years in the Sacramento region, and then voyaged back home via a circuitous route traversing the eastern Pacific Ocean, Central America, Gulf of Mexico, and eastern United States. The account of Nichols permits study of the spatial and temporal patterns in weather and climate across North America during the mid-nineteenth century. My work, therefore, expands upon previous historical climatological studies by taking a spatiotemporal approach in qualitatively analyzing the observations of Nichols throughout his travels to track weather systems and elucidate understanding of regional climates and their comparison to the present day. To do so, I present a series of case studies which highlight the most intriguing of his observations.

9.2 Case Studies

Three years of Gideon Nichols' diary were transcribed, from February 1849 through February 1852. His account chronicles his transcontinental journey from Hempstead, Long Island, New York to California to participate in the gold rush from late 1849 through mid-1851, and his return travel via the Pacific Ocean, Central America, and Caribbean Sea. In the remainder of this chapter, I apply a historical geographic approach to the diary of Gideon Nichols, expanding upon previous work to jointly analyze spatial and temporal patterns in weather and climate. Specifically, I present a series of case studies which capitalize on Nichols' unique insight to further our understanding of past climate and recent change.

9.2.1 The Journey Westward: Tracking Weather Systems, South Pass, Wyoming Snow Cover, and the "100th Meridian" Dividing Line

When Nichols began his journey westward in early 1849, winter conditions still dominated. However, as he traveled via train across Pennsylvania, a weather pattern more common for early spring developed (Fig. 9.1). A series of highly amplified "troughs" and "ridges" in the

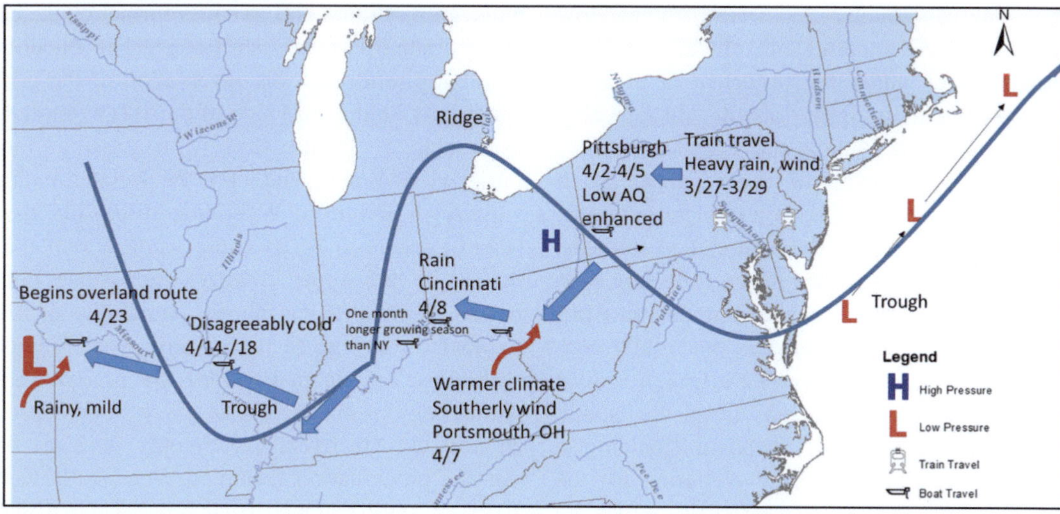

Fig. 9.1 The travels of Gideon Nichols and weather events encountered, March–April 1849

atmosphere, or dips and bumps in the jet stream, allowed for unsettled conditions and considerable temperature swings. While riding from Philadelphia to Harrisburg in late March, Nichols experienced heavy rain and wind as part of a trough, accompanied by a strong coastal low pressure system. Had cooler temperatures been present, snow may have fallen, especially in interior sections of Pennsylvania. When Nichols reached Pittsburgh a few days later on April 2, that area was under the influence of the ridge and surface high pressure, allowing for clear and calm conditions. In Pittsburgh, he noted poor air quality, unsurprising given the industrial activity (i.e., burning coal to produce iron; Tarr 2003), but it was likely exacerbated by the ridge, as rain and winds can disperse atmospheric pollutants, but the stagnant air present during his visit likely helped to keep the pollution in place. After a few days in Pittsburgh, Nichols boarded a boat on April 5th, which traveled southwest down the Ohio River. In the Northern Hemisphere, surface winds circulate in a clockwise direction around high pressure systems, so as Nichols traversed the Ohio-Kentucky border, he experienced the western periphery of the high, and observed a "warmer climate" with southerly winds upon stopping at Portsmouth, Ohio, during the evening of April 7. The following day, he proceeded further west, into the next trough and associated low pressure system, and noted rain in Cincinnati.

While travelling past Cincinnati, and then further southwest to Madison, Indiana, on April 10th, Gideon Nichols remarked that he could easily observe the "spring advance," and in Madison, the growing season appeared to be one month ahead of what it would be on Long Island. Such an observation is curious, as current data indicate that southern Indiana and western Long Island have a growing season of similar length (i.e., comparable USDA Plant Hardiness Zone). In 1849, however, both Madison and Hempstead were rural landscapes, contemporarily, the latter is now urbanized. Nichols' account may be read as further evidence that urbanization and the resultant enhancement of the urban heat island effect has lengthened the growing season on Long Island at a much quicker rate than global climate change has increased it at other locations where urbanization has not developed with as great and intensity (i.e., Madison, Indiana.) On April 14th, Nichols arrived in St. Louis via the Mississippi River, and witnessed another shift in the weather conditions, as he reached the other side of the low-pressure trough which had initially ensured milder weather as he traversed the Midwest. Opposite to high pressure systems, winds move counter-clockwise around low pressure in the Northern Hemisphere, so in Saint Louis, he was under the influence of a northwesterly wind, bringing about "disagreeably cold" conditions. Nichols did take heart, however, in the "coldness of the weather," as it limited the onset and risk of a cholera outbreak. Just north of St. Louis, the boat shifted from the Mississippi to the Missouri River, travelling west across Missouri, until reaching Camden on April 23rd. There, he noted unsettled but milder conditions, as the next low-pressure trough moved in, transporting warmer air from the south. Nichols disembarked in Camden, spending a week and a half in the town preparing for the long overland journey to California, which he commenced as part of a traveling party on May 2.

As Nichols and his party traveled westward, a rather serendipitous notation was made in his diary on June 9, 1849, while his party followed the Platte River in Nebraska. They encamped near where the south fork of the Platte meets the main river, around the present day city of North Platte. Nichols described the soil character shifting markedly at that point—to a sand and gravel mix, "not so excessively washed by water as before." Such an observation is notable because he was located between the 100th and 101st meridian—a significant demarcation line for United States physical geography. East of the line, a wetter climate dominates, with soil moist throughout much of the year. West of the line, most locations experience an arid climate, with either distinct wet and dry seasons, or low precipitation year round. The idea of the 100th meridian as a climatic dividing line was first articulated by John Wesley Powell in a report to the United States Department of Agriculture in 1878 (Powell 1878). The 100th meridian has received increased attention as of late due to anthropogenic climate change, as Seager et al. (2018) concluded that the "effective 100th meridian" is shifting eastward due to anthropogenic climate change. The observation of Nichols reinforces the notion that 100th meridian (or even a location slightly west of it) was an effective demarcation of hydroclimatology in the nineteenth century, and provides further evidence of its gradual eastward shift at present. Nichols frequently reinforced his observation of the arid conditions on the west side of the 100th parallel, expressing a climatic shift while passing Fort Laramie, Wyoming, and a few weeks later, in the vicinity of Riverton, Wyoming, disclosing that "there appears to be an entire want of rain this season in this upper country, none since Fort Laramie." Despite the inherent variability of the weather on any given day in any given region, such observations match well the present day climatology of this region, as Riverton possesses an arid climate overall, with only 8.42 in. of average precipitation (National Weather Service 2018) in stark contrast to Nichols' native Long Island, which experiences in excess of 40 in. per year.

As Gideon Nichols and his party traversed the rugged terrain of the Rocky Mountains, he frequently mentioned snow being visible on the higher mountain peaks—a phenomena unfamiliar to him while living on Long Island. In Wyoming, they reached what was likely the highest elevation of their journey while crossing the Continental Divide at a mountain pass in western Wyoming referred to as South Pass, at an elevation of approximately 7400–7500 feet. Nichols wrote in his diary that snow was still present in "banks," "fields," and "patches" as they travelled through on July 19. Despite its relatively high elevation, was the presence of snow cover notable? According to recent data—yes. Snowpack, precipitation, and other conditions at remote, high elevation sites in the western United States are monitored by automated weather stations as part of the Snow Telemetry (SNOTEL) Network (Natural Resources Conservation Service 2018). A SNOTEL site is located in the higher terrain just north of South Pass, with data collected the past 15 years (2004–2018). According to those data, snow typically disappears for the season during May or June, with a median date of June 3rd, and latest date of June 21. Thus, Nichols' observation of snow in the pass and surrounding mountains nearly a month later would be exceedingly rare given current climatic conditions there. Thus, this evidence from his diary presents, for one location in the Rocky Mountains, further compelling evidence of amplified climatic change across global mountains, and the rapid loss of snow and ice mass in those places (e.g., Rangwala and Miller 2012).

9.2.2 The Sacramento Flood of 1850

Nichols' description of the Sacramento Flood of 1850, during early January, and the storms that preceded it, allow for additional insight into the physical causes of the flood. Precipitation in California and much of the Pacific coast of the United States primarily falls during the cool season, mostly as part of heavy events. Annual precipitation, and resultant streamflow in hydrologic systems, varies considerably from year to year. During cool seasons with several storms, rivers and streams may flood, both during the event itself, and later, in the spring, due to snowmelt from their mountain sources. However, other winters may be devoid of any notable storms, resulting in reduced water levels and no flooding. It has long been understood that precipitation variability in the Western United States has been directly linked to "teleconnection" patterns over the Pacific Ocean, such as the El Niño-Southern Oscillation and Pacific-North American pattern (Cayan et al 1998). However, more recent work asserts that a prime driver of extreme precipitation and flooding are atmospheric river (AR) events, whose frequency may or may not be influenced by so called "teleconnection" patterns including El Niño. Research has shown that for central and northern California in particular, AR events may account for 30–40% of annual precipitation and explain the magnitude of variation in streamflow patterns (Dettinger et al. 2011).

In the case of the winter of 1849–1850 and January flood, Nichols noted several storms, both during his final approach to Sacramento, and while he settled into the city. Those storms included rain and snow on October 30 and 31, cool rain on November 2 and 3, mountain snow and valley rain during his travels of November 5 and 6, rain and stormy conditions the day after arriving in Sacramento on November 16, and another rainstorm on December 20th. Unsurprisingly, in his entry on the 22, Nichols warningly noted water overflowing the river, then described it overtaking lower elevation portions of Sacramento on the 29, and finally mentioned the city being "nearly submerged" on January 1—a prelude to the major flooding to come several days later. The extreme of the wet season, rising waters, and ultimate inundation of the city was described in dramatic detail by Nichols. Upon arrival in Sacramento, he noted that the storms to date had already filled the streets of the city with mud, filth, and dead carcasses. During his first full day in town, November 16, Nichols attempted to find work in the gold mines outside of Sacramento. However, after travelling only a few miles through the American River valley in increasingly dire conditions, he was advised by veteran workers that mining would not be possible due to the "unusual severity with which the rains had set in." Thus, he was forced to seek alternative employment, and after nearly going bankrupt, eventually found work chopping wood. He also boarded with his boss, working out an arrangement to pay reduced rent while assisting with housework. This arrangement persisted from November 30 until early January when "the flood came and expelled us all." As described by Nichols, water rapidly rose and flooded the city on January 9, inundating streets under two feet of water, and submerging the ground floors of many buildings. Transportation was only possible by canoe, and rations were short. Although the water level began to recede after the initial torrent, the waters did not fully recede until January 31.

Nichols' insight suggests that the flood event was most likely caused by a series of AR events in November and December. This hypothesis is strengthened by the findings of Gergis and Fowler (2009), who used proxy sources to determine that 1849 and 1850 were most likely weak to moderate *La Niña* years. Further, the deep inland penetration of this strong AR flow of moisture likely originating from the tropics is strongly supported by Nichols recording severe

Storm Date(s) 1849	Diary Description	San Francisco Precipitation (inches)
10/30-10/31	Rain (30th) to snow (31st), 2 to 3 inches of snow accumulation	1.69
11/2-11/3	Cold driving rain (2nd), day cold and rainy (3rd)	1.84
11/4-11/6	Severe snow storm (5th), day very stormy, windy, rainy (6th)	2.88
11/15-11/16	Evening rain (15th), rainy and stormy (16th)	0.84
12/19-12/21	Rainy day (20th and 21st)	2.12

Fig. 9.2 *Table 1*. Nichols diary descriptions of 1849 California storms

impacts from the early November storms despite still being well east of Sacramento. Just how anomalous were these antecedent conditions which caused the flooding? Reliable precipitation records for Sacramento are only available beginning in 1877. However, data are available, accessed from the Northeast Regional Climate Center (http://climod.nrcc.cornell.edu/climod) from a weather station in downtown San Francisco, which is often impacted by the same AR events as Sacramento. Unsurprisingly, the months of October through November 1849 are among the wettest in San Franciscan history. Eighteen inches of precipitation were recorded, the fourth highest total in the station's 168 year history, and well above the mean of 7.82 in. Moreover, the October, November, and December storms mentioned by Nichols are also reflected in the daily data, as shown in Fig. 9.2, Table 1.

9.3 Description of Local Climates

A particularly compelling aspect of the diary of Gideon Nichols is his vivid description of the many unique local and regional climates encountered during the journey. As a farmer, Nichols was keenly aware of the physical geography of his native Long Island, and thus provided detailed observations of the contrasting conditions experienced in California, Nicaragua, and Jamaica.

9.3.1 Sacramento

Nichols spent nearly 2 years in Sacramento and adjacent areas of central California, primarily Salmon Falls and Mormon Island (Fig. 9.3). He resided in Sacramento upon arriving in California in November 1849 through March 1850. From March through August, he mined for gold on Mormon Island, located northeast of Sacramento and surrounded by the South Fork of the American River. After experiencing only modest financial success, he decided to relocate to Salmon Falls, seven miles further upstream on the South Fork of the American River. However, in subsequent months, Nichols continued to turn a minimal profit, and opted to supplement his income with a more reliable flow of cash by farming the land. After purchasing the necessary seeds, tools, and equipment in Sacramento, he moved back to Salmon Island, setting up the garden in December, and planting in January, during the height of the typical wet season. He continued to plant and work in the garden over the ensuing months, along with some digging for gold. By May, Nichols began to harvest and sell smaller crops such as lettuce, with other vegetables, primarily potatoes, maturing by July, which he spent the summer selling to other miners, before closing the garden in late August and September. Much like the gold mining, the gardening allowed Nichols to earn sufficient income to subsist and save for passage home, but not enough to warrant staying in California any

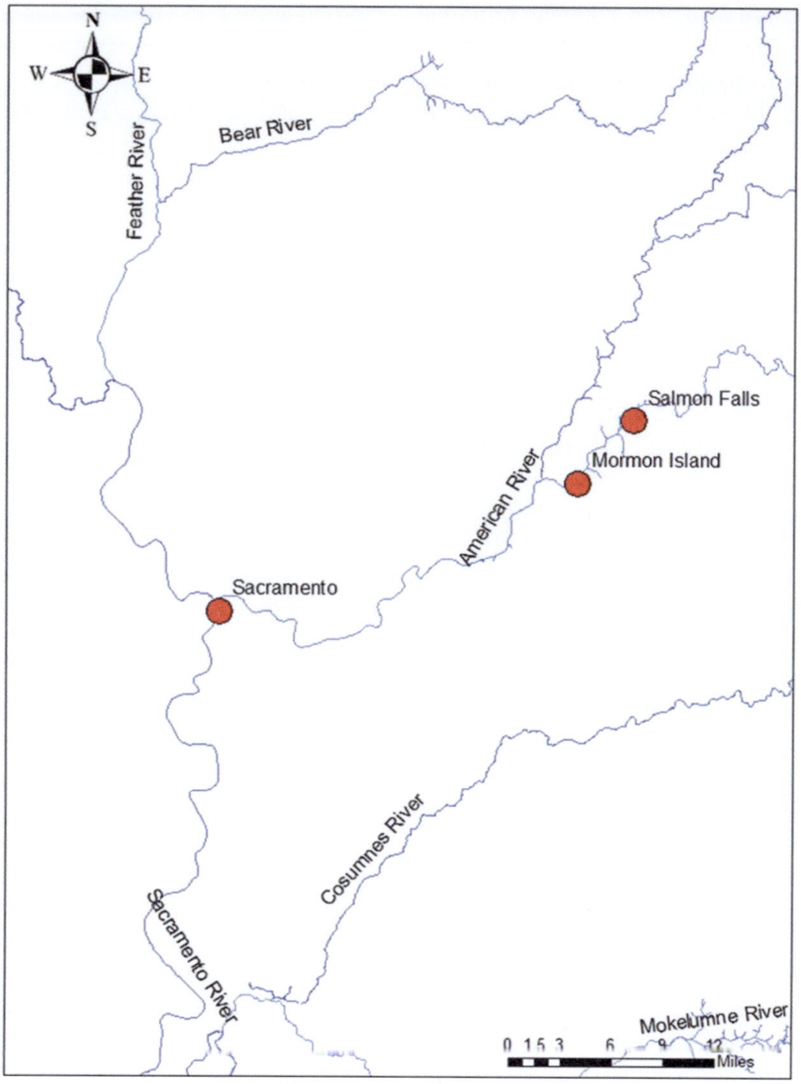

Fig. 9.3 Sacramento area places of residence, Gideon Nichols, 1849–1851

longer, especially with many of the most profitable mines being exhausted, and increased farming lowering the prices of vegetables (e.g., Chan 1989).

The decision of Nichols to farm the land motivated him to closely monitor the climate of 1851, and to maximize his output, compare it to both the previous year in California and conditions back home, in order to adjust his gardening strategy. Aforementioned in the previous subsection, the 1849–1850 cool season was exceptionally wet in California, with an especially stormy start from late October through December. Conversely, the 1850–1851 wet season began tepidly, as Nichols noted the "rains held off finely," until an initial storm on November 19. By that time in the previous year, by contrast, there had already been four precipitation events of consequence. Abnormally dry conditions prevailed over the following two months as well, and Nichols stated that through January 25 only one additional "decidedly rainy day" had occurred. Nevertheless, he made the decision in December to begin gardening, and the unusually dry weather proved favorable, allowing him time to prepare the fields for planting. The dearth of rainfall persisted, though, for another month, and in late February, Nichols once again lamented the landscape suffering from the "want of rain" during the wet season. He did, add, however, that light rain had begun to fall, and hoped that more would fall during March, the final month of the wet season.

Fortuitously for Nichols, precisely that transpired. He recounted in late April that instead of occurring during the normal peak of the wet season (i.e., the winter months), the rains had been "deferred till spring," falling in copious amounts in March and April. The late surge of precipitation allowed his primary cash crop, the potato, to "grow stronger and more thriftily than any I see elsewhere." Although he hypothesized that the dry season would quickly set in at that juncture in late April, there were additional rain events of consequence during the first few weeks of May, and by late that month, his garden was "flourishing in luxuriant beauty"—setting the stage for a reasonably profitable summer selling vegetables. Having only experienced two years in central California, how exactly did Gideon Nichols come to understand the local climate and its precipitation seasonality so well? His notes during the 1850–1851 wet season implied that climatologically, it starts in November and ends in April, with the winter months, December through February, the wettest, which identically matches the modern climatology for Sacramento (Bevan and Cline 2005). It is possible, though not explicitly mentioned, that he directly or indirectly (i.e., from other miners) conversed with indigenous peoples in the region, as he mentions occasional sightings of and contact with Native Americans. Those individuals would have possessed Traditional Ecological Knowledge (TEK), that is, deep knowledge of the local climate, from generations of living off the land. Nevertheless, the only source of scientific information specifically referenced by Nichols is the Bible. He compares the Mediterranean climate of Sacramento and surrounding areas to that of Judea, specifically mentioning he had been "often struck with a force and beauty of appropriateness in many of the passages and expressiveness of the Scripture which I have never perceived in them before." Indeed, Jerusalem and Sacramento bear strikingly similar climates, with that of the former regularly described in biblical readings and associated prayers. For example, one of the primary Jewish prayers, the *Amidah*, specifically contains a blessing for rain to be said seasonally, typically from late September through April (Kern-Ulmer 1995). Whether it through interaction with veteran miners and local natives, or close observation of the landscape and weather, Nichols was able to readily connect the climate he faced in California to that described in Biblical times.

9.3.2 Regional Climates of the Return Journey

Nichols took a circuitous route home after concluding his gold mining (Fig. 9.4), experiencing a variety of weather systems and regional climates along the way. Notable weather events included a tropical cyclone in the Eastern Pacific Ocean (described further below), heavy wind and waves in the Caribbean Sea off the east coast of Jamaica, and a strong winter frontal passage in the Gulf of Mexico. In September 1851, Nichols wrapped up his mining and farming in central California, sold off his assets, and returned to nearby Sacramento. At the end of the month, he took a ferry from Sacramento to San Francisco, and departed from there aboard a brig in early October. The most noteworthy weather event reported by Nichols while at sea on the return journey was storminess during late October. After departing San Francisco on October 7 and navigating through thick coastal fog, the craft travelled in a general southerly direction, paralleling the coast several hundred miles offshore, and experiencing mostly fair conditions between the 35th and 20th parallels. Such conditions are not atypical of the subtropical eastern Pacific Ocean during October, as winds often circulate from northeast to southwest, centered around the large subtropical high pressure system frequently present in the subtropics (i.e., 30° latitude) over

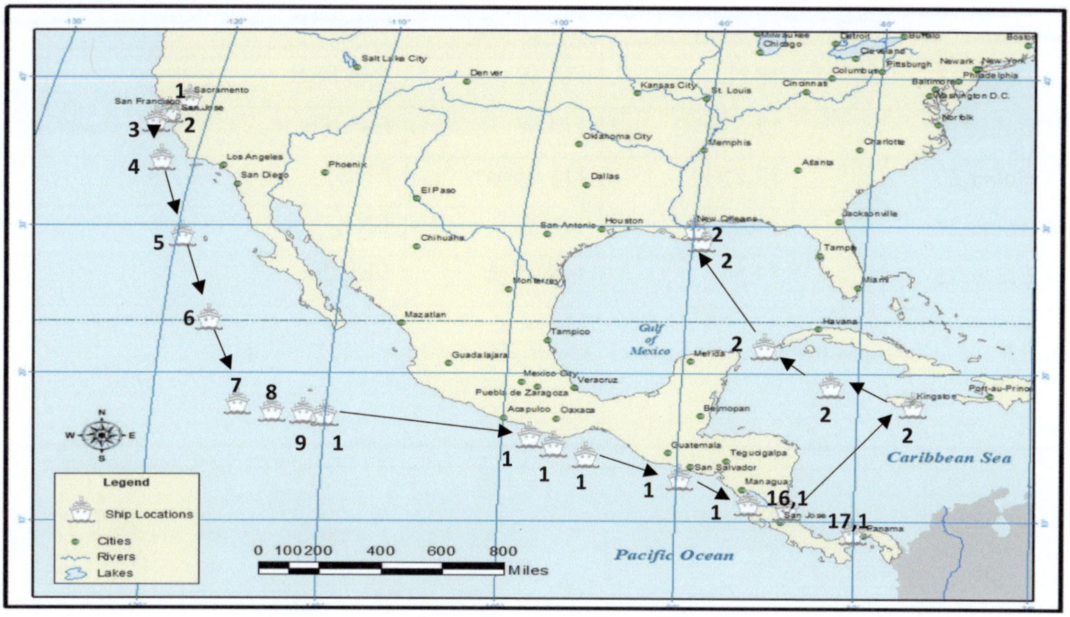

Fig. 9.4 Return journey of Gideon Nichols by sea, September 1851–January 1852. The ship locations are based on specific (but sporadically recorded) latitude and longitude coordinates noted in Nichols' diary entries, or descriptions of key features (e.g., the Mississippi River Delta) and where applicable, the port of departure or arrival mentioned. The locations are chronologically numbered, with dates and additional details found in the Appendix. This procedure is similar to that used in other analyses of travel diaries, such as Milson (2017). While he makes no mention of the exact source of these measurements, it is assumed that Nichols gleaned the coordinates from instrumentation onboard

the open ocean. Moreover, the subtropical high often promotes sinking air and dry conditions. However, by October 18, the ship had reached the 18th parallel, and Nichols observed a shift towards breezier and cloudier conditions. By the 20, as the ship approached the 17 parallel, he reported rising wind, heavy seas, squalls, and some rain, with the wind shifting to the east-northeast. The following day, the heavy seas, winds, and squalls persisted, which "caused the brig to pitch and plunge violently." By the 22, the worst of the storm had passed, with Nichols describing "light, changing and baffling winds and some of the time calm," with no mention of precipitation. Given the nature of these observations, latitude, and time of year, it is likely Nichols and his fellow passengers were impacted by a tropical cyclone (i.e., hurricane). As shown in Fig. 9.5, Table 2, strong but shifting winds were observed as the boat moved slowly east-southeast. Based on those observations, and data from three studies, the ship likely traversed the far western periphery of one of two hurricanes to form in the Eastern Pacific during October 1851. Visher (1922), using data from Redfield (1857), lists a hurricane first reported on October 21, and also mentions a second storm during the month, at an unspecified date, with details attributed to Seewarte (1897). The latter two papers also provide track maps of each storm. Thus, Fig. 9.6 contains the tracks of those two hurricanes, georeferenced from figures in Deutsche Seewarte and Redfield, along with the ship's path according to Nichols. Initially, the cyclone was located at a lower latitude than Nichols, resulting in a southerly component of the wind. Soon after, however, the storm moved to a higher latitude, resulting in a shift towards the east-northeast. Moreover, as depicted in Fig. 9.6, the Redfield-tracked storm was considerably further from the boat, and would not have resulted in noticeable impacts. Nevertheless,

Date	Latitude	Longitude	Wind Direction
October 18	18 N	115 W	E to ESE
October 19	17.5 N	113 W	E to NE
October 20	17.25 N	111.25 W	ENE
October 21	17.25 N	110 W	ENE
October 22	17 N	108.25 W	Variable

Fig. 9.5 *Table 2*. Gideon Nichols observations of October 1851 Eastern Pacific hurricane

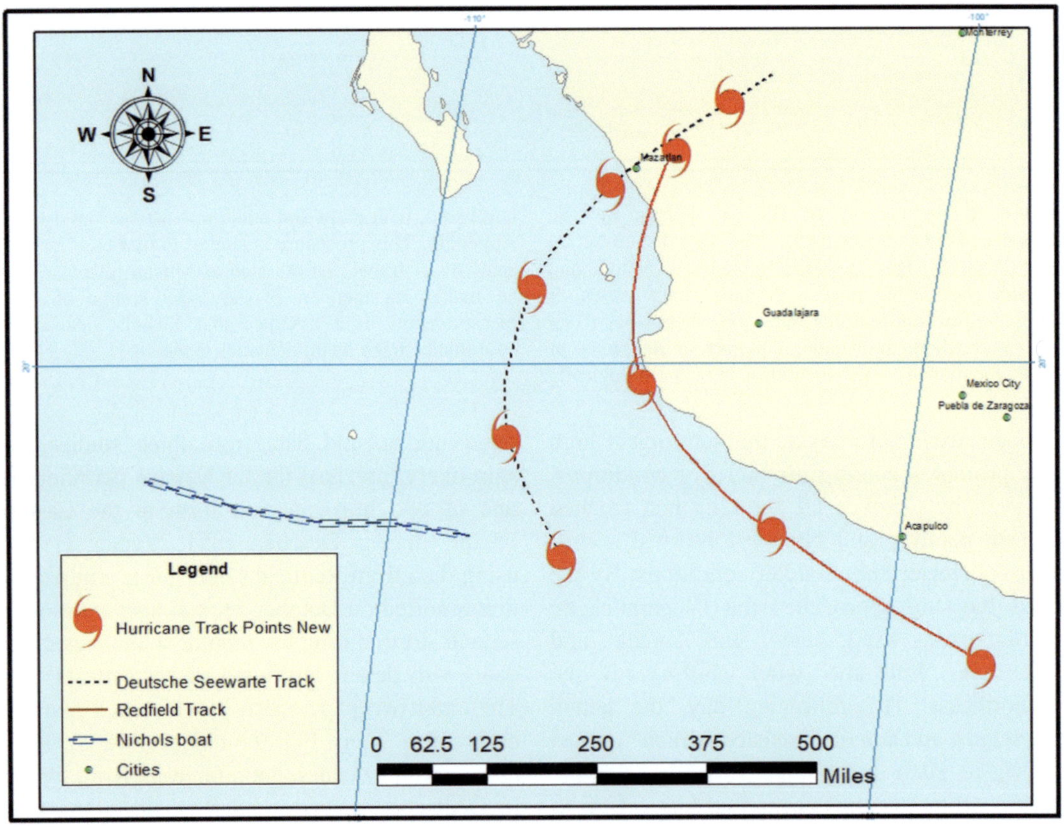

Fig. 9.6 October 1849 hurricane and path of Nichols' ship. As in the previous figure, the ship locations are determined from coordinates specified in Nichols' diary entries. The two proposed hurricane tracks are based on tables and figures from three sources, detailed in the text below and contained in full in the Appendix. For the Seewarte (1897) track, the coordinates in their Table 7 (p. 269) and approximate track in Fig. 19 (p. 266) were used, with the latter being georeferenced. Regarding the Redfield (1857) track, the approximate track from their unnumbered Figure was georeferenced, and corroborated with coordinates from an unnumbered Table in Visher (1922)

according to the Deutsche Seewarte track, the hurricane eventually moved far north and east, towards the Mexican coast, and away from the boat. The cyclone probably made landfall sometime on the 22nd and quickly weakened due to rugged terrain just inland, explaining why Nichols noted suddenly slackening winds coming from variable directions that day. Thus, this diary account serves as important corroborating evidence for Seewarte's (1897) approximated track of the storm, and adds important details such as the probable date of landfall. Soon after the hurricane, Nichols arrived in San Juan, on the Pacific (west) coast of Nicaragua. After a night's rest, he proceeded on foot several miles inland to Rivas, and immediately noted "luxuriant vegetation" and "exuberantly fertile soil," while the oppressive heat caused him to feel ill and fatigued. Western Nicaragua features a monsoonal climate, and Nichols arrived towards the end of the wet season, which dictated the hot and still-sultry conditions. He eventually proceeded to Lake Nicaragua, taking a vessel across it, before transferring to a large, open canoe to traverse the San Juan River across southern Nicaragua to Greytown, on the Caribbean (east) coast. Nichols noted different conditions in the eastern half of the country, still hot, but accompanied by frequent showers. Moreover, the vegetation was even denser, "so impervious to the eye that the ground is scarcely to be seen." Indeed, eastern Nicaragua is part of a tropical rainforest, with intense, year-round precipitation, and hot, humid conditions. The health hazards Nichols faced in Nicaragua also varied by climatic region. In the west, where the dry season was setting in, he complained of sunburn so severe that he could hardly walk for three or four days, and feared for his overall welfare. Conversely, in Greytown, on the humid east coast, he noted that the English individuals present took on the "color of aching under the king of Mosquito." Moreover, Nicaragua's location within the northeasterly trade wind belt severely delayed his journey and nearly cost Gideon Nichols his life.

Upon reaching Greytown in late November, he desired to take passage on a ship to New Orleans, in order to hasten his return to the United States. Against his better judgement, Nichols decided to purchase a ticket for the next available vessel, a brig, instead of a steam-powered craft. Constant showers and the northeasterly trade winds, however, provided a stiff challenge to the vessel in attempting to navigate to the northeast, away from the Nicaraguan coast. After several false starts, and increasing passenger sickness due to the rough seas, the brig finally attempted to set sail, only to be pushed off course, back to the coast, but below their initial location. With one of the ship's sails damaged, their only recourse was to proceed southeasterly along the Central American coast, in hopes of locating a safe natural harbor to land in. However, strong winds and low visibility from the constant storms made this impossible, and the captain eventually decided to crash the boat ashore near Chagres in present day Panama. Fortunately, Nichols was physically unharmed and even retrieved his possessions, but he failed to recover all of his ticket purchase, and saw his journey home delayed a few weeks. He finally secured passage on a steamship, which departed Chagres, only to return to Greytown for cargo, and then once again become stranded in the harbor due to unfavorable weather conditions. Then, to Nichols' great dismay, the vessel traveled back to Chagres to drop off the freight picked up in Greytown, and again had to wait for more tranquil conditions to depart. Finally, on December 29, the craft departed for Kingston, Jamaica—over a month after he initially arrived in Greytown on November 24.

Nichols arrived at Kingston on January 1, 1852, and immediately observed a more tranquil scene than Nicaragua, particularly calm winds and sea being "intercepted" by the land. That was an apt description of Kingston, as it sits in a protected natural harbor, and further, lies on the leeward side of the Blue Mountains in eastern Jamaica, favoring relatively dry conditions compared to other locations at its latitude. After some time in Kingston, Nichols left aboard a vessel bound for New Orleans on January 11. The craft was equipped with a thermometer, and he read temperatures in the lower 80s as they moved away from Jamaica. However, as they passed Grand Cayman Island and approached Cuba, the conditions changed, courtesy of a strong cold frontal passage. Strong cold frontal passages and accompanying air masses of polar origin are rare in the Caribbean Sea, but most common during the depths of winter, exactly the time Nichols was passing through. Just after midnight on January 14, between Grand Cayman Island and Cuba, Nichols provided evidence of such an occurrence. He described a wind shift to the north, and during the ensuing daytime hours, it "blowed with violence," while temperatures decreased 10° from the low 80s the previous day to the low 70s. Then, as the vessel progressed northward into the Gulf of Mexico, an even stronger cold front moved through. On January 18th, Nichols' entry noted daytime temperatures in the mid 70s, but at 9 PM, a "furious norther burst suddenly upon us," resulting in "an unpleasant and anxious night." The following day, temperatures were 20° cooler, and heavy gales continued—enough to cause Nichols and other passengers to stay below deck, along with damage to the ship's thermometer. Although the storminess abated the following day, a cold air mass had been ushered in, as Nichols detailed howling northwest winds on the 21st, "cool and chilly" weather on the 22, and continued cold conditions as they reached the mouth of the Mississippi River on the 23. Finally, on the 24, he once again set foot on United States soil, landing at New Orleans in the evening. Nichols completed the remainder of his journey over the following week and a half, using a combination of stagecoach, rail, and boat to travel from New Orleans to Atlanta and then up the east coast to New York, finally returning home on February 6. Along this final leg of his journey, Nichols did not mention any major weather systems of consequence, besides the expected cold conditions.

9.4 Conclusion

After completing his 3-year expedition, Gideon Nichols returned to his homestead in Hempstead and resumed his life as a farmer. Although he did not amass a great fortune in the Gold Rush, it is clear from his diary that the long trip left an indelible mark on Nichols. Moreover, his detailed account of the weather and climate conditions encountered along the way allow useful data to be extracted, despite the minimal presence of quantitative data. I feel confident in my interpolation of his written records, and historical climatologists with an in-depth knowledge of regional climates can offer a skilled reading of sources provided by an individual with a well-founded understanding of the environment, such as Gideon Nichols. Readily accessible primary source accounts such as diaries will become more ubiquitous as records become increasingly transcribed and disseminated. The field of historical climatology will therefore continue to serve the critical purpose of allowing geographers and others to broaden interpretations of past climates.

Acknowledgements I am indebted to several individuals for making the work described in this chapter possible. Thank you to Geri Solomon, Hofstra University archivist, for first bringing the transcribed diary to my attention and providing access. I am also grateful to Kristen Aloysius and Jethel Bayawa for their initial analysis of the diary as part of the 2018 Hofstra University Summer Science Research Program. Last, thank you to Deryck Holdsworth for his review of preliminary drafts and valuable feedback, and the editors of this volume for providing useful guidance on the final draft.

Appendix

See Figs. 9.7, 9.8 and 9.9.

#	Date(s)	Notes
1	9/29/1851	Departed Sacramento
2	9/30 - 10/6	San Francisco
3	10/8	
4	10/9	
5	10/12	
6	10/15	
7	10/18	Impacted by tropical cyclone
8	10/19	Impacted by tropical cyclone
9	10/20	Impacted by tropical cyclone
10	10/21	Impacted by tropical cyclone
11	10/30	Mexican coast in sight
12	10/31	
13	11/1	
14	11/6	
15	11/9	Landed at San Juan, Nicaragua
16	11/28	Departed Greytown, Nicaragua
17	12/8 - 12/12	Chagres, Panama
18	12/13	Greytown, Nicaragua
19	12/23	Chagres, Panama
20	1/1 - 1/11 1852	Kingston, Jamaica
21	1/12	
22	1/14	Cape San Antonio (Cuba) in sight
23	1/23	Landed at Mississippi River Delta
24	1/25	Reached New Orleans

Fig. 9.7 Dates and details of Nichols' return journey depicted in Fig. 9.4

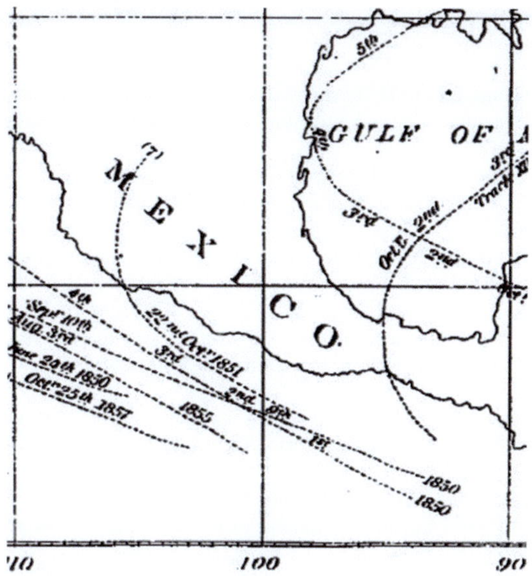

Fig. 9.9 Snippet of Redfield (1857) Eastern Pacific Ocean tropical cyclone track map; the track labeled 22nd October 1851 (or #7) was georeferenced in Fig. 9.6

Fig. 9.8 Sample image of the georeferencing process for the Seewarte (1897) tropical cyclone track depicted in Fig. 9.6

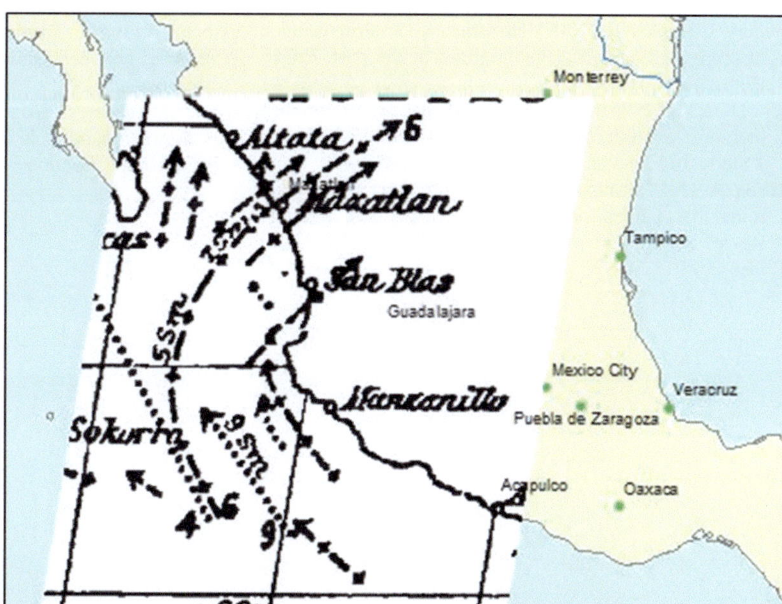

References

Adamson G (2015) Private diaries as information sources in climate research. Wiley Interdiscip Rev Clim Chang 6:599–611

Bernhardt J (2015) Determining regional weather patterns from a historical diary. Weather Clim Soc 7:295–308

Bevan L, Cline G (2005) Climate of Sacramento, California. NOAA Technical Memorandum NWS WR-272, 87 pp

Businger S, Nogelmeier M, Chinn P, Schroeder T (2018) Hurricane with a history: Hawaiian newspapers illuminate an 1871 storm. Bull Am Meteor Soc 99:137–147

Butzer K, Endfield G (2012) Critical perspectives on historical collapse. Proc Natl Acad Sci 109:3628–3631

Cayan D, Dettinger M, Diaz H et al (1 add. co-author) (1998) Decadal variability of precipitation over western North America. J Clim 11:3148–3166

Chan S (1989) This bittersweet soil: The Chinese in California agriculture, 1860–1910. University of California Press, Berkley

Chenoweth M, Mock C (2013) Hurricane "Amanda": rediscovery of a forgotten US Civil War Florida Hurricane. Bull Am Meteor Soc 94:1735–1742

Dettinger M, Ralph F, Das T et al (2 add. co-authors) (2011) Atmospheric rivers, floods and the water resources of California. Water 3:445–478

Fyfe D, Holdsworth D, Weaver C (2009) Historical GIS and visualization: insights from three hotel guest registers in Central Pennsylvania, 1888–1897. Soc Sci Comput Rev 27:348–362

Gergis J, Fowler A (2009) A history of ENSO events since AD 1525: implications for future climate change. Clim Chang 92:343–387

Gregory I, Healey R (2007) Historical GIS: structuring, mapping and analysing geographies of the past. Prog Hum Geogr 31:638–653

Kern-Ulmer B (1995) Consistency and change in rabbinic literature as reflected in the terms Rain and Dew. J Study Jud Persian Hell Rom Period 26:55–75

Milson A (2017) Mapping travelers' cultural and environmental perceptions: Thomas Nuttall and Henry Rowe Schoolcraft in Arkansas, 1818–1819. Hist Geogr 45:172–187

Moodie D, Catchpole A (1976) Valid climatological data from historical sources by content analysis. Science 193:51–53

National Weather Service (2018) 1981–2010 Normals for Wyoming Station. https://www.weather.gov/riw/cms_wyclimatenormals. Accessed 1 Dec 2018

Natural Resources Conservation Service (2018) Automated snow monitoring. https://www.wcc.nrcs.usda.gov/about/mon_automate.html. Accessed 1 Dec 2018

Pfister C (2006) Climatic extremes, recurrent crises and witch hunts: strategies of European societies in coping with exogenous shocks in the late sixteenth and early seventeenth centuries. Mediev Hist J 10:33–73

Powell J (1878) Report on the lands of the Arid Region of the United States: with a more detailed account of the lands of Utah. Gov Print Off, Washington DC, U.S

Rangwala I, Miller J (2012) Climate change in mountains: a review of elevation-dependent warming and its possible causes. Clim Change 114:527–547

Raymond A (2011) Denny regrade, 1893–2008: A Case Study in Historical GIS. Soc Sci Hist 35:571–597

Redfield W (1857) On the cyclones or typhoons of the North Pacific ocean, with a chart showing their courses of progression. Am J Sci Arts 1820–1879 (24):21–40

Seager R, Lis N, Feldman J et al (5 add. co-authors) (2018) Whither the 100th meridian? The once and future physical and human geography of America's Arid–Humid Divide. Part I: the story so far. Earth Interact 22:1–22

Seewarte D (1897) Segelhandbuch für den Stillen Ozean: Mit einem Atlas von 31 Karten. L. Friederichsen and Co., Hamburg

Tarr J (ed) (2003) Introduction: some thoughts about the Pittsburgh environment. Devastation and renewal, an environmental history of Pittsburgh and its region. University of Pittsburgh Press, Pittsburgh

Visher S (1922) Tropical cyclones in the Northeast Pacific, between Hawaii and Mexico. Mon Weather Rev 50:295–297

White S, Pfister C, Mauelshagen F (eds) (2018) The Palgrave handbook of climate history. Palgrave MacMillan, London

10 Unmappable Variables: GIS and the Complicated Historical Geography of Water in the Rio Grande Project

Daniel R. Beene and K. Maria D. Lane

Abstract

At the turn of the twentieth century, the U.S. Bureau of Reclamation's Rio Grande Project radically changed irrigation practices and the legal structure of water deliveries along the entire reach of the Rio Grande. The construction of Elephant Butte Dam and other modern water infrastructure in the Lower Rio Grande Basin of New Mexico set the stage for a complicated relationship between water governance, agricultural landscapes, and water users. Geospatial technologies are uniquely poised to quantify the observable effects of spatial change, but are limited in their ability to explain the human causes or repercussions of such phenomena. Using a sequential exploratory research design, pairing GIScience analysis of contemporary agrarian change and models of observed landscape transformation with historical-critical physical geography, we explore how water governance has precipitated ecological change and the loss of cultural knowledge, political change and the exacerbation of inequalities, and legal conflict leading to political battles downstream of the dam. We conclude with a discussion of the limits of geospatial models to capture the complexity and nuance of agrarian patterns and demonstrate the value of a mixed-methods approach for analyzing the formative and lasting ramifications of water governance in the Lower Rio Grande Basin.

Keywords

Rio Grande Project · Water governance · Sequential exploratory design · GIS · Historical-Critical Physical Geography

10.1 Introduction

Along the banks of the Rio Grande in southern New Mexico, tensions over water have been rising since the turn of the current century. Nearly two decades of drought and diminished river water have prompted farmers to engage in intensive groundwater pumping to ensure the viability of their crops. This practice has ecological implications and has become part of the legal basis for an interstate dispute over New Mexico's Rio Grande water deliveries to Texas. Historical agreements about water sharing and water deliveries are today complicated by climate change, as New Mexico and Texas vigorously engage the legal system to protect and manage their share of dwindling water supplies for irrigation and other needs. Most recently, this

D. R. Beene (✉) · K. M. D. Lane
University of New Mexico, Albuquerque, NM, USA
e-mail: dbeene@unm.edu

K. M. D. Lane
e-mail: mdlane@unm.edu

C. Travis et al. (eds.), *Historical Geography, GIScience and Textual Analysis*,
Historical Geography and Geosciences, https://doi.org/10.1007/978-3-030-37569-0_10

conflict has gone to the U.S. Supreme Court, with New Mexico and Colorado defending against charges from Texas that they have failed to deliver promised water supplies as a result of farmers' groundwater pumping in southern New Mexico (Texas v. New Mexico & Colorado, 138 S. Ct. 954 [2018]).

In the middle of these tensions stands Elephant Butte Dam, the southernmost gauging station along the Rio Grande in New Mexico, about 100 miles north of Texas and the US/Mexico border. The dam's completion in 1916 initiated a wave of settlement, as homesteaders purchased federal land grants by the hundreds and thousands of acres for as low as forty dollars an acre (Schönfeld LaMar 1984). The influx of new settlers converted the Lower Rio Grande's floodplain into a significant agricultural region, while simultaneously changing the political landscape during the early years of New Mexico statehood. Today, the Rincon and Mesilla valleys, between Elephant Butte Dam and the state line, support some of New Mexico's largest agricultural enterprises, including one of the most prolific pecan orchards in the world. Although cotton dominated the region from the 1920s until the 1960s, pecans are the most influential single crop in the region today, and New Mexico is now the top producer of pecans in the United States (Autobee 1994; Bustillos and Bautista 2018; Hedden 2019; Schönfeld LaMar 1984). And despite the fact that pecans are a water-intensive crop not naturally suited to arid lands cultivation, pecan production is actually expanding in southern New Mexico at the same time water conflict is increasing.

This paradox prompted us to ask what might seem like a straightforward question: Why do farmers in the Rincon and Mesilla valleys choose to grow pecans despite the long-term and water-intensive investments these orchards require? Our efforts to find an answer have followed a few different paths. First, we interviewed farmers downstream of the dam, asking them to narrate recent ecological change and explain how they decided what to plant. Second, we used a GIS analysis to render these changes and drivers in an explicitly spatial way, analyzing how specific variables relate to and influence a farmer's decisions to plant one crop versus another. Finally, we dug into the historical geographies of material, ecological, and political changes in the valleys, chronicling the legal and economic contexts that set the stage for modern agricultural decisions and issues. In our research process, we found geospatial methods both enriching and limited. GIS augments our understanding of complex physical geographies but cannot fully replace qualitative modes of analysis and storytelling.

This chapter illustrates our mixed-methods approach, intertwining geospatial modeling with social-science analysis and historical narrative. The first section reviews the basic historical geography of irrigation in southern New Mexico, followed by a discussion of recent legal issues. The central sections of the chapter present our GIScience work, which quantifies landscape change in the last decade and analyzes the spatial relationship between pecan cropping and a variety of economic and ecological variables. The final section pairs spatial results with historical context, tracing the complicated relationship between water governance, irrigators, and landscapes. Throughout, we consider the way that these different methodologies encounter "unmappable variables" that can be addressed only by stitching together different methodological frameworks. In the end, these multiple methods all point toward a similar conclusion, showing that water governance has perpetuated power inequalities among different classes of agriculturalists while also promoting specific types of landscape impact.

10.2 Study Area: Historical Geographies of Irrigation in an Arid Valley

All irrigable farmland along the Rio Grande between Elephant Butte Dam and the Texas state line is served by the Elephant Butte Irrigation District (EBID), which spans two New Mexico counties and encompasses both the Mesilla and Rincon valleys (Fig. 10.1). The governing body of EBID is made up of nine farmers who manage

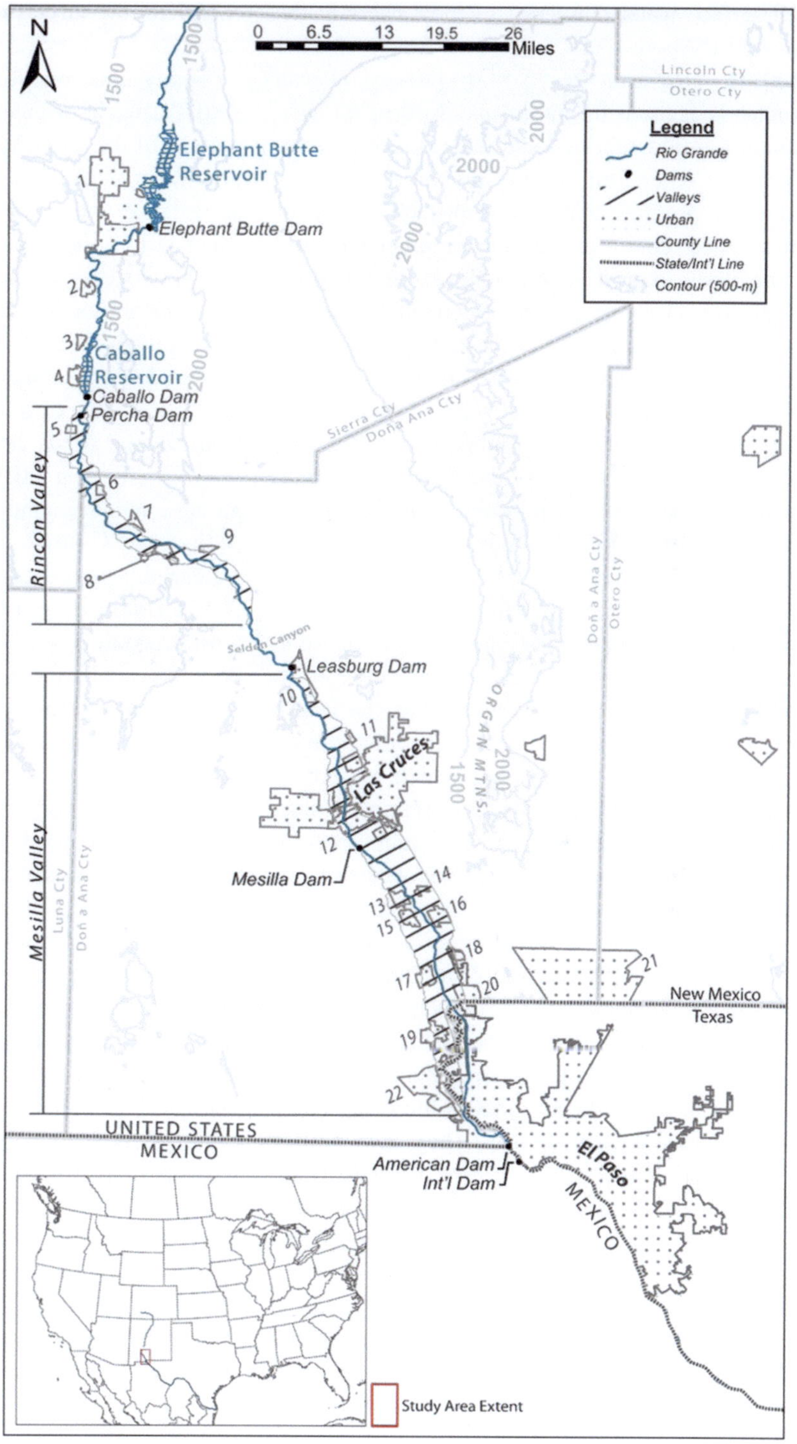

Fig. 10.1 Study Area Map showing political and physical features. Numbered cities and towns: 1. Truth or Consequences; 2. Las Palomas; 3. Caballo; 4. Oasis; 5. Arrey; 6. Garfield; 7. Salem; 8. Hatch; 9. Rincon; 10. Radium Springs; 11. Doña Ana; 12. Mesilla; 13. San Miguel; 14. Mesquite; 15. La Mesa; 16. Vado; 17. Chamberino; 18. Berino; 19. La Union; 20. Anthony; 21. Chaparral; 22. Santa Teresa

all surface water deliveries to their peers in the district ("About EBID" n.d.; NM Stat § 73-10-1 [2014]; NM Stat § 73-11-1 [2014]). Pecan orchards account for about one-third of all agricultural production in these valleys, with a scattering of other fruit and nut orchards producing apricots, peaches, apples, and pistachios. As for row crops: cotton and alfalfa or other hays are ubiquitous, and it is common to see diverse vegetable crops of red or green chile, onions, lettuce, and melons. All of these crops grow well in the Mesilla and Rincon valleys, which are generally flat and have an elevation around 3,900 feet (1,100 m). Climatic conditions here are well suited for both orchards and row crops, with warm temperatures year round ranging from a mean low of 41.1 °F to a mean high of 76.6 °F. This portion of the state receives an average of just under 10 inches of rain per year, however, meaning that the key constraint for agriculture in these valleys is water. The federal government and multiple irrigation districts offset the dearth of precipitation by capturing and redistributing water from the Rio Grande, which is fed by snowmelt originating in its mountainous Colorado headwaters region. These systems require extensive infrastructure to get this water out of the river (or its associated aquifers) and into the fields. In the early twentieth century, the federal government constructed Elephant Butte Dam to ensure a reliable and regular delivery of water to farmers, building a massive concrete structure just south of Hot Springs, New Mexico[1] to hold back up to 2 million acre feet of the Rio Grande in a 40-mile long reservoir. They also built a much smaller secondary facility, Caballo Dam, 28 miles to the south. Two miles below that, Percha Dam redirects waters through the narrow canyons separating the lowlands to the northern Rincon Valley. South of Selden Canyon, water is diverted to the Mesilla Valley from the Leasburg Dam and the Mesilla Dam 20 miles below that. Hundreds of miles of canals and smaller ditches complete the diversion of irrigation water to fields.

Today, this infrastructure supports more than 90,000 acres of irrigated farmland in EBID, a large portion of the federal Rio Grande Project. Irrigation water in this area is apportioned to farmers in accordance with a set of agreements that clarify water-sharing between Texas and New Mexico as well as between the United States and the Republic of Mexico. Elephant Butte Dam is the infrastructural cornerstone for these accords, and its impounded waters have long been a point of contention. Along with the current U.S. Supreme Court battle, *Texas versus New Mexico and Colorado,* farmers in EBID are simultaneously navigating a lengthy legal process in New Mexico to "adjudicate"—prove and quantify—all historic water rights for the state government (Utton Center 2010). Both legal processes inhibit farmers' economic certainty, incentivize certain agricultural decisions, and promote political alliances focused on protecting farmers' livelihoods and access to resources (Boelens 2009; Gallardo et al. 2017; Warner 2016).

The long history of legal tension and altercation over irrigation water in the Mesilla and Rincon valleys stems in large part from the original decision about where to build a dam. The U.S. Reclamation Service[2] constructed Elephant Butte Dam between 1911 and 1916 on the exact site that a private company had previously selected for its own dam project to serve farmers in southern New Mexico (Littlefield and Douglas 2008). When the United States government heard of the private company's intentions, it sued for an injunction and began machinations to develop the 1902 Reclamation Act, which essentially trumped the plans of private land and water speculators throughout the arid West (Pisani 1992, 2002; Worster 1985).

Using eminent domain and eventually the Homestead Act, the federal government commissioned the Rio Grande Project, which brought

[1]Hot Springs was renamed "Truth or Consequences" in 1950 and is labeled on the study area map with its modern name.

[2]The U.S. Reclamation Service was renamed the "Bureau of Reclamation" in 1923, and we use both names in this paper, depending on which was in effect at the time referred to.

Elephant Butte Dam to life as a federal project instead of a private operation (Baxter 1927; Littlefield 2008). The complex historical relationship between private water speculators, individual irrigators, and government resource managers persists in modern times, animating legal maneuvers and ill will both in the interstate dispute cited above and in the Lower Rio Grande Adjudication (Utton Center 2013).

A compounding factor in much of the ongoing animosity is the fact that the 1938 Rio Grande Compact (hereinafter "Compact") designated Elephant Butte Dam as the site where New Mexico would deliver its water share to Texas and where the U.S. would deliver its water share to Mexico (Fig. 10.2). This large-capacity reservoir thus functions now primarily as a holding location for water destined for downstream water users. Yet its location well north of New Mexico's southern border raises the question of how irrigation water should be delivered to and shared by New Mexico's southernmost farms and irrigators in the Mesilla and Rincon valleys. In these two valleys, agricultural lands are subject to New Mexico law and New Mexico water planning, yet they lie downstream of the delivery point for Texas' and Mexico's water. In this complex political landscape, many New Mexican irrigators are considered part of Texas for purposes of water allocation and management (Ray 2018).

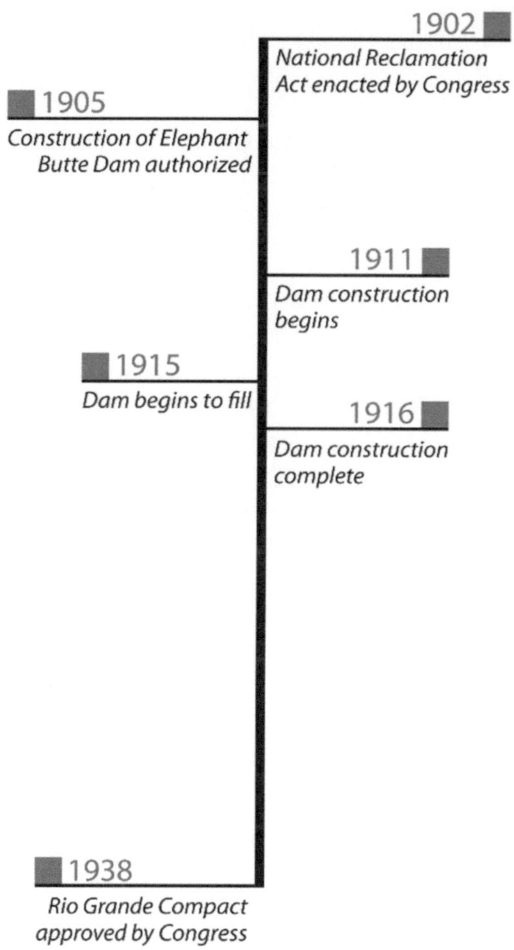

Fig. 10.2 Timeline of major water governance events in the Lower Rio Grande Basin of New Mexico. The National Reclamation Act is also commonly referred to as the Federal Reclamation Act or the Reclamation Act

10.3 Recent History: "Proving up" to Pecans

Landowners in the Mesilla and Rincon valleys are in the midst of an ongoing legal process to quantify and prove their New Mexico water rights in court. Although numerous agreements were established in the early twentieth century to clarify how water should be shared among the many users of the Rio Grande, demand often outstrips the river's supply, leading to litigation threats against irrigation districts and states suspected of "over-appropriation" beyond their legally allowed share. No one knows the true extent of over-appropriation in New Mexico, a problem which gave rise to the Lower Rio Grande Adjudication (New Mexico v. EBID, et al., 96-CV-888 [1996]) in New Mexico's third judicial district court as a mechanism to formally define the number and scope of all water rights in the river basin. The adjudication began in 1996 and is ongoing as of 2019.

The adjudication process defines the principal parameters in which today's farmers make decisions with economic and environmental impacts. In 2008, the adjudication court produced an intermediate judgment—Stream System Issue 97-101 (SS-97-101)—meant to simplify these parameters by defining the amount of water associated with individual rights. Stream system issues are legal clarifications of tangential points that must be addressed in order for the broad adjudication to proceed. SS-97-101 makes two important clarifications. First, it defines the farm delivery requirement (FDR) of every crop in the basin as 4.5 acre feet per acre per year (afay), regardless of the crop's specific water needs. Under the purview of SS-97-101, however, pecan farmers were given the opportunity to demonstrate a higher FDR, or "prove up" to 5.5 afay on any orchard planted before March 31, 2009. Given the high value of water rights, it's likely that all pecan farmers claimed this right to extra water. SS-97-101 defines each rightsholder's FDR of 4.5 or 5.5 afay as a combination of surface water (1.476 afay) and groundwater (3.024 to 4.024 afay), but this relationship changes in years when surface water deliveries are "short," or less than 1.476 afay. Every year since 2008 has seen short deliveries of surface water, which means irrigators are regularly pumping supplemental groundwater to satisfy their allowable FDR. Second, and perhaps more importantly, the stream system issue clarifies that farmers are allowed to "stack" their water rights by leasing them from other farmers. There is no limit to how much an irrigator may legally acquire through financial arrangements with other rightsholders.

As the same time that southern New Mexico's irrigators were implementing the SS-97-101 ruling and pecan farmers were proving up their FDR, water managers began to realize the magnitude of claims over Rio Grande water in New Mexico. The state's inability to adhere to Compact obligations by delivering required water to Texas became a growing concern. To address this crisis, Texas and New Mexico entered a Reclamation-brokered agreement in 2008 that balanced New Mexico farmers' need for groundwater (as a supplement for insufficient surface waters) with Texas farmers' need for Rio Grande water (to dilute salty groundwater drawn from the Ogallala aquifer). The 2008 Operating Agreement settled disputes over water delivery from Elephant Butte Dam to the Texas state line by balancing the diversion ratio—how much water is released from the dam and how much water actually reaches the border—with annual debits and credits of stored water behind the dam, as defined by the Compact.

Farmers on both sides of the Texas/New Mexico border were generally happy with this arrangement. However, in a lawsuit that has been called "politicized hydrology" and as having a "sinister motive", New Mexico's then-attorney general, Gary King, sued the Bureau of Reclamation for allegedly giving away the state's water to Texas (Esslinger 2011; Haussamen 2012; New Mexico v. United States, et al., D.N. M. 11-CV-691 [2011]). This lawsuit exacerbated interstate tensions, with Texas responding in kind by suing New Mexico in the U.S. Supreme Court in 2013, arguing that excessive groundwater pumping in the Rincon and Mesilla valleys had violated the Compact for decades due to the hydrologic connection between groundwater and surface water in the river channel (Holden 2014). Although the case has not yet been decided, any change it may introduce to current water sharing agreements would likely decrease New Mexican irrigators' water rights and would have large-scale economic impacts on the state (Griggs 2017).

10.4 The Puzzle of Contemporary Crop Change

Land change models of satellite-based classifications of cropland show a marked increase in pecan orchards since 2008, the year that SS-97-101 was decided (Fig. 10.3). Pecans, like nearly all of the crops in the Rincon and Mesilla valleys, are irrigated through controlled flooding with both surface water and groundwater between late spring and late fall. At the time of this writing in 2019, the American Southwest is nearing the end of a drought that began in the late 1990s, though climate projections are still dire. Temperatures are expected to rise by up to 5.5 °F by 2070, contributing to persistent drought conditions in this region of the United States (Meyer 2018; Riganti 2019). The most pronounced impacts of the warming are expected to manifest as summer heatwaves and as longer intervals between annual freezes, which will be

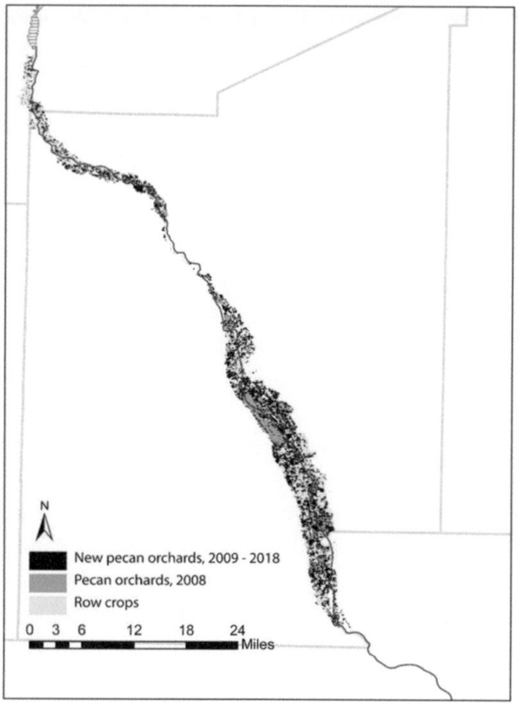

Fig. 10.3 GIS analysis of increase in pecan orchards between 2008 and 2018

detrimental to the yield and productivity of regional crops like established fruit and nut trees (Garfin et al. 2014). These climate projections do not bode well for New Mexico's famed pecan crop. Today, pecan orchards account for 32% of all diverse crops in EBID—an increase of 6% from ten years earlier in 2008 (Fig. 10.3). National-level market superiority makes pecans seem like a safe bet for New Mexican farmers who are considering a transition from diverse row crops, and in some areas dark green trees span the valley as far as the naked eye can see. Yet, this popular crop choice might seem counterintuitive to the casual observer. Pecan orchards require a large amount of water to survive and thus are not naturally suited to a region that receives less than a foot of rain annually. Furthermore, pecan orchards are a significant and long-term investment for farmers. The trees take about 7 years to bear their first fruit and therefore cannot respond to perturbations in water supplies as readily as crops with a shorter lifespan, like cotton. So why have so many farmers planted pecan trees in the Mesilla and Rincon valleys over the last decade? And how do modern cropping choices relate to the historical decisions of farmers whose residence within EBID was made possible only by government infrastructure and policy?

As critical physical geographers know, these questions are not as straightforward as they might appear. To better understand the links between water infrastructure, resource governance, environmental change, and socioecological relations, we must use multiple methods and research approaches that combine scientific analysis of landscape change with social-scientific analysis of economic and political change (Grabowski et al. 2017; Lave et al. 2014). This work is further enriched by additionally grappling with historical contexts and stories that link the decisions and livelihoods of today's farmers to those of the past (Greer et al. 2018). Sequential exploratory research design occurs in three general phases proceeding from qualitative to quantitative data collection and culminates with holistic analysis informed by these complimentary datasets. This structure satisfies what contemporary critical geographers see as tangible and systemic limitations of any one epistemology to model or understand drivers of socioecological change. (For modern calls for interdisciplinarity in socioecological research see Kinzig 2001; Kull and Rangan 2016; Miller et al. 2008; Schoolman et al. 2012; Widgren 2012.) Moreover, the effects of epistemological approaches have been shown to have pronounced impacts on conclusions drawn in studies of agrarian and landscape change (Gallardo et al. 2017). Regarding modeling SES, Roy Chowdhury and Turner (2006) demonstrate that incorporating wide-ranging structure models and individualized agent-based models of change are generative of more holistic explanatory power.

In order to effectively quantify the extent of land use change resulting from these recent developments, we started by talking to farmers. We met with 15 farm managers in EBID, ranging from a pecan grower on a 2,000-acre orchard to a diverse organic row cropper on 20 acres, to discuss the historic, socioeconomic, environmental, and institutional drivers they consider when deciding what to grow (Campos et al. 2014). Each farmer also specifically addressed why they have or have not transitioned any of their farmland to pecans. We then used these semi-structured interviews to build GIS models of observed landscape transitions. Interviewees identified 24 salient factors as driving or influencing farm management decisions, which we then translated into ten annual robust GIS models of a sample set of 1,430 farms covering 8,000 acres per year from 2008 to 2017. The dependent variable in these models is specifically "transition to pecan orchards" from some other crop or land use. Our statistical analyses clarified exactly which factors from the broad dataset of farm management factors best explain this specific transition.

First, to determine the spatial relationships between the different factors farmers had identified, we conducted a series of correlation tests using Pearson's coefficient—and found significant collinearity between variables. However, because pairwise correlations do not adequately capture the overall influence of variable interplay, we also conducted principal components analysis to examine the degree to which total model variance is explained by each variable. Results showed more than 60% of the variance in the initial model of 24 variables was attributable to seven variables grouped into three components: total farm acreage, property value, and depth to groundwater.

Grouping seven of the 24 criteria into three distinct variables reduced the dataset to (a still substantial) 20 variables. We then used linear and binary logistic regression models to explain transition to pecan orchards to pare down the 20 explanatory variables to eight. In all, the eight most important factors that explain transition to pecans are as follows:

- Total farm acreage
- Property value
- Depth to groundwater
- Whether the farm is in the Rincon or Mesilla Valley
- Distance to urban zones
- The areal trend of transition to pecan orchards
- Whether there are already established pecans on the farm
- Distance to fallow or idle cropland

All 24 variables are listed in Table 10.1, with those eight that best explain this land use change noted.

10.5 Regression Modeling of Transition to Pecans

Linear logistic regressions had better model strength and were more mathematically defensible than binary logistic regressions for these data. However, despite a strong R^2 value of higher than 0.7 when all eight variables were included in the model, the odds ratios of each variable to explain change were essentially meaningless, suggesting some unaccounted interplay between variables. While it is possible that the interplay is encapsulated wholly by "unmappable variables"—likely stemming from historic and cultural context, global regression models do not account for spatial heterogeneity of explanatory variables. Using GIS software, we then conducted a geographically weighted regression (GWR) to account for dissimilarity by using the relative spatial distribution of farms to fit a regression line.

Comparing a global ordinary least squares regression (OLS) to a GWR of the same input dataset conclusively demonstrated that a legitimate unidentified relationship between variables is their spatial distribution (Fig. 10.4). Though the overall model strength in terms of the R^2 value was lower for each of the 10 annual models —between 0.5 and 0.6—the odds ratios of each variable demonstrated significant explanatory power contributing to the transition to pecan orchard from some other crop. Furthermore, the Aikike Information Criterion (AIC) and variance inflation factors (VIF) were consistently higher for the GWR, indicating better overall goodness of fit (Table 10.1).

Using correlation analyses, we developed regression models that explore the degree to

Table 10.1 Mappable variables gleaned from semi-structured interviews. An * denotes variables used to explain transition to pecans

Valley (Mesilla or Rincon)*	
Distance to urban*	
Trend of transition (hot spots)*	
Total Acreage (Grouped variable)*	Total acreage of farm
	Irrigable acreage
	Total water diverted
	Surface water allotment
Property value (Grouped variable)*	Land value
	Total value (w/structures)
Percentage of pecans on farm before transition*	
Depth to Groundwater (Grouped variable)*	Depth to water in well
	Distance to river
Distance to fallow/idle cropland *	
Soils (Irrigation interpretation rating, USDA)	
Crop transitioned from (ranked by relative value)	
Fallow acreage within farm	
Crop variety	
Ownership (private, trust, bank, company, municipal)	
Wells on farm	
Adjudication status	
Source (surface only, groundwater only, conjunctive)	
Depth of well	
Priority date	
Well drill date	

which an array of factors relating to farm management decisions can specifically explain a transition to pecan orchards. While many of the factors we initially identified are indeed relevant to best management practices in farming, they do not directly explain tangible land use change in the form of switching to pecan growing.

In the realm of pecan production within EBID, each of these eight variables reveal underlying associations with ecological and political change. The limit of the explanatory power of the models is explicit in the numbers. Looking at the power of our GWR model, we see that it only explains or otherwise accounts for 50% of the drivers of transition to pecans. To augment this, we will need to turn to historical-critical physical geography to contextualize and wrestle with the unmappable variables of irrigated agriculture in EBID (Table 10.2).

10.6 Initial GIS Evaluation and Unmappable Variables

A cursory exploration of the eight confounding drivers of the transition in question helps us set the stage for deeper qualitative analysis. The valley in which a farm is situated carries with it specific biophysical, social, and traditional factors. As recently as half a century ago, soil differences between the Rincon and Mesilla valleys played a key role in what crops would grow

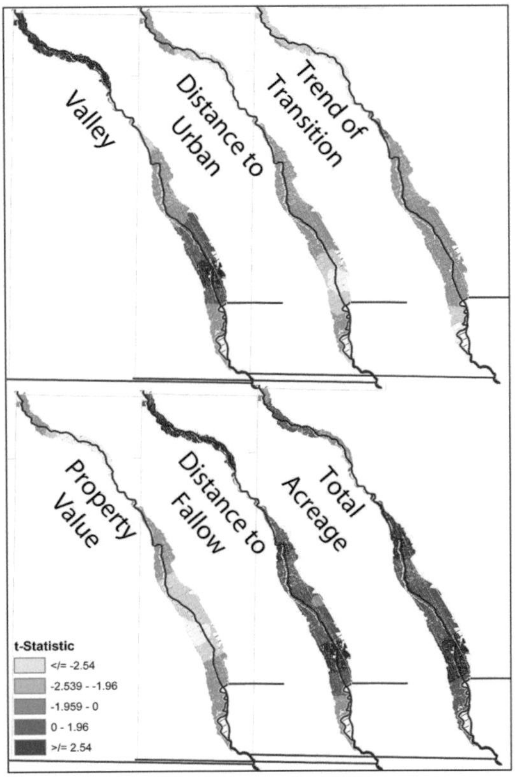

Fig. 10.4 Significance of drivers of agrarian change. Original figure. Refer to Fig. 10.1 to see distribution of political boundaries. (Not pictured: depth to water, which is uniformly the most consistent variable, and existing pecans, which is the least consistent among the eight)

(Childs and Hanks 1975; Cominelli et al. 2013; Schönfeld LaMar 1984). More irrigable land in the Mesilla Valley meant more room for larger and more profitable farms (hence the importance of total acreage and property value). Knowledge bases and traditions are deeply connected to the historic trajectory of crops grown. From the results, we see that more importance is placed on distinguishing between valleys in the smaller Rincon Valley.

It follows that the relationship to urban land is a confounding factor. Again, the results suggest a significant difference between the Mesilla and Rincon valleys. Figure 10.1 illustrates the geographic location of all urban centers in both valleys. Cities and towns represent centers of local power where the proximity and reach of political influence are necessarily interconnected (Allen and Cochrane 2010). There is a statistical indifference to the proximity to Las Cruces, the power center where sprawl is the norm, but a statistically positive relationship between smaller towns and transition to pecans.

This makes sense. Some of the oldest pecan orchards are closest to Las Cruces, and we see that established orchards influence continued transition nearby. The idea of a trend of transition to pecans borrows from agent-based modeling, which seeks to understand or even predict environmental transition and human response. The assumption in this technique is that behavior and innovations diffuse across space (Lybbert and Sumner 2012; Padel 2001; Rogers 2003). However, this criterion looks only at other farms that are *entering* the pecan industry, while it neglects *intensification* of pecan cultivation on farms that were once diversified between pecans and other crops. The positive shift in the market is making crop diversification—generally a safe farming practice—less necessary as a physical insurance policy. Looking at these results, we can see that pockets of diverse crops are filling in with a pecan monocrop.

A general rule of thumb in EBID is that fallowed land indicates that water is being put to beneficial use for irrigation elsewhere. The distance of a farm to fallowed land does not explicitly tell us if that farm's manager has leased water, but as a whole we begin to see regions where that distance has a positive relationship with new pecan orchards, suggesting that these monocrops provide less valuable unused water rights in immediate proximity.

Finally, the depth to groundwater is the only consistently positive explanatory variable for transition across the study area. Drawing from this, it is apparent that the political changes that have justified a large-scale transition to pecans have not entirely insulated irrigators from an awareness of the ecological burden placed on the aquifer of the Lower Rio Grande basin.

Ongoing adjudication in the basin has formalized these effects by establishing a mechanism to ensure water availability for individual irrigators with sufficient capital to accumulate

Table 10.2 Outputs of OLS and GWR models

Variable	2009	2010	2011	2012	2013	2014	2015	2016	2017	2018
Valley	NS	NS	NS	NS	NS	NS	NS	NS	NS	NS
Distance to urban	1.012**	1.011**	1.005**	1.007**	1.023**	1.012**	1.013*	1.021**	1.013*	NS
Trend	NS	NS	NS	NS	1.059*	1.02*	NS	1.022*	NS	NS
Pecans before transition	1.045**	1.02**	1.011**	1.022*	1.047**	1.06**	1.018**	1.008**	1.03**	1.023**
Distance to fallow	1.013*	1.025**	1.012**	NS	1.076*8	NS	1.019*	1.018**	NS	NS
Acreage	1.034**	1.049**	1.038**	1.035**	1.049**	1.048**	1.048**	1.034**	1.053**	1.042**
Property value	1.024*	1.022**	1.022**	1.025**	1.022*	NS	1.02**	1.021**	1.021*	1.021*
Groundwater level	NS	NS	NS	NS	NS	1.057**	NS	NS	NS	1.031*
n	1430	1430	1430	1430	1430	1430	1430	1430	1430	1430
R^2 for OLS	0.428	0.371	0.333	0.321	0.327	0.489	0.459	0.340	0.378	0.443
R^2 for GWR	0.543	0.543	0.537	0.545	0.533	0.650	0.533	0.534	0.533	0.535
AIC for OLS	8006.2	8138.8	8223.5	8249.0	8235.4	7842.9	7924.8	8208.7	8123.3	7966.2
AIC for GWR	7773.3	7778.9	7791.1	7766.7	7804.6	7384.8	7805.7	7801.3	7799.5	7794.6
Moran's *I* for OLS model residuals	0.026	0.027*	0.026	0.024*	0.035*	0.037**	0.020*	0.025*	0.021*	0.017*
Moran's *I* for GWR model residuals	0.005*	0.003**	0.006**	0.010**	0.009**	0.020**	0.009*	0.002**	0.011*	0.011*

NS Relationship is Not Significant
* Significant at $p < 0.05$
** Significant at $p < 0.01$
AIC Akaike's Information Criterion

excess rights. Within this framework, environmental resilience is secured through the exacerbation of inequality. In the face of uncertain water availability and limited economic resilience, powerful actors have an outsized ability to remain insulated from hydrologic uncertainties and to exploit agricultural market opportunities at the expense of less wealthy farmers. The results of our geospatial analysis indicate that the ruling in SS-97-101 has given unequal power to farmers on larger and more valuable parcels closer to urban areas, since they now have the ability to compete in the dominant agricultural economy for New Mexico: pecans. This simultaneously results in the diminished opportunity of smaller-scale farmers to react to changing markets or to simply maintain their water rights by leasing them to pecan farmers.

By interrogating the historic context of political and ecological change, mechanisms of resilience building, and quantifiable land use transitions, we can begin to paint a better picture of the critical physical and human geographies of irrigation and land use in EBID. For example, the transformation or loss of specific forms of agricultural knowledge, cultural practices, or familial traditions may not be easily detected in a GIS analysis, despite the certain influence of such loss or transformation on subsequent land management decisions. Preparation for parlous environmental or economic regime shifts can likewise be informed largely by "unmappable" practices that will not show up in a GIScience analysis. In this project, we sought to confront the methodological limits of understanding how people prepare and react to long-term drought and legal uncertainty by fleshing out the historic context to augment more positivist results. This final section thus extends the sequential exploratory research framework.

In evaluating our model's explanation of the current shift to pecan farming in EBID, we must consider the possibility that "unmappable variables" drive part of the uncertainty in the model. No single methodology can provide a perfect analysis; thus we need multiple methodologies to identify and address factors that would otherwise escape detection.

10.7 The Long Context of Ecological and Political Change

Three years after the first controlled water deliveries from Elephant Butte Reservoir were released in 1915 to the canals and ditches of southern New Mexico, farmers struggled to turn a profit on crops that simply would not grow. Land that had been productive a few years previously was left idle, and land clearance for farm expansion had come to a halt across both valleys. Where vineyards and alfalfa had briefly ensured a steady return on investment, farmers found the roots of these crops had rotted away. The Rio Grande Project, which had been expected to vastly expand farmers' prospects, was instead causing catastrophic landscape impacts (Schönfeld LaMar 1984).[3]

These impacts stemmed from the ways that new infrastructure impacted existing environmental knowledge and practice. Before the Project was authorized in 1905, both subsistence and cash crops had been irrigated mainly via communal irrigation ditches that conveyed rich, sediment-filled waters from the river to individual farms. Irrigators were accustomed to these muddy waters and the specific soil moisture conditions they created, with communal irrigation norms contributing to the widespread use of standard field flooding techniques to achieve optimal crop growth. After the dam was complete, however, the naturally suspended sediments of the Rio Grande became trapped behind the concrete structure, leaving only clear water to be sent to the farms downstream. When farmers implemented their standard techniques, they soon discovered that it caused significant waterlogging in the upper soil horizons. Just one year after the dam's completion, six percent of all irrigable lands in the Mesilla Valley were covered by standing water, and 70% were classified by the

[3]Schönfeld LaMar (1984) provides a detailed discussion of changes introduced to the Mesilla and Rincon valleys after the building of Elephant Butte Dam. Her fine-grained analysis of property records and oral histories provides the basis for this section's discussion of historical context.

Reclamation Service as having "poor drainage" (United States Reclamation Service 1917). Farmers used trial and error to pilot new irrigation techniques on compromised low-lying fields, but not before alkali salt deposits had rendered the soil useless for growing most of their common crops. Eventually, farmers learned to leach the soil of salts and adopted new farming practices more suitable for sediment-poor waters, leaving behind traditional communal irrigation techniques.

In those early years between the disastrous waterlogging of entire valleys and farmers' transition to new irrigation techniques, however, the instability in ecological conditions was mirrored by a destabilization of other community norms. The Mesilla and Rincon valleys contained mixed rural settlements at the turn of the twentieth century, with both Anglo and Hispanic farmers using communal irrigation practices to raise primarily subsistence crops. In this pre-dam era, few farmers had any reason to sell their lands, since there was plenty of public land available for homestead claims and no need of large cash sums in what was then a barter-based economy. The new Reclamation project, however, upended the social order on which agricultural life had been based, introducing taxes and maintenance fees that incentivized farmers to switch to cash crops and to put uncleared lands into production or sell them off. From 1904 to 1915, homestead claims picked up significantly in southern New Mexico, as did speculative subdivision and re-sale of quarter section land parcels.

Many of the new settlers were Anglos who brought experience with cash economies from other, more eastern areas of the U.S. They obtained land first through homestead claims and later by purchasing it from existing landowners, many of whom were struggling to meet new tax levies that kept increasing as settlement and speculation caused land values to increase. The dam's authorization thus began a chain of events that created financial pressures on existing landowners, many of whom were Hispanic, to sell part, or all of, their land. In this period while the dam was being constructed and its reservoir filled, a detailed analysis of landownership records shows that changes in landownership increased dramatically throughout southern New Mexico: "By 1915, only one small portion of one of the original homesteads in the sample areas was still owned by a relative of an original grantee. All other homestead lands had changed hands, some as often as six or seven times since their appropriation" (Schönfeld LaMar 1984, 153). Although land ownership changes are complex and based in individual decisions about whether and why to sell individual plots of land, there was a clear pattern at the time of Anglo residents increasing their percentage of land ownership while Hispanic residents' land ownership decreased, despite the prospect of turning handsome profits once the irrigation infrastructure was complete. In this way, an economic stratification developed between those who were able to hold on to their lands and those who were not. Hispanic farmers were disproportionately unable to hold on to their land, thus decreasing their economic and political powers.

10.8 Conclusion

This brief historical review confirms the long context in which water governance changes produce specific types of ecological, economic, and political change. A hundred years ago, waterlogging was the most significant ecological change caused by new water infrastructure. Today, the transition from row crops to orchards throughout the valley is of similar ecological magnitude and is also caused by specific material infrastructures for water management. In economic and political terms, the historical consequences of reclamation policy are nearly identical to those occurring today in southern New Mexico. Historically, increased inequality and class stratification originated from changes in land ownership. Current changes in water rights ownership and leasing accentuate and widen these existing inequalities, also as a direct result of water governance policy.

In this project, GIS helped represent and spatialize modern conditions of land and water use in EBID, analyzing the nature and significance of variables that impact farm management decisions. The importance of each variable to determine the behaviors of irrigators and the resulting land use transitions spatially radiates outward from the irrigation district's economic hub, Las Cruces. Generally, the farms adjacent to this city are larger and more profitable, meaning that economic and ecological uncertainty plays a vastly different role there than around the smaller communities of the Rincon Valley or the southern reaches of the Mesilla Valley. Therefore, the ability of farmers to interact with the dominant pecan market—an ostensibly rational decision—is preferential to those with outsized capital.

Within EBID, farmers who respond to drivers of agrarian change by transitioning to pecan orchards are making economically sound decisions in the absence of water as a confounding factor. Overall water availability is always a consideration in crop selection, but it is not always manifested uniformly in EBID because conjunctive use has allowed a full water supply, even in years in which minimal water is released from upstream reservoirs. However, between the neoclassical economic concerns of externalities and the political ecology of agrarian change lies this middle ground of understanding how wealthier farmers are folding in capital accumulation (Moore 2015) by increasing their appropriation of water rights when water scarcity is a confounding factor. As in the early days of EBID, inequality manifests through capital and political influence. Water rights may theoretically have been created equally, but access to those rights was quickly stratified, with spatial patterns that still dominate today's landscape.

References

"About EBID" n.d. Elephant Butte Irrigation District. Accessed May 28, 2019. https://ebid-nm.org/index.php/about-ebid/

Autobee R (1994) "Rio Grande Project." Bureau of reclamation. Accessed May 28, 2019. https://www.usbr.gov/projects/pdf.php?id=179

Baxter JO (1927–1997) Dividing New Mexico's Waters, 1700–1912. Albuquerque: University of New Mexico Press. http://agris.fao.org/agris-search/search.do?recordID=US201300014344

Boelens Rutgerd (2009) The politics of disciplining water rights. Develop Change 40(2):307–31. https://doi.org/10.1111/j.1467-7660.2009.01516.x

Bustillos L, Maria B (2018) New Mexico agricultural statistics 2017. Annual Bulletin. New Mexico Annual Bulletin. Las Cruces, NM: United States Department of Agriculture, National Agricultural Statistics Service

Campos M, Velázquez A, McCall M (2014) Adaptation strategies to climatic variability: a case study of small-scale farmers in rural Mexico. Land Use Policy 38(May):533–40. https://doi.org/10.1016/j.landusepol.2013.12.017

Utton Center (2010) "Basic Water Law Concepts." 2. Water Matters! Accessed March 10, 2019. http://uttoncenter.unm.edu/resources/research-resources/basic-water-law-concepts.pdf

Utton Center (2013) Water litigation in the lower Rio Grande. 24. Water Matters! Accessed March 10, 2019. http://uttoncenter.unm.edu/resources/ombudsman/pdfs/24-water-litigation-lrg.pdf

Childs SW, Hanks RJ (1975) Model of soil salinity effects on crop growth. Soil Sci Soc Am J 39:617–22

Cominelli E, Conti L, Tonelli C, Galbiati M (2013) Challenges and perspectives to improve crop drought and salinity tolerance. New Biotechnol 30(4):355–61. https://doi.org/10.1016/j.nbt.2012.11.001

Esslinger G (2011) AG plays fast and loose with southern NM water. NMPolitics.net. September 21, 2011. Accessed March 11, 2019. https://nmpolitics.net/index/2011/09/ag-plays-fast-and-loose-with-southern-nm-water/

Gallardo G, Saunders F, Sokolova T, Börebäck K, van Laerhoven FS, Kokko S, Tuvendal M (2017) We adapt… but is it good or bad? Locating the political ecology and social-ecological systems debate in reindeer herding in the Swedish Sub-Arctic. J Politic Ecol. http://dspace.library.uu.nl/handle/1874/358057

Grabowski ZJ, Matsler AM, Thiel C, McPhillips L, Hum R, Bradshaw A, Miller T, Redman C (2017) Infrastructures as socio-eco-technical systems: five considerations for interdisciplinary dialogue. J Infrastruct Syst 23(4):02517002. https://doi.org/10.1061/(ASCE)IS.1943-555X.0000383

Greer K, Hemsworth K, Csank A, Calvert K (2018) Interdisciplinary research on past environments through the lens of historical-critical physical geographies. Historical Geogr 46(1):32–47. https://doi.org/10.1353/hgo.2018.0024

Griggs BW (2017) The political cultures of irrigation and the proxy battles of interstate water litigation. Nat Res J 57(1):1–73

Haussamen, H (2012) "EBID Accuses AG of 'Sinister' Motive for Expanding Water Lawsuit| NMPolitics. Net." NMPolitics.Net. Accessed March 9, 2012. https://nmpolitics.net/index/2012/03/ebid-accuses-king-of-sinister-motive-for-expanding-water-lawsuit/

Hedden A (2019) "NM Leads Nation in Pecan Production." Las Cruces Sun-News, April 14, 2019, sec. News. Accessed May 28, 2019 from https://infoweb.newsbank.com/apps/news/document-view?p=NewsBank&t=state%3ANM%21USA%20-%20New%20Mexico&sort=YMD_date%3AD&fld-base-0=alltext&maxresults=20&val-base-0=pecan&docref=news/172CC0FA183AE180

Holden J (2014) Water Resources: an integrated approach. Routledge, London, p 2014

Allen J, Cochrane A (2010) Assemblages of state power: topological shifts in the organization of government and politics. Antipode 42(5):1071–1089. https://doi.org/10.1111/j.1467-8330.2010.00794

Kinzig AP (2001) Bridging disciplinary divides to address environmental and intellectual challenges. Ecosystems 4 (8):709–15. https://doi.org/10.1007/s10021-001-0039-7

Kull CA, Rangan H (2016) Political ecology and resilience: competing interdisciplinarities? In: Hubert B, Mathieu N (eds) Interdisciplinarités Entre Natures et Sociétés: Colloque de Cerisy. Peter Lang, P.I.E, pp 71–87

Lave R, Wilson MW, Barron ES, Biermann C, Carey MA, Duvall CS, Johnson L et al (2014) Intervention: critical physical geography. Canadian Geographer Le Géographe Canadien 58(1):1–10. https://doi.org/10.1111/cag.12061

Littlefield DR (2008) Conflict on the Rio Grande : water and the Law, 1879–1939. Norman : University of Oklahoma Press, ©2008

Lybbert TJ, Sumner DA (2012) Agricultural technologies for climate change in developing countries: policy options for innovation and technology diffusion. Food Policy 37(1):114–23. https://doi.org/10.1016/j.foodpol.2011.11.001

Meyer R (2018) The Southwest May be deep into a climate-changed mega-drought. The Atlantic. December 18, 2018. https://www.theatlantic.com/science/archive/2018/12/us-southwest-already-mega-drought/578248/. Accessed March 14, 2019

Miller TR, Baird TD, Littlefield CM, Kofinas G, Chapin III FS, Redman CL (2008) Epistemological pluralism: reorganizing interdisciplinary research. Ecol Soc 13(2). https://doi.org/10.5751/ES-02671-130246

Moore JW (2015) Capitalism in the web of life: ecology and the accumulation of capital. Verso Books

Padel S (2001) Conversion to organic farming: a typical example of the diffusion of an innovation? Sociologia Ruralis 41(1):40–61. https://doi.org/10.1111/1467-9523.00169

Pisani DJ (1992) To reclaim a divided west: water, law, and public policy, 1848–1902. University of New Mexico Press, Albuquerque

Pisani DJ (2002) Water and American government: the reclamation bureau, national water policy, and the west, 1902–1935. University of California Press, Oakland

Pryor SC, Scavia D, Downer C, Gaden M, Iverson L, Nordstrom R, Patz J, Robertson GP (2014) Southwest climate change impacts in the United States: the third national climate assessment. In: Melillo JM, Richmond TC, Yohe GW (eds) Highlights of Climate Change Impacts in the United States: The Third National Climate Assessment (Low Resolution-PDF EBook) (003–300-00012-9). U.S. Global Change Research Program. https://doi.org/10.7930/J08G8HMN

Ray K (2018) TX v. NM SCOTUS Decision. March 15, 2018. https://ebid-nm.org/index.php/2018/03/15/tx-v-nm-scotus-decision/. Accessed May 28, 2019

Riganti C (2019) "U.S. drought monitor." National Drought Mitigation Center. Droughtmonitor.unl.edu. Accessed May 28, 2019

Rogers EM (2003) Diffusion of innovations. Free Press, New York, p 2003

Roy Chowdhury R, Turner BL (2006) Reconciling agency and structure in empirical analysis: smallholder land use in the southern Yucatán, Mexico. Ann Assoc Am Geogr 96(2):302–322. https://doi.org/10.1111/j.1467-8306.2006.00479.x

Schönfeld LaMar B (1984) "Water and land in the Mesilla Valley, New Mexico." Ph.D. Geography, Eugene, OR: University of Oregon

Schoolman ED, Guest JS, Bush KF, Bell AR (2012) How interdisciplinary is sustainability research? Analyzing the structure of an emerging scientific field. Sustain Sci 7(1):67–80. https://doi.org/10.1007/s11625-011-0139-z

United States Reclamation Service (1917) "[Annual] Project History." Unpublished Manuscript at United States Bureau of Reclamation Office. El Paso, TX: United States Reclamation Service

Warner BP (2016) Understanding actor-centered adaptation limits in smallholder agriculture in the central american dry tropics. Agric Hum Values 33(4):785–97. https://doi.org/10.1007/s10460-015-9661-4

Widgren M (2012) Resilience thinking versus political ecology: understanding the dynamics of small-scale, labour-intensive farming landscapes. In: Pleninger Thomas, Bieling Claudia (eds) Resilience and the cultural landscape: understanding and managing change in human-shaped environments. Cambridge University Press, Cambridge

Worster D (1985) Rivers of Empire: water, aridity, and the growth of the American West. Oxford University Press

Supplying the Conquest: A Geospatial Visualization and Interpretation of Available Environmental Resources at the Battle of Hastings in 1066

11

C. M. Hewitt

Abstract

This chapter, will present a never before considered geographic interpretation on the environmental resources available to English and Norman combatants at the Battle of Hastings in 1066. In the volume of literature on this pivotal battle in English history, such resources have only been discussed briefly, with historians questioning some of the findings. First, the battle's historical and geographical context will be provided, followed by a discussion of the local *fieldscape* uses. Next, comments regarding what resources were available to the two combating armies are explored. These findings contribute to a detailed discussion of the volume and location of abundant resources including food, water, iron, salt and timber available in the area. The environmental framework of this study acts as a GIS template analysis for other similar investigations of medieval military campaigns. By considering the Middle Ages through environmental studies, we can obtain a richer and more comprehensive view of the resources and logistics available during battles than has been previously presented.

Keywords

Battle of Hastings · Environmental resources · Crops · Animals · Water · GIS inverse distance weighting (IDW) interpolation

11.1 Introduction

Early medieval armies were comprised of people and animals which lived in environments, and required food, water, weapons, and tools. In historical accounts of the world's great battles, however, such aspects of warfare are rarely discussed from a geographical or environmental perspective in terms of geographical information systems (GIS) visualization and analysis. Certainly, this is true of the battle of Hastings, which brought together English and Norman forces in a battle in 1066 from which the origins of modern Britain can be traced. A possible exception is Bernard Bachrach's (1985) article on the military administration of the 1066 campaign which

C. M. Hewitt (✉)
Department of Geography, The University of Western Ontario, London, Canada
e-mail: chewitt4@uwo.ca

C. Travis et al. (eds.), *Historical Geography, GIScience and Textual Analysis*,
Historical Geography and Geosciences, https://doi.org/10.1007/978-3-030-37569-0_11

examined the challenges of supplying the Norman army at Dives, prior to launching for England.[1]

As the documentary sources on the battle do not provide sufficient detail on the resources as would be acceptable in standard GIS studies, this chapter presents an in-depth geographic interpretation through visualization and analysis of the local resources likely available at the time of the battle. It is anticipated this chapter will be a partial corrective to a gap in the literature on the battle of Hastings.[2] Following a discussion of the historical context, the study will commence with a consideration of the local population in the southern region of England where the armies fought. Next, based upon documentary, cartographic and environmental evidence, novel interpretations regarding what resources (e.g. food, water and industry) were available to the English and Norman armies will be explored through a GIS reading and interpretation. The chapter will conclude with a discussion of the implications of this research on the historiography of the Battle of Hastings in addition to its impact on military studies in historical geography and medieval history.

[1]Bernard S. Bachrach, "Some Observations on the Military Administration of the Norman Conquest," *ANS VIII* (1985): 1–25; Based on horse depictions in the *Bayeux Tapestry*, horse shoes and seals, the horses in use at the time of Hastings were considered small in comparison to later medieval warhorses in the fourteenth century. R. H. C. Davis, *The Medieval Warhorse: Origin, Development and Redevelopment* (London: Thames and Hudson Ltd, 1989), 21–22; Ann Hyland, *The Medieval Warhorse from Byzantium to the Crusades* (Stroud, UK: Alan Sutton Publishing Limited, 1994), 85–86; Ann Hyland, *The Warhorse: 1250–1600* (Stroud, UK: Sutton Publishing Limited, 1998).

[2]Bachrach, *Ibid*., R. H. C. Davis, "The Warhorses of the Normans," *ANS X* (1987): 80; M. K. Lawson, *The Battle of Hastings: 1066* (Stroud, UK: Tempus Publishing Ltd, 2003), 180–81; Stephen Morillo, *Warfare Under the Anglo-Norman Kings, 1066–1135* (Woodbridge, UK: The Boydell Press, 1994), 126–27.

11.2 Context

The Battle of Hastings was the climax of a succession crisis to the English throne. In early January 1066, King Edward of England died, and his successor was Harold II Godwineson. However, there were two other claimants. One, Harald Hardrada was king of Norway; the other William, duke of Normandy. On September 25, 1066, Harold defeated Harald who was killed at the battle of Stamford Bridge in Yorkshire. Almost immediately afterward, on September 28 or 29, William landed on the southern coast of England at Pevensey and camped at Hastings. Harold subsequently rushed south to engage William, arriving the evening before the battle.[3] The next day, October 14, 1066 Harold and the English faced William on a hill just north of Hastings. In the ensuing battle, Harold was killed, and the English were defeated, marking the last time England was successfully invaded by a foreign power.[4]

11.2.1 Armies and Provisioning in the Literature

There are several important historical studies on military logistics. One is Jonathan P. Roth's work on Roman military logistics to A.D. 235, which undertakes important discussions related to food, timber and water resources. Logistics of the Crusades are similarly covered in John H. Pryor's edited volume, *Logistics of Warfare in*

[3]Frank Barlow, *The Godwines: The Rise and Fall of a Noble Dynasty* (Harlow, UK: Pearson Education Limited, 2002), 139; R. Huscroft, *The Norman Conquest: A New Introduction* (Harlow, UK: Pearson Education Limited, 2009), 123–24.

[4]J. Bradbury, *The Battle of Hastings* (Stroud, UK: Sutton Publishing Limited, 1998); R. Allen Brown, 1996. "The Battle of Hastings." In *The Battle of Hastings: Sources and Interpretations*, by Stephen Morillo, 196–218. (Woodbridge, UK: The Boydell Press, 1996); Lawson, *The Battle of Hastings*: 1066; Stephen Morillo, *The Battle of Hastings: Sources and Interpretations* (Woodbridge, UK: The Boydell Press, 1996); Matthew Strickland, *Anglo-Norman Warfare: Studies in Late Anglo-Saxon and Anglo-Norman Military Organization and Warfare* (Woodbridge, UK: The Boydell Press, 1992).

the Age of the Crusades. This study covers a number of important topics such as food, water and travel during the period.[5] Although these studies mention resource availability, they have yet to analyze this issue within a geographical or environmental dimension. A number of environmental studies have employed visualizations (maps, graphs and tables), focusing on economics and supplies to demonstrate the state of peacetime resource production within the medieval English context.[6] By linking such approaches to the study of military resources, findings can be displayed in a similar fashion, as typical peacetime production would produce an ideal amount of resources, from an army's perspective. Regarding military campaigns, Bachrach's studies provide much of the information regarding specific relations between army logistics and provisioning. For example, in his *Logistics in Pre-Crusade Europe* (1993) study, Bachrach suggests that logistics strategies employed by the combatants in the eleventh century largely followed practices derived from the Roman Empire.[7] Bachrach also examined in detail the administration of the Norman army camp at Dives in the late summer of 1066. His discussion focused on the army size, the local environment, supplies at Duke William's expense, and the administrative leadership of the camp. In a later medieval example, an article by David Bachrach covers the requisition of supplies in Edward I's reign (1272–1307). Bernard and David also recently co-authored a book on medieval warfare with a chapter specifically on logistics. Additionally, there is the edited book, *Landscapes for Defence in Early Medieval Europe* (2013) which includes several maps relating topography to defense.[8] What these studies do not include, however, is any comprehensive analysis or visualizations, including maps of the local environment of a military campaign, as would be applied in the environmental studies referenced above. By engaging GIS, this chapter will consider the relevant factors for interpreting the Battle of Hastings in such a fashion, including the local population, food, water and industry. GIS has been engaged in re-analyzing historical geography, battlefields and text. In the literature, this is known as historical GIS or HGIS.[9] An important examination of the philosophy behind the use of GIS in historical studies is provided by Alexander von Lünen's examination of the origins and developments of different philosophies such as positivism and postmodernism. In addition his work explores different types of reasoning including induction, deduction and abduction in relation to geographic and historical thought.[10] The value of GIS to history is detailed in an article by Venkaka Ratnadeep Suri and Hmid R.

[5]John H. Pryor ed. *Logistics in the Age of the Crusades* (Aldershot, UK: Ashgate Publishing Limited, 2006); Jonathan P. Roth, *The Logistics of the Roman Army at War* (264 B.C.–A.D. 235) (Boston: Brill, 1999).

[6]Bruce M. S. Campbell, *English Seigniorial Agriculture, 1250–1450* (Cambridge, UK: Cambridge University Press, 2000); Christopher Dyer, *Making a Living in the Middle Ages: The People of Britain 850–1520* (New Haven and London: Yale University Press, 2002); Christopher Dyer, *Standards of Living in the Later Middle Ages: Social Change in England, c. 1200–1520* (Cambridge, UK: Cambridge University Press, 1989); Peter Fowler, *Farming in the First Millennium AD: British Agriculture between Julius Caesar and William the Conqueror* (Cambridge, UK: Cambridge University Press, 2002); Kathy L. Pearson, "Nutrition and the Early-Medieval Diet," *Speculum* 72 (1) (1997): 1–32; Derek Yalden, *The History of British Mammals* (London: T & A D Poyser Ltd., 1999); Derek Yalden and Umberto Albarella, *The History of British Birds* (Oxford, UK: Oxford University Press, 2009).

[7]Bernard S. Bachrach, "Logistics in Pre-Crusade Europe." In *Feeding Mars: Logistics in Western Warfare from the Middle Ages to the Present*, by John A. Lynn (Oxford: Westview Press, 1993), 58.

[8]Bachrach, "Some Observations on the Military Administration of the Norman Conquest"; David Stewart Bachrach, "Military Logistics during the Reign of Edward I of England, 1272–1307," *War in History* 13 (4) (2006): 423–44; Bernard S. Bachrach and David S. Bachrach, *Warfare in Medieval Europe c. 400–1453* (New York, Routledge, 2017); John Baker, Stuart Brookes, and Andrew Reynolds eds., *Landscapes for Defence in Early Medieval Europe* (Brepolis, 2013).

[9]I. N. Gregory & P. S. Ell, *Historical GIS: Technologies, Methodologies and Scholarship* (Cambridge: Cambridge University Press, 2007).

[10]Alexander von Lünen, "Tracking in a New Territory: Re-imagining GIS for History" in A. von Lünen and C. Travis (eds.), *History and GIS: Epistemologies, Considerations and Reflections*, https://doi.org/10.1007/978-94-007-5009-8 14, 211–239.

Ekbia on the Budapest Ghettos in the Second World War.[11] In the study, Suri and Ekbia analyze spatial changes to the Jewish Ghetto in June 1944 through a cartographic framework, underscoring that GIS mapping is "both the process and product of historical reasoning."[12] Eugene Ch'ng's GIS study examined experimental archaeology and virtual reality to digitally recreate historical environments. Such virtual environments could then be explored by situating the public or researchers in the time period being studied.[13] The future directions of GIS, mapping and history include "'deep mapping' . . . [which] aims to evoke the richness of small places and different kinds of experiences and emotions they can engender."[14] In terms of sources, the primary sources in HGIS projects are historical maps. Historical maps generally depict the environment as it existed in a slice of time. Unlike satellite images or aerial photographs, there is an inherent blurriness in historical maps, as they often incorporate accrued knowledge, which can become outdated, before the date of a map's creation. These maps can be scanned and geo-referenced to their location in GIS. Other sources include archaeological findings and written records, all of which can be combined in a GIS to map and analyse a historical topic (battlefield) in question.[15]

Several authors have conducted extensive research on battlefields and battlefield archaeology from a number of different periods. Many tend to focus on more recent conflicts since the French Revolutionary and Napoleonic Wars (1793–1815) and in particular the American Civil War (1861–1865), although a selection of older battlefields has been studied. Glenn Foard has examined the Battle of Bosworth (1485) in detail as well as English and Flemish medieval and early modern battlefields.[16] Towton (1461) was investigated from an archaeological perspective by T. Sutherland and A. Schmidt.[17] The weather patterns at the time of the battle of Trafalgar (1805) were reconstructed using historical documents.[18] Anne Knowles and others in a landmark study (2008) drew conclusions about General Lee's perspective and decisions at the battle of Gettysburg (1863) by visualizing his viewshed across the terrain of the battlefield. This analysis led to the conclusion that Lee was unable to see a number of important areas of the battlefield and that this in turn contributed to a number of poor decisions on his part during the battle.[19] Steven Smith, Christopher Clement and Stephen Wise employed Global Positioning System (GPS) technology to investigate a Civil War battlefield in South Carolina and its associated earthworks.[20] Scott Hippensteel reconstructed the entrance to Charleston Harbor during the Civil War, by using sediment cores to locate and map tidal areas, natural dunes and artillery

[11]Venkaka Ratnadeep Suri and Hmid R. Ekbia, "Spatial Mediations in Historical Understanding: GIS and Epistemic Practices of History" *Journal of the Association for Information Science and Technology*, 67(9): 2296–2306.

[12]Ibid., 2299; 2304.

[13]Eugene Ch'ng, "Experimental Archaeology: Is virtual time travel possible?" *Journal of Cultural Heritage* 10 (2009): 458–470.

[14]M. Dodge, 2017. "Cartography I: Mapping deeply, mapping the past," *Progress in Human Geography* 41, 1 (2017): 91.

[15]Glenn Foard & Tracey Partida, "The archaeology of medieval and early modern battlefields in Flanders," *Journal of Conflict Archaeology* 13, 1 (2018): 19.

[16]Foard, G., and A. Curry, Bosworth, *1485: A Battlefield Rediscovered* (Oxford: Oxbow Books, 2013); Foard, G., and R. Morris, *The Archaeology of English Battlefields: Conflict in the Pre-Industrial Landscape* (York: Council for British Archaeology, 2012); Foard & Partida, "The archaeology of medieval and early modern battlefields in Flanders"; Glenn Foard, T. L. Sutherland, A. Schmidt, Tony Pollard & Mark Newman, "Battlefields", *Landscapes* 4, 2 (2003): 5–43.

[17]Foard, Sutherland, Schmidt, Pollard & Newman, "Battlefields".

[18]Dennis Wheeler, "The Weather of the European Atlantic Seaboard during October 1805: An Exercise in Historical Climatology," *Climate Change* 48, (2001): 361–385.

[19]A. K. Knowles, W. Roush, C. Abshere, L. Farrell, A. Feinberg, T. Humber, G. Kuzzy, and C. Wirene, "What could Lee see at Gettysburg?" in *Placing history: How maps, spatial data, and GIS are changing historical scholarship*, eds. A. Hillier & A. K Knowles (Redlands, CA: ESRI Press, 2008), 235–266.

[20]Steven D. Smith, Christopher Ohm Clement and Stephen R. Wise, "GPS, GIS and the Civil War Battlefield Landscape: A South Carolina Example," *Historical Archaeology* 37, 3 (2003): 14–30.

sites. Hippensteel also visualized his findings with maps and considered the evidence from historical maps.[21] Thomas Nolan used GIS to visualize data and re-examine the military engagement which led to the awarding of the Medal of Honor to Alvin York in the First World War.[22] Recently, Anne Knowles and others investigated the Holocaust through a geographic framework that incorporated traditional and nontraditional mapping techniques to visualize collective and individual experiences through maps and diagrams.[23] From these studies, it can be observed how changes in data quality and quantity influence the reconstructions and conclusions to be drawn about battles through GIS investigations. The knowledge gained from such studies shaped the methods employed in this chapter by providing theory and detailed examples upon which to base this new research approach. At a broader level, HGIS mapping projects can aid in the conservation and preservation of the locations of historical events and battlefields. Examples include Tony Pollard's work on Civil War battlefields in the United States in general and the battle of Culloden (1746) in Britain; in addition, Mark Newman's study provided a rational for methods of battlefield conservation.[24]

11.2.2 Land Use

The land uses discussed in this chapter were constructed back to 1066 through a process known as map regression, whereby modern land use is regressed back in time using historical maps, documents and information from the *East Sussex Historical Environmental Record* (ESHER) GIS-database.[25] The amount of crop land in 1066 was estimated from agricultural information per manor listed in the *Domesday Book* (*DB*).[26] The value of the *DB* as a source is that it records detailed information about "every manor in Norman England in 1086. From the wealth of manorial data it is possible to reconstruct, with considerable precision, the economic system of [the time]".[27] Agricultural information was derived from the recorded number of ploughlands and plough-teams. A ploughland was "the taxable amount of land that can be ploughed by a [plough-] team of eight oxen".[28] It has been estimated by Frederic William Maitland that for twelve counties, there were 120 acres on average per ploughland. This statistic assumes one ploughland per plough-team.[29] Alternatively, R. V. Lennard's analysis, based on 28 counties, estimated there was an average of 100 acres per plough-team.[30] Both estimates were based on the *fieldscapes* of southern England, delineated as south of the river Trent.[31] As the study area included 374.5

[21]Scott P. Hippensteel, "Reconstruction of a Civil War Landscape: Little Folly Island, South Carolina," *Geoarchaeology: An International Journal* 23, 6 (2008): 824–841.

[22]Thomas J. Nolan, "Geographic Information Science as a Method of Integrating History and Archaeology for Battlefield Interpretation," *Journal of Conflict Archaeology* 5, 1 (2009): 81–104.

[23]A. K. Knowles, T. Cole, and A. Giordano eds., *Geographies of the Holocaust* (Bloomington, IN: Indiana University Press, 2014); A. K. Knowles, L. Westerveld, and L. Strom, "Inductive visualization: A humanistic alternative to GIS" *GeoHumanities* 1, 2 (2015): 233–265.

[24]Foard, Sutherland, Schmidt, Pollard & Newman, "Battlefields".

[25]Christopher Macdonald Hewitt, "How did East Sussex Really Appear in 1066? The Cartographic Evidence," *Digital Studies/Le champ numérique*, 8(1) (2018) http://doi.org/10.16995/dscn.278; East Sussex County Council, *East Sussex Historic Environmental Record, Computer File* (Lewes, East Sussex: East Sussex County Council, September, 2013): https://www.eastsussex.gov.uk/environment/archaeology/her.

[26]David Roffe, *Decoding Domesday* (Woodbridge, UK: The Boydell Press, 2007), 230.

[27]John McDonald & G. D. Snooks, *Domesday Economy: A New Approach to Anglo-Norman History* (Oxford: Clarendon Press, 1986), 3.

[28]National Archives, *Domesday Book*, Retrieved March 2019, from Domesday Book: Glossary: https://www.nationalarchives.gov.uk/domesday/glossary/.

[29]Campbell, *English Seigniorial Agriculture*, 386–87; Frederic William Maitland, *Domesday Book and Beyond: Three Essays in the Early History of England* 2nd. (Cambridge: Cambridge University Press, 1987), 435.

[30]Campbell, *English Seigniorial Agriculture*, 387; R. V. Lennard, *Rural England 1086–1135: A Study of Social and Agrarian Conditions* (London: Oxford University Press, 1959), 393.

[31]Campbell, *English Seigniorial Agriculture*, 387.

ploughlands or 474.875 plough-teams, then the rough extent of arable land was between 44940 and 47487.5 acres. However, not all of the *fieldscapes* would have been agricultural, pastoral or left to fallow. Natural unused grassland and trees beyond the woodland (see below) existed during the period, therefore designated land uses in this chapter are informed estimates for 1066.[32]

11.3 Population

This section examines the local population and their ways of life. Understanding the local population allows us to conceptualize who was present and what they could have provided to the armies. The following sub-sections will explore the population size and distribution in the Hastings area, as well as relevant aspects of the daily life of the local inhabitants.

11.3.1 Population Size

Figure 11.1 presents the manors recorded in the *DB* for the Hastings area. This map was derived from two DB datasets. The first is a simplified *DB* GIS dataset to display ownership, amount of land and land values that is based at King's College London.[33] The second is a non-GIS complete tabular *DB* dataset from the UK Data Archive.[34] These data were joined together as tables in ArcMap using unique ID codes and then summarized by code and location. Based on the records for the DB sites, the recorded population consisted of 700 villagers, 14 bordars (unfree peasant), 157 cottagers (unfree peasant), 95 smallholders, 3 slaves, and 202 burgesses (urban resident), totaling 1,171 people.[35] Based upon H. C. Darby's analysis of *Domesday*, the recorded population only represented households. For total population a multiplier was required. Considering a multiplier of 4.5 and 5 as well as a ten per cent error estimation above the estimated figures, the local population in the Hastings area may have ranged from 5,270 to 6,441 people.[36]

The life expectancy at the time was approximately 25 years of age.[37] Demographic models by Ansley J. Coale, Paul Demeny and Barbara Vaughan estimate the number of people per age category for a series of life expectancies. Using their calculations for a "west" model level 3, with a female life expectancy at birth of 25 years and 22.852 for males, -as well as assuming a gender ratio of 1:1, -a demographic profile was determined for the local population.[38] This was achieved by summing the column of the "[n]umber of survivors at age x out of an original population of 100,000"and determining the percent per age of that total population.[39] Next, these percentages

[32]See Hewitt, "How did East Sussex Really Appear in 1066," for a detailed land use reconstruction.

[33]King's College, London, *Prosopography of Anglo-Saxon England*, Retrieved September 2012, from Prosopography of Anglo-Saxon England: http://www.pase.ac.uk/index.html.

[34]J. Palmer, 2010. *Electronic Edition of Domesday Book: Translation, Databases and Scholarly Commentary*, 1086; Second Edition. Computer File. Colchester, Essex: UK Data Archive, September. http://dx.doi.org/10.5255/UKDA-SN-5694–1.

[35]Ibid.

[36]H. C. Darby discusses the use of multipliers on the Domesday counts. In this study a range of multipliers will be used. Campbell, *English Seigniorial Agriculture*, 389; H. C. Darby, *Domesday England* (Cambridge: Cambridge University Press, 1977): 87–91; Samuli Helle, Samuli Helama and Kalle Lertola "Evolutionary ecology of human birth sex ratio under the compound influence of climate change, famine, economic crises and wars," *Journal of Animal Ecology* 78 (2009): 1226–1233; S. H. King, "Chapter IX Sussex," In *The Domesday Geography of South-East England*, by H. C. Darby and Eila M. J. Campbell, (Cambridge: Cambridge University Press, 1962), 438; Palmer, *Electronic Edition of Domesday Book*.

[37]Ian Mortimer, *The Time Traveller's Guide to Medieval England: A Handbook for Visitors to the Fourteenth Century* (New York: Touchstone, 2008), 36.

[38]A west model was selected because it is based on historical populations for England, Wales and France among others. Model 3 was selected because it has the same life expectancy as was estimated for the time. A study by Karen Cokayne on ancient Roman populations used the same model. Karen Cokayne, *Experiencing Old Age in Ancient Rome* (London: Routledge, 2003), 2–3; Ansley J. Coale, Paul Demeny and Barbara Vaughan, *Regional Model Life Tables and Stable Populations* (New York: Academic Press, 1983), 12.

[39]Coale, Demeny and Vaughan, *Regional Model Life Tables and Stable Populations*, 34.

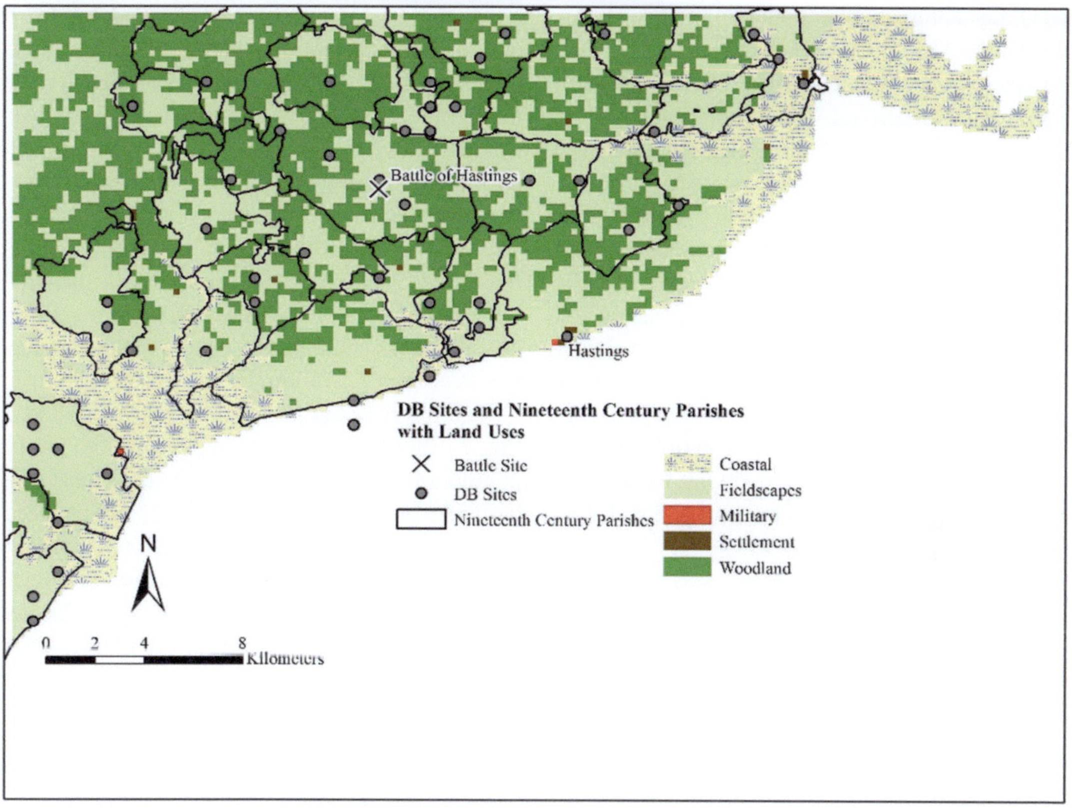

Fig. 11.1 Hastings area *DB* sites and nineteenth century parishes (This map includes data from a selection of sources (N. Burton, J. Westwood, and P. Carter, "GIS of the Ancient Parishes of England and Wales, 1500–1850," (Colchester, Essex: UK Data Archive, March, 2004); Palmer, *Electronic Edition of Domesday Book*.))

were multiplied by the number of males or females, which returned the number of people per age. To determine the average age, the number of people per age was multiplied by their age and then averaged. This analysis for both males and females was combined to report a value for the entire population. For the number of people per age bracket, the number of people within the age bracket was summed. The results revealed that the average age would have been between 20 and 22 with 44% of the population aged 0–15 and 52% between the age of 15 and 60.[40]

11.3.2 Population Distribution

In terms of the local population, the *DB* databases were compared to S. H. King's population maps of Sussex (1962).[41] King's maps were part of a larger study on mapping the *DB* by H. C. Darby which is considered a standard work on the subject.[42] Based on visual inspection, King's maps were based on nineteenth-century parish boundaries. Using a nineteenth-century parish dataset, the local parishes (Fig. 11.1) were aggregated

[40]Coale, Demeny and Vaughan, Ibid., 43; Mortimer, *The Time Traveller's Guide to Medieval England*, 36–37.

[41]King, "Chapter IX Sussex," 436.

[42]Darby, Domesday England; Andrew Lowerre, "Mapping Domesday Book using GIS," *Research News: Newsletter of the English Heritage Research Department* February, (2008), 3; King, "Chapter IX Sussex," 436.

based on King's maps and supplied with his population data.[43] Through this process, a rough estimate of the spatial distribution of the local population was derived. The population based on the *DB* sites in Fig. 11.1 were then aggregated to King's boundaries. When the two sets of values were compared, King's estimation for some areas was either higher or lower than the total number of people reported at the village level. Since the *DB* sites database was organized by manor and village, it would be more realistic to consider those values as they have not been aggregated to an area. Non-aggregated data are in fact preferred since the method of aggregation can influence the results (see the Modifiable Areal Unit Problem or the ecological fallacy).[44] The village level data would also be relevant because at this time "agrarian society in England was [. . .] being localised within thousands of such estates."[45]

Using population data at the village level, it is possible to determine the spatial distribution and density of the local population at a finer scale than was presented by King. This information is important because it indicates with more precision where the local population was living at the time. The revised distribution is visualized through an inverse distance weighting (IDW) interpolation. The interpolation of the *DB* population was based on an example provided by English Heritage for several counties (Bedfordshire, Cambridgeshire, Huntingdonshire and Northamptonshire).[46] Before the interpolation, the *DB* counts had to be modified to obtain workable values. First, the range of the interpolation was determined by first finding the average area of every parish in the study area with a *DB* record within its boundary (Fig. 11.1) and then estimating the radius as if it were a circle. This exercise produced an average area of 14.363 km^2 and a radius of 2.138 km. Next, the *DB* counts were multiplied by 4.5 to obtain the total population in the area. Subsequently, to calculate the population density, the total population values were divided by the average parish size. Then, the density values were interpolated using IDW over the study area. However, since the densities were for a specific area, it was decided that the values should only be interpolated to that area. Therefore, in the IDW interpolation dialog box in ArcMap, the maximum distance was specified as the radius of the average parish area or 2.138 km.[47]

The interpolated population map presented in Fig. 11.2 indicates that the areas with the largest population density were along the coast and particularly in the Pevensey area. When compared to the land use map, this area contains extensive *fieldscapes* and marshland. The areas with lower population densities were located further inland which corresponds to large areas of woodland where few people would have been living.

11.3.3 Daily Life and Employment

The population at Hastings is described in detail in the *Chronicle of Battle Abbey* (*BA*), including some general characteristics that can be inferred from the record.[48] Local trade jobs such as bakers, brewers, carpenters, farmers and various labourers were probably present in many of the local villages at the time.[49] The list in the *BA* provides some limited insight into who may have had the valuable skills necessary to provision the armies at Hastings.[50] Another source that is particularly useful for its information on typical daily life and employment at the village level is

[43] Burton, Westwood and Carter, "GIS of the Ancient Parishes of England and Wales"; King, "Chapter IX Sussex," 436.

[44] Gregory & Ell, *Historical GIS*, 163–66.

[45] Burton, Westwood and Carter, "GIS of the Ancient Parishes of England and Wales"; Darby, *Domesday England*, 92; Fowler, *Farming in the First Millennium AD*, 269; Palmer, *Electronic Edition of Domesday Book.*

[46] Lowerre, "Mapping Domesday Book using GIS," 4.

[47] Ibid.

[48] Eleanor Searle, *Lordship and Community: Battle Abbey and its Banlieu, 1066–1538* (Toronto: Pontifical Institute of Mediaeval Studies, 1974), 70; *The Chronicle of Battle Abbey*, trans. E. Searle, (Oxford: Oxford University Press, 1980).

[49] Judith M. Bennett and C. Warren Hollister *Medieval Europe: A Short History.* 10th. (New York: McGraw-Hill Higher Education, 2006), 161.

[50] *The Chronicle of Battle Abbey.*

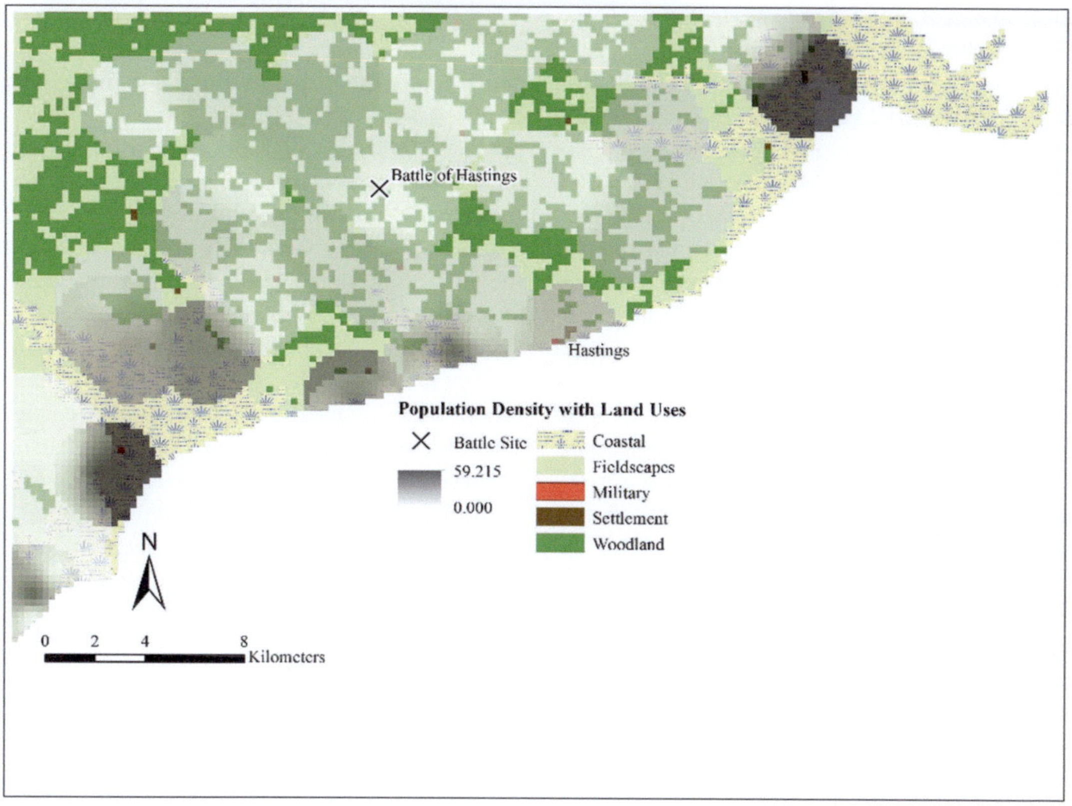

Fig. 11.2 Estimated Hastings area population density in c. 1066 (The calculations behind this map are based upon a selection of sources. Burton, Westwood and Carter, "GIS of the Ancient Parishes of England and Wales"; Lowerre, "Mapping Domesday Book using GIS"; Palmer, *Electronic Edition of Domesday Book*.)

the *Rectitudines Singularum Personarum* (*RSP*). This document describes the various positions in the archetypical village and their employment contributions.[51]

11.4 Food (Crops, and Animals)

One of the most important considerations for a military commander was the supply of food for their men and horses. As William was plundering the area to force Harold into a battle, he could have supplied his army as well.[52] Given the sizes of the army, considerable provisions would have needed to be procured. This section will investigate crop yields and animals.

11.4.1 Crops

Table 11.1 indicates the potential number of people the Hastings area could have supported, based upon crops, and was calculated based on a method described in Bruce M. S. Campbell's

[51]Fowler, *Farming in the First Millennium AD*, 269; P. D. A. Harvey, 1993. "Rectitudines Singularum Personarum and Gerefa." *English Historical Review* 108 (426): 1–22; D. C. Douglas & G. Greenaway, eds., *English Historical Documents, 1042–1189*, 2nd ed., Vol. II. (London: Taylor & Francis, 1981), 875–78.

[52]Morillo goes into detail about the various goals of plundering by armies Morillo, *Warfare Under the Anglo-Norman Kings*, 98–102.

book, *English Seigniorial Agriculture, 1250–1450* (2000).[53] The arable area is derived from the arable estimates presented in the land use section, above. Key variables as discussed below were manipulated to provide a series of six estimates of the population potentially supported by the arable land available. Campbell indicated the grain area was estimated to range between 50 and 56 per cent of the total arable area based on data from the fourteenth to nineteenth centuries.[54] Campbell selected 56% as the grain area for 1086. The value was calculated by multiplying the arable land estimate by a factor of 0.56. The harvested grain for a typical year was deduced by multiplying the grain area by the number of kilocalories per acre. The amount of processed grain was obtained by multiplying the harvested grain by the processing rate as indicated in the note to Table 11.1. This value was then converted into daily supply by dividing it by 365 days. The potential population was calculated by dividing the daily supply by 1500 calories.[55] The results indicate the area could have fed between 13,561 and 18,226 people. As the local population is estimated to be between 5,270 and 6,441 people, there would have been grain supplies remaining for both armies.

[53]P. F. Brandon, "Demesne Arable Farming in Coastal Sussex during the Later Middle Ages," *The Agricultural History Review* 19, (1971): 127. In the Hastings area, farming was done through enclosed fields as opposed to an open-field the two or three crop rotation system. Alan R. H. Baker, 1973. "Field Systems of Southeast England." In *Studies of Field Systems in the British Isles*, by Alan R. H. Baker and R. A. Butlin, 377–429. (Cambridge: Cambridge University Press, 1973), 423–24; Graeme J. White, *The English Medieval Landscape: 1000–1540* (London: Bloomsbury Academic, 2012), 61. The crop rotation system simply indicates how much of the field will be pasture. In the eleventh century, it has been estimated that 56 per cent of the arable was farmed at any given time (Campbell, *English Seigniorial Agriculture*, 396).

[54]Campbell, *English Seigniorial Agriculture*, 398–99.

[55]Ibid., 397.

11.4.2 Animals

More than just grain, William's fighting men required meat as well. The *Bayeux Tapestry* provides evidence of food acquisition associated with conflict, including a scene depicting men taking farm animals for use by the army.[56] Local animals that would have been available to the armies included poultry (chickens, ducks and geese), cattle, pigs, ponies and sheep. Cattle, pigs and sheep were an important commodity in the twelfth century with their remains identified in archaeological surveys of castles, towns and villages. According to Pearson, French society (of which Normandy was a part) preferred cattle and pigs.[57] Therefore, the local domestic animals would have been an additional resource for the competing armies in 1066.[58]

11.5 Water

Accessibility to fresh water was important for all medieval armies. In this section, the local hydrology is discussed in detail specifically with respect to the stream network, discharge and importance to the army.

11.5.1 Stream Networks

The current Ordnance Survey data provides a detailed hydrological network for the study region. However, there have been human modifications to these channels since 1066. These include channel straightening, and the creation of

[56]Stenton, *The Bayeux Tapestry*, Plate 47; Britain's Bayeux Tapestry, *Britain's Bayeux Tapestry at Reading Museum*, accessed March 2019. http://www.bayeuxtapestry.org.uk/bayeux20.htm.

[57]Pearson, "Nutrition and the Early-Medieval Diet," 7.

[58]Annie Grant, "Animal Resources." in *The Countryside of Medieval England*, eds Grenville Astill and Annie Grant, 149–87 (Oxford: Basil Blackwell Ltd, 1988), 152; Pearson, "Nutrition and the Early-Medieval Diet".

Table 11.1 Estimating the grain yields in 1066. Estimate 1 is based upon kilocalories per acre in 1300 and a processing rate (with wastage) of 0.56 while estimate 2 uses a processing rate with wastage of 0.5. Estimate 3 reduces the kilocalories per acre by 20% to provide "a 20 per cent improvement in net kilocalorie yield per acre" and uses the processing rate from estimate 2 (Bruce M. S. Campbell, *English Seigniorial Agriculture*, 397)

	Estimate 1		Estimate 2		Estimate 3	
Variable	A	B	A	B	A	B
Arable area	44940	47487.5	44940	47487.5	44940	47487.5
Grain area (56% of arable area)	25166	26593	25166	26593	25166	26593
Kilocalories per acre	670075	670075	670075	670075	558396	558396
Harvested grain (kcal—$\times 10^{10}$)	1.686	1.782	1.686	1.782	1.405	1.485
Processed grain including wastage (kcal —$\times 10^{9}$)	9.443	9.979	8.432	8.910	7.026	7.425
Daily supply (kcal—$\times 10^{7}$)	2.587	2.734	2.310	2.441	1.925	2.034
Potential population	17248	18226	15400	16273	12834	13561

ponds and reservoirs.[59] Figure 11.3 presents the streams with the local topography while Fig. 11.4 presents the streams with the local land uses.

11.5.2 Stream Discharge

River discharge indicates the amount of water flowing past a single location known as a gauging station over time (Figs. 11.3 and 11.4). For the Combe Haven River, which drains the western part of the battle area (Figs. 11.3 and 11.4 depict the extent of the Combe Haven River basin), a flow duration curve is reported by the Natural Environment Research Council (NERC)'s Centre of Ecology and Hydrology and has been replicated in Fig. 11.5.[60] The Figure displays the river's discharge versus the percent of time a discharge is exceeded for September to October as well as annually. Based on the curve for September to October (1963 to 2003), approximately 95 times out of 100, the flow exceeded 0.01 m^3/s which equals 228 thousand gallons per day. In the absence of any indication of severe drought in 1066 (although one existed from 1064 to 1065), such amounts would have been more than enough for William's horses (Bachrach estimates they required 20,000 to 30,000 gallons per day).[61] Additionally, the fact that this drought did occur actually

[59]Ordnance Survey, Maps of Britain from Ordnance Survey, accessed February 2012. http://www.ordnancesurvey.co.uk/oswebsite/index.html; E. Rhodes, "Identifying Human Modification of River" In *Waterways and Canal-Building in Medieval England*, by J. Blair, (Oxford: Oxford University Press, 2007).

[60]Natural Environment Research Council, "Centre for Ecology and Hydrology," accessed January 2013, https://www.ceh.ac.uk/.

[61]See the following article for data to depict drought: E. R. Cook, R. Seager, Y. Kushnir, K. R. Briffa, U. Buntgen, D. Frank, P. J. Krusic, W. Tegel, G. van der Schrier, L. Andreu-Hayles, M. Baillie, C. Baittinger, N. Bleicher, N. Bonde, D. Brown, M. Carrer, R. Cooper, K. Cufar, C. Dittmar, J. Esper, C. Griggs, B. Gunnarson, B. Gunther, E. Gutierrez, K. Haneca, S. Helama, F. Herzig, K-U. Heussner, J. Hofmann, P. Janda, R. Kontic, N. Kose, T. Kyncl, T. Levanic, H. Linderholm, S. Manning, T. M. Melvin, D. Miles, B. Neuwirth, K. Nicolussi, P. Nola, M. Panayotov, I. Popa, A. Rothe, K. Seftigen, A. Seim, H. Svarva, M. Svoboda, T. Thun, M. Timonen, R. Touchan, V. Trotsiuk, V. Trouet, F. Walder, T. Wazny, R. Wilson, and C. Zang, "Old World megadroughts and pluvials during the Common Era," *Science Advances* 1 (2015) https://doi.org/10.1126/sciadv.1500561; Old World Drought Atlas, "Old World Drought Atlas," accessed March 2019, http://drought.memphis.edu/OWDA/Default.aspx. See Christopher Macdonald Hewitt, "The Historical Topography of the Battle of Hastings: A Geographical Reconstruction", *Transactions in GIS* (Forthcoming) for an annual precipitation graph back to 1066. Bachrach, "Some Observations on the Military Administration of the Norman Conquest," 12. The river discharge for 20,000 gallons a day would be 0.0009 m^3/s which is less than the lowest discharge on the graph.

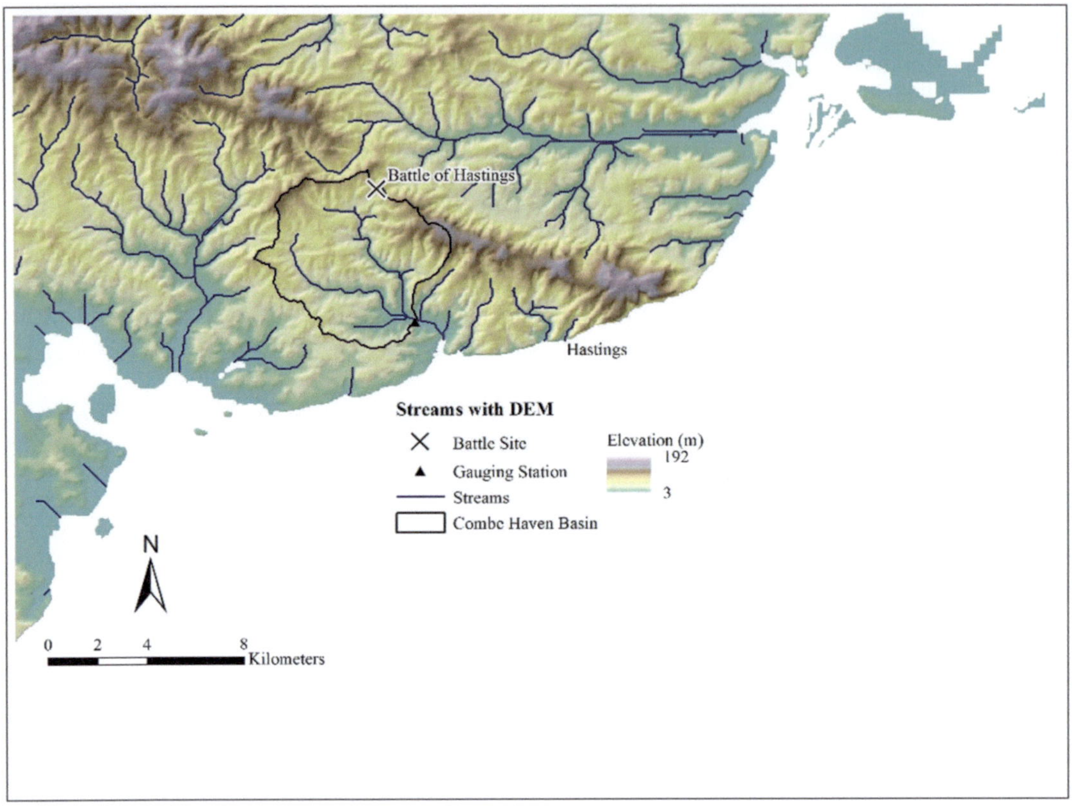

Fig. 11.3 Estimated Hastings area streams with *Digital Elevation Model* (*DEM*) for c. 1066 (This figure was developed by first calculating the stream network and a hillshade map from the *DEM*. Next, the *DEM* raster was set to transparent and layered on top for the 3-D visual effect known as *relief shading*. A. Jon Kimerling, Aileen R. Buckley, Phillip C. Muehrcke and Juliana O. Muehrcke, *Map Use: Reading and Analysis* (Redlands, CA: ESRI Press, 2009), 117; A. Mitchell, *The ESRI Guide to GIS Analysis, Volume 3: Modeling Suitability, Movement and Interaction*, (Redlands, CA: ESRI Press, 2012), 269–87; Ordnance Survey, Maps of Britain from Ordnance Survey.)

makes the analysis of the river flow levels even more significant as it shows that the river in question would be able to supply a large volume of water even under very severe drought.

11.6 Industry

This section provides an overview of the local industries which existed in the area. These include iron sites (including mines and forges), grain mills, salt houses and wood suppliers. These resources were important in providing armies with armour repair, food processing and preservation, as well as cooking and construction of defenses. Each type of industry in the Hastings area is mapped and analyzed in turn in the sub-sections which follow.

11.6.1 Iron Sites

According to Darby, "[i]ron-ware must have been in widespread use in the eleventh century, but we hear little in Domesday."[62] Therefore, other sources must be consulted with respect to local metal use. Peter Brandon has noted that there was an active mining industry in Sussex.

[62]Darby, *Domesday England*, 266.

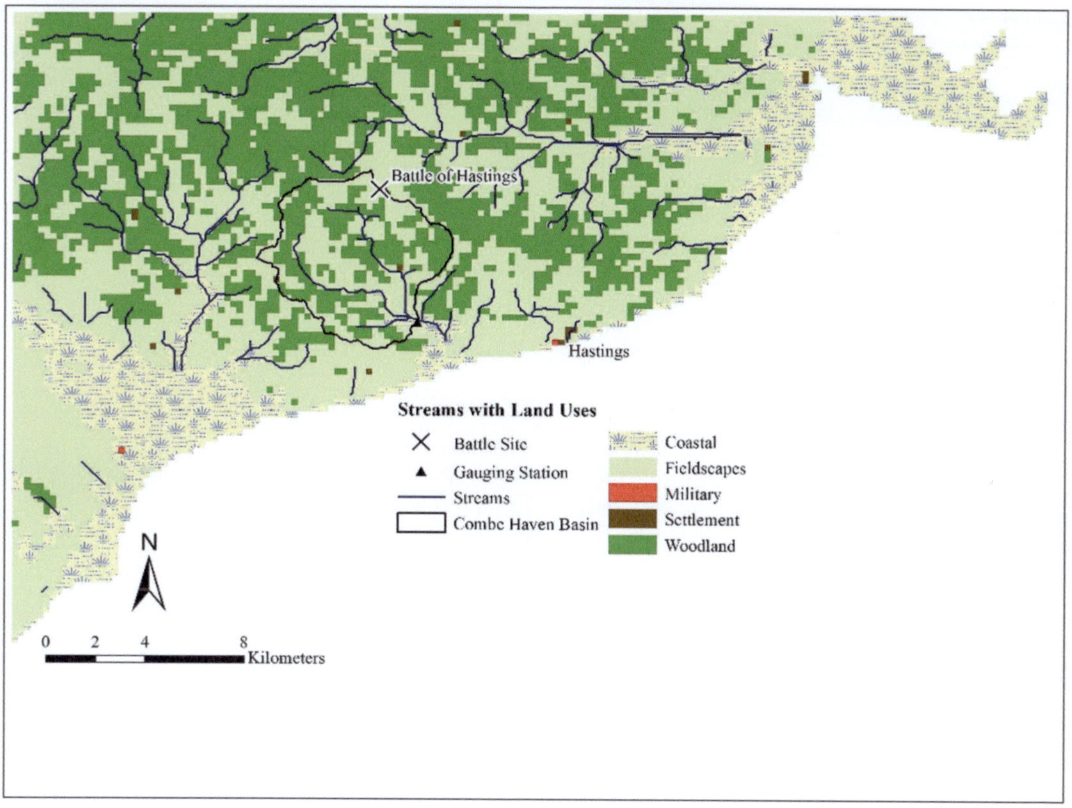

Fig. 11.4 Estimated Hastings area streams with land use for c. 1066

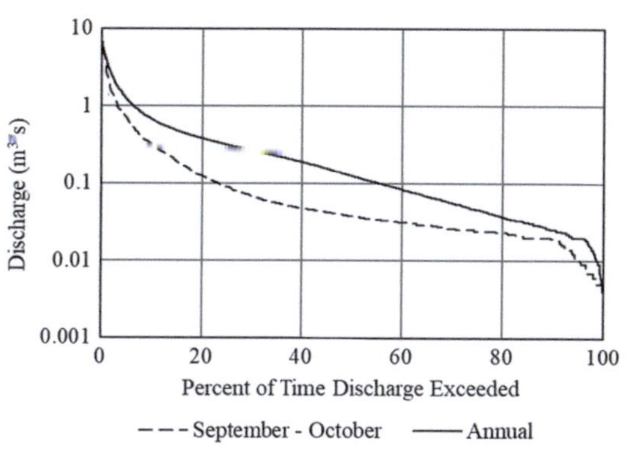

Fig. 11.5 Flow duration curve for discharge from Combe Haven gauging station for 1969 to 2003 (Natural Environment Research Council, "Centre for Ecology and Hydrology," accessed January 2013, https://www.ceh.ac.uk/.)

He also indicates how the local iron was extracted from the rock and that these sites can be approximately dated. These locations can be accessed from a database maintained by the Wealden Iron Research Group.[63] However, it is also noted that the extent of pre-Conquest English mining is not known and thus many sites which were active in Roman times may not have been active at the time of the Norman Conquest.

[63]Peter Brandon, *The Sussex Landscape* (London: Hodder and Stoughton, 1974), 148–50; Wealden Iron Research Group, "Wealden Iron Research Group," accessed August 2014, www.wealdeniron.org.uk. Based on a map in an environmental history textbook, South East England was predominantly used for iron production and smelting. R. C. Hoffmann, *An Environmental History of Medieval Europe* (Cambridge: Cambridge University Press, 2014), 216–17. In the eleventh century, iron smelting was done through a bloomery. A bloomery was a contained furnace where the iron was heated so the impurities would be separated from the iron. The product was known as a bloom and if kept hot could have further impurities removed manually. Hoffmann, *An Environmental History of Medieval Europe*, 223; H. Cleere, D. W. Crossley & B. C. Worssam, *The Iron Industry of the Weald*, (Cardiff: Merton Priory Press Ltd, 1995), 85–86.

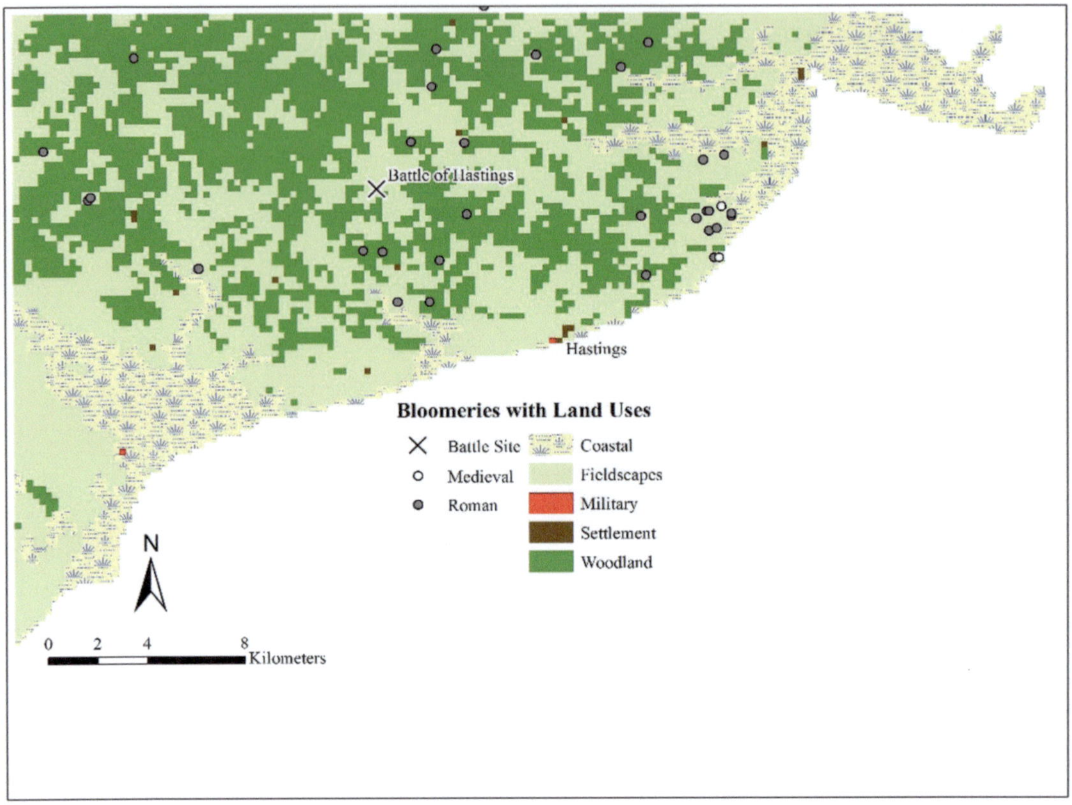

Fig. 11.6 Estimated Hastings area iron sites in c. 1066 (The locations of the iron sites were accessed from the Wealden Iron Research Group, "Wealden Iron Research Group".)

Nonetheless, Brandon does point out that it is likely settlement continued around the old mines into the Anglo-Saxon period. Therefore, William's men could have potentially accessed iron to repair their weapons and shoe the horses during their stay in the Hastings area.[64] Figure 11.6 presents the local iron sites with the time period indicated as well. Note the many sites which were classified as Roman versus medieval. This trend indicates the growth of iron mining in the Roman period and its possible continued existence at those sites in 1066—therefore, potentially providing both sides with a source of fresh iron for armaments and armament repair.

11.6.2 Mills

According to the *DB* survey, there were around 5–6000 mills in England.[65] It has been assumed that these mills were water-mills, the technology of which had existed in England since the mid-eighth century, but the *DB* surveyors did not record the type of mill. Roffe has indicated the *DB* surveyors could have been reporting the presence of millstones.[66] The locations of the mills at the time of Hastings is important because (as was indicated earlier) bread was a staple of the armies and a mill would have been required to process the wheat to produce flour for bread. As William could not have known when Harold would have arrived, he could have fed his army

[64]Bachrach, "Logistics in Pre-Crusade Europe", 72; Brandon, *The Sussex Landscape*, 76–77.

[65]Dyer, *Making a Living in the Middle Ages*, 25–26; Fowler, *Farming in the First Millennium AD*, 178.

[66]Roffe, *Decoding Domesday*, 230.

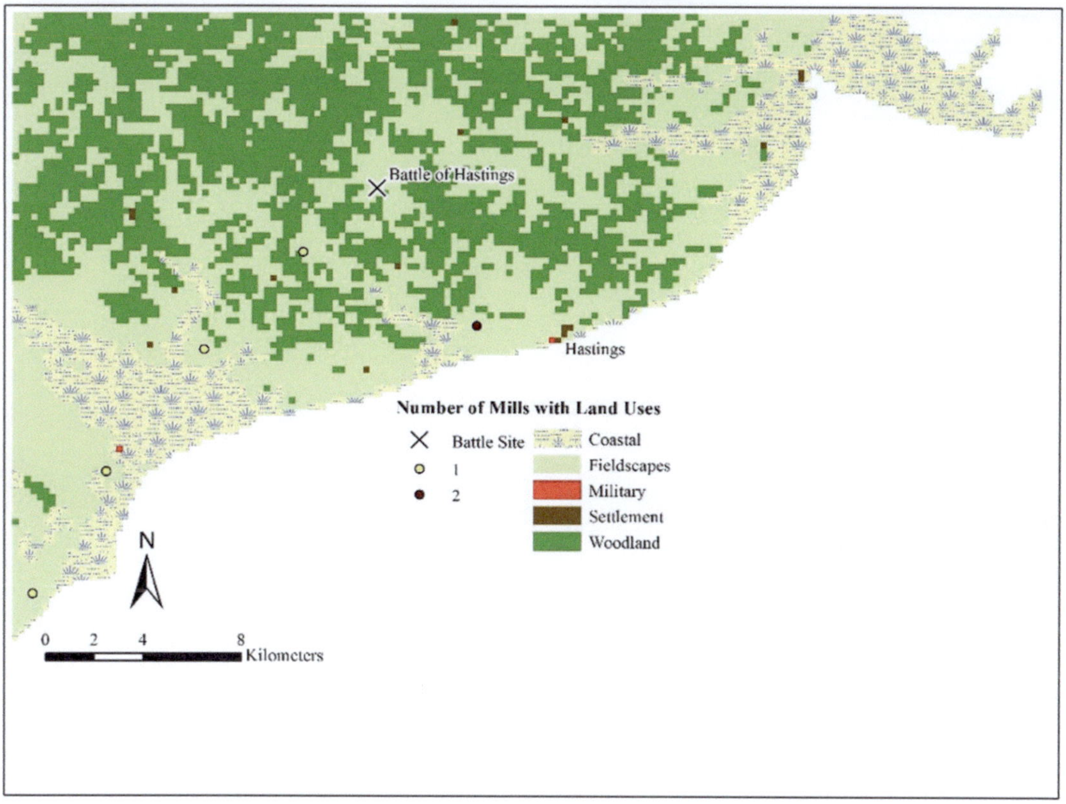

Fig. 11.7 Estimated Hastings area mills in c. 1066 (The number of mills was recorded in the *DB*. King, Ibid., 459; Palmer, *Electronic Edition of Domesday Book*.)

with the local resources. Therefore, having a number of mills in close proximity to the encampments would have been ideal.[67] Based on evidence in the *DB*, Fig. 11.7 was developed to illustrate the locations of the known mills in the Hastings area. The majority of the mills in Sussex were to the west of the area under study, with few in the Hastings area. This suggests that soldiers may have ground the unprocessed wheat manually with hand-mills, a well-developed technology, or processed their food supplies elsewhere.[68]

11.6.3 Salt

The use of salt was important in the Middle Ages particularly as a meat or fish preservative. Salt is of interest to the battle of Hastings because it would have been used by both sides to preserve food into the late fall and winter. There are several studies which cover the use of salt and salt production at the time of the Norman Conquest. The *Domesday* studies presented by H. C. Darby also map salt production.[69] Figure 11.8 presents data showing the number of salt houses per village. What is interesting about this map is that some salt houses appear to be far inland. The

[67]Darby, *Domesday England*, 270–72; Fowler, Ibid., 174–78.

[68]Darby, Ibid., 270; Dyer, *Making a Living in the Middle Ages*, 26; Fowler, Ibid., 172; Gardiner, "Shipping and Trade between England and the Continent during the Eleventh Century,"; King, "Chapter IX Sussex," 459. For evidence of hand-mills on military campaigns, see: *Calendar of Documents relating to Ireland Preserved in Her Majesty's Public Record Office*, London: 1171–1251, H. S. Sweetman (Ed.). (London: Longman, 1875), 1–2.

[69]Darby, *Domesday England*, 260; Laurence Keen, "Coastal Salt Production in Norman England." *ANS XI* (1989): 134; King, "Chapter IX Sussex," 457.

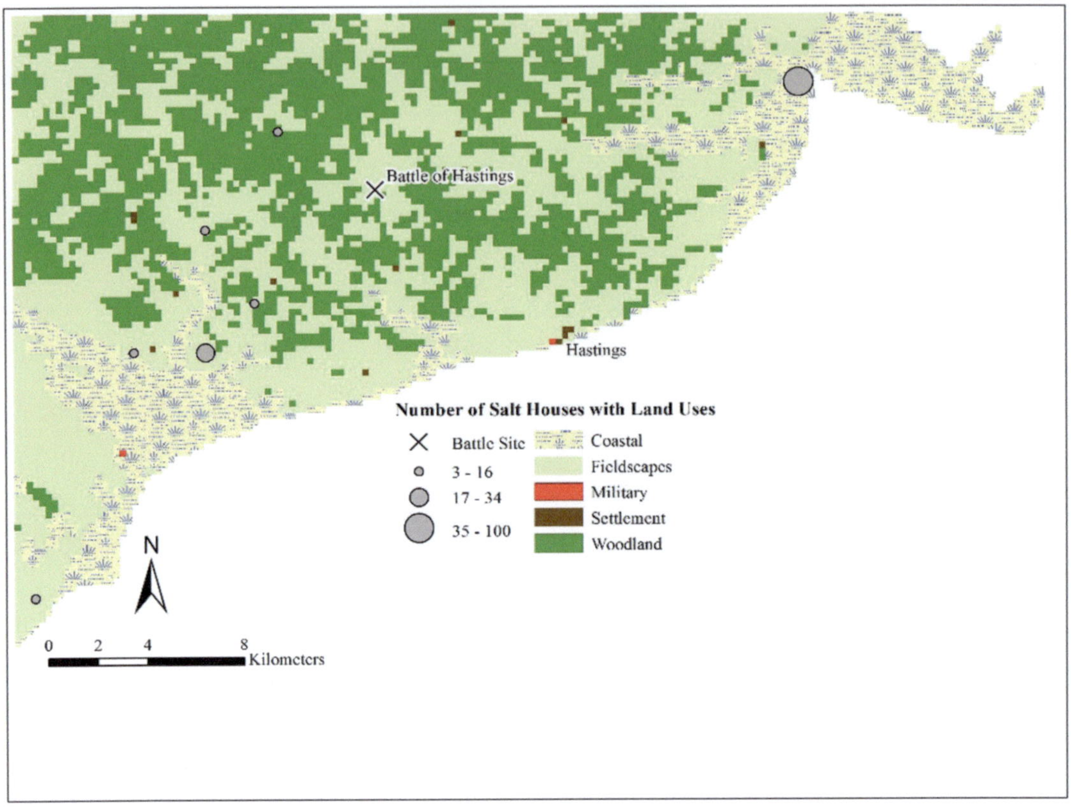

Fig. 11.8 Estimated Hastings area salt houses in c. 1066 (The number of salt houses was recorded in the *DB*. King, "Chapter IX Sussex," 457; Palmer, *Electronic Edition of Domesday Book*.)

best explanation, proposed by Darby, is that "the pans entered for some inland manors must have been physically situated elsewhere."[70] What is more important, however, is that the map indicates that the Norman landing site and camp were ideally situated to access the local salt reserves required for meat preservation.

11.6.4 Wood

"Woods are very stable features of the landscape; like roads and churches, they have outlasted many changes in the societies which they serve."[71] The distribution of wood resources potentially utilized by either side can be obtained from the *DB*. However, in the *DB*, the woodland is listed as "x wood for y swine" which is misleading because it does not indicate the extent of the woodland. There are locational concerns as well, as some of the woodland may not have been next to the village it belonged to, but some distance away (see previous comment pertaining to salt). Consequently, the distribution of woodland as presented above in the figures relies upon evidence from the East Sussex Historical Environmental Record database which displays the spatial extent of woodland. As the figures reveal,

[70]Darby, *Domesday England*, 265; King, "Chapter IX Sussex," 457. For locations of salt works in the Pevensey area, see Stephen Rippon, *The Transformation of Coastal Wetlands: Exploitation and Management of Marshland Landscapes in North West Europe during the Roman and Medieval Periods*, (Oxford: Oxford University Press, 2000), 187–88.

[71]Oliver Rackham, *Ancient Woodland: Its History, Vegetation and Uses in England* (London: Edward Arnold (Publishers) Ltd, 1980), 137.

woodland was readily available in the area.[72] Woodland would have been important to both sides for a number of reasons. The most prominent would have been for firewood to cook meals, building castles, ships and other defensive works in case of attack.[73] Oliver Rackham confirms this interpretation by discussing in detail the uses of wood in the Middle Ages including building material, charcoal, fuelwood, pasture and shipbuilding from underwood and timber.[74]

11.7 Conclusion

This applied historical geography and GIS research on environmental resource influence on the Battle of Hastings presents a number of interesting findings. It is estimated that there were between 5,270 and 6,441 people in the area with an average age of between 20 and 22. The population density was calculated to range between close to 0 and 59.215 people per square kilometre with a mean of 7.755 people. Additionally, 52% of the population was estimated to be between the age of 15 and 60. The examination of the food sources indicated that after the local population was fed, there would have been extra provisions available for the armies during their presence in the area. A local stream discharge graph suggests that the streams draining the area would have provided sufficient water as required. Mapping of iron and salt sites as well as the extent of woodland indicated abundant supplies of those resources. At the same time, milling facilities were less readily available as these were scattered and few in number. All in all, then, this study has provided previously unexplored contextual details regarding the local population and their activities in relation to the provisioning needs of the two armies at the Battle of Hastings. As a result, we can conclude with reasonable assurance that both sides would have been well supplied in the region -even had the length of their conflict extended beyond 14 October 1066. This would have been particularly important for Duke William's army as this would have alleviated the costly need to transport supplies across the English Channel from France.

In terms of methodologies, through its development of GIS cartographic representations and analyses, this chapter provides new insight into the volume and type of resources likely available to both armies at the time of the Norman Conquest in the Hastings area. The value of this GIS visualization and analysis approach is that medievalists can gain new knowledge and perspectives about a topic that would not have been accessible without GIS applications. For example, in this study, the population data was visualized at a village level, -only undertaken once before for another region of England (Bedfordshire, Cambridgeshire, Huntingdonshire and Northamptonshire).[75] Subsequently, the findings in this chapter present estimates of the East Sussex population at the village level in 1066 for the first time. In addition, to the author's knowledge, iron sites in the Weald have not been visualized simultaneously representing their geographical spread and periodization. Further research is suggested on the length of time these iron sites were active, to confirm the accessibility of iron in 1066. Other industrial factors were also visualized and analyzed from various environmental frameworks, facilitating a more informed discussion on the battle of Hastings and providing a richer context for historians to discuss and

[72]Darby, Domesday England, 172–73; East Sussex County Council, *East Sussex Historic Environmental Record.*

[73]Hoffmann, *An Environmental History of Medieval Europe*, 201–02; Stenton, *The Bayeux Tapestry*, Plates 38 and 51. Lavelle discusses fortifications including signal beacons in his book on Anglo-Saxon warfare. Ryan Lavelle, *Alfred's Wars: Sources and Interpretations of Anglo-Saxon Warfare in the Viking Age* (Woodbridge, UK: The Boydell Press, 2010), 209–63.

[74]Rackham, *Ancient Woodland*, 137–47 and 153–56. The importance of wood to medieval society is discussed by Hoffmann in his book on the medieval environment. Hoffmann, *An Environmental History of Medieval Europe*, 181–88.

[75]Lowerre, *Mapping Domesday Book using GIS.*

explore the parameters of the battle. In summary, this study demonstrates the value of interdisciplinary research on historical military topics. In the examples above, without research on demographics, iron archaeology, environmental research and GIS visualization, this study and its specific contributions would not have been possible. In addition, the study contributes to the historical GIS literature by using medieval data sources to extensively estimate the resources available for a battle from a period when there were few extant written sources. The study also illustrates the means to reconstruct an environmental picture from an era in human history, where little data can be gleaned.[76] The chapter also emphasizes the value of a geographic or environmental perspective on the role of logistics in military history. Moving beyond the period and place of the Battle of Hastings, the geographical and environmental framework discussed in this chapter is applicable to other battles, events and locations, within and beyond the scope of the medieval period. By using geographical analysis, in fact, a much richer and more comprehensive portrait of military campaigns and their environments may be obtained to both interrogate and enhance existing historical studies.

[76]M. T. Clanchy, *From Memory to Written Record: England 1066–1307*, 3rd ed, (Chichester, UK: Blackwell Publishing Ltd., 2013), 1–2.

12 Mapping the Irish *Rath* (Ringfort): Landscape and Settlement Patterns in the Early Medieval Period

Robert Legg, Francis Ludlow and Charles Travis

Abstract

The landscape motif of the Irish *rath*, or ringfort, is closely associated with early medieval settlement in Ireland (400–1100 AD). Ringforts were circular or near-circular enclosures, typically constructed by digging a surrounding ditch, the material from which would be used to construct an adjacent earthen bank. The majority of ringfort sites feature a single ditch and bank (known as a *univallate* site), but double or triple banked sites are also known, and are described as *bivallate* and *trivallate*, respectively. Variations in ringfort morphology and in the physical geographical characteristics of their site location provides a valuable field record that is indicative of the diversity of socioeconomic roles played by these sites and of status of individuals and clans responsible for their construction and occupation. Ringfort locations generally represented the habitation centre of the farmstead, comprising one or more dwellings. Petal-shaped fields would surround the site. This chapter discusses a geographical information science (GIScience) and geostatistical study of ringfort locations in the Irish Midlands, utilizing data sourced from the Irish Government's Department of the Environment, Heritage and Local Government archaeological site spatial database. This study also draws on archaeological and historical database records and fieldwork analysis in order to consider the relations between physical environmental context and ringfort settlement patterns.

Keywords

Early Medieval Ireland Midlands · Ringforts · Typology · Settlement · Environment GIScience · Cluster analysis · *ANOVA*

R. Legg (✉)
Northern Michigan University, Marquette, USAe-mail: rlegg@nmu.edu

R. Legg · F. Ludlow
Trinity Centre for Environmental Humanities, Trinity College Dublin, Dublin, Ireland

F. Ludlow
Department of History, Trinity College Dublin, Dublin, Ireland
e-mail: ludlowf@tcd.ie

C. Travis
Trinity Centre for Environmental Humanities, Trinity College Dublin, Dublin, Ireland

University of Texas, Arlington, USA
e-mail: charles.travis@uta.edu

12.1 Introduction

The Early Medieval Period (c. 400–1100 AD) represents a transitional period in Ireland's history. Coinciding with the spread of Christianity, these seven centuries saw marked climatic fluctuations between higher and lower precipitation

C. Travis et al. (eds.), *Historical Geography, GIScience and Textual Analysis*,
Historical Geography and Geosciences, https://doi.org/10.1007/978-3-030-37569-0_12

and temperatures, extreme weather events, advances in agricultural technology and food production (McManus et al. 1998; Edwards-Charles 2000; McDermott and Mattey 2001; Cróinín 2005; Kiely et al. 2010; Ludlow et al. 2013; Cook et al. 2015). The early medieval period in Ireland is also closely associated with the landscape motif of the Irish *rath*, or ringfort. Representing a dominant settlement type during the period, ringfort sites have left an indelible imprint on the island's landscape and associated archaeological record, with tens of thousands still visible or known from cartographical sources, and many more thought to have existed before their erasure from the landscape by modern agriculture (Warner 2002; Stout 2017). In the Irish context, settlements during this period can be defined as the way in which people were situated on the landscape with respect to the type, arrangement, and location of their dwellings.

Typically featuring a circular or near-circular enclosure built of earth (and more occasionally of stone, particularly in western parts of Ireland, in which case the term *cashel* is preferred), ringforts generally represented the centre of a farmstead, with one or more dwellings clustered within the banks (O'Connor 1998; Stout 2017). While some ringforts exhibited ongoing usage into the high and late medieval period, particular in western regions of Ireland (Warner 2002; Fitzpatrick 2009), the vast majority appear to date to the early medieval period (Stout 2017). The majority of ringforts feature a single earthen bank and ditch (known as a *univallate* site). Sites with double or triple banked ringforts are described as *bivallate* and *trivallate*, respectively (Fig. 12.1). Morphological and other variations in ringfort sites offer a valuable field record that can reveal the diversity of social roles and the varying status of individuals and clans responsible for the construction and occupation of the ringforts in question (Patterson 1994; Cody 2007). It is often inferred that the early medieval construction of ringforts across the Irish landscape is associated with increased prosperity and possibly a dramatic increase in population during the period (e.g., Stout 2017).

Early medieval Irish society was rural and employed a client-based hierarchical organisation (Binchy 1954; Patterson 1994; Kelly 1997a; Lyttleton and Monk 2007). Members typically secured social ranking through heredity, but some mobility could also be attained by increasing personal wealth during an individual's lifetime. Ringforts, as products of considerable manual labour, were a visible symbol of social rank. The complexity of design and location of these ringforts within the landscape of a given territory can thus be read (then and now) as a symbol of the social status and wealth of the settlement's owners. The labour necessary to build a ringfort was obtained primarily through clientship, a practice involving the lending of cattle by social elites as well as providing a degree of physical and legal protection, in return for a negotiated form of repayment (McCormick 1995). Repayment often consisted of manual labour and could also include the provision of food and military service (Patterson 1994).

As cattle acted as a critical store of wealth and medium of exchange in the early medieval period (Kelly 1997a), securing territories from the ravages of cattle raiding, wild animals and hostile clan invasions was critical. The earthen enclosures around most ringforts likely played some protective role in this regard, but are generally considered insufficient to play a more significant military function (Warner 2002), being a plausible casual deterrent, but not sufficient to withstand a prolonged attack or siege (Stout 2017). Between 45,000 and 47,000 documented ringforts, varying in size and shape, remain visible across the Irish landscape (Barrett 2002; Stout 1997, 2017), standing as lonely earthen and stone sentinels to a by-gone but still iconic era in Irish settlement history. This chapter employs geographical information science (GIScience) and geostatistical approaches to studying ringfort landscapes in the Irish Midlands. Drawing on archaeological and historical database records and fieldwork analysis, the chapter considers the relations between environmental context and ringfort settlement patterns.

Fig. 12.1 Kite aerial photograph of the *multivallate* ringfort at Rathrá, Co Roscommon, Ireland *Source* West Lothian Archaeological Trust (Wikimedia Commons)

12.1.1 Ringfort Geo-Database

In a general sense, ringfort sites across the Irish landscape appear to offer a simple and relatively consistent morphology. Most often, sites are often overlooked or studied only individually rather than as part of a potentially larger pattern (Monk 1998). In this chapter, we adopt a broader landscape approach, examining ringfort location attributes as a means of providing further insights into the imperatives of the complex and dynamic communities responsible for their construction. Ringfort coordinates are available from the Irish Government's Department of the Environment, Heritage and Local Government archaeological site spatial database. This records the locations of individual sites using the Irish national grid and tags each site with a *Record of Monument and Place* (RMP) identifier and other class descriptors (e.g., site type such as ringfort or monastic settlement). This data can be complemented by further information on, for example, ringfort morphology as detailed in secondary sources.

The overall scope of the work discussed in this chapter begins with 1312 ringforts identified in three areas in the Irish Midlands. Statistical classification techniques, including Cluster Analysis, offer frameworks for parsing large datasets into smaller and more manageable geo-relational groupings (e.g., Staeheli and Clarke 2003; Papageorgiou et al. 2001; Poon et al. 2004; Rice and Saffer 1982; Sutton 1998). This approach was thus employed to group similar ringfort types according to their morphology (*univallate, bivallate, trivallate*), and to highlight and associate dominant site characteristics with dominant settlement patterns. A geo-dataset with over 1000 ringfort sites was constructed as a result, and its structure suggests a potential framework for an island-wide

GIScience collation, visualization and analysis of other extensive archaeological and historical datasets.

12.2 Ringfort Study Region

Figure 12.2 shows the distribution of known ringforts in three adjacent study areas: the Blackwater Valley, the Inny River catchment and the Lough Ramor catchment. These areas variously comprise parts of counties Cavan, Meath, Westmeath and Longford. In the early medieval period, this region comprised the wealthy extended Gaelic kingdoms of Brega and Mide, dominated by the powerful Ui Neill clan dynasty. The kingdoms were located in a strongly contested territory in proximity to the Hill of Tara, the symbolic seat of the high-kingship of Ireland. The Blackwater Valley study area covers an area of 345 sq. km and contains 59 known ringfort sites, representing a density of one site per 5.8 km^2. The extent of this study area was defined through a field survey and literature review that examined its topographic boundaries (Newman et al. 2007; Meehan and Warren 1999; Legg 2006). The Inny River and Lough Ramor catchments were defined by reference to watershed boundaries in lands adjoining the important Blackwater Valley (Geological Survey Ireland 1962). The Inny River catchment study area is the largest of the three, comprising 1,282 km^2 and containing 1,067 known ringfort sites (a density of approximately one site per 1.2 km^2). The Lough Ramor catchment covers 352 km^2 and contains 186 known ringfort locations (a density of approximately one site per 1.9 km^2).

The three study areas are separate but contiguous and accommodate a complex range of environments, ringfort densities, distributions and historical settings. The Blackwater Valley, for instance, offers fertile agricultural soils and holds a prominent position in Ireland's settlement history. Situated near several important historical sites, the area encompasses prehistoric monuments at Lough Crew, the Hill of Tara complex, and the Newgrange and the Boyne River Valley. The Donnaghpatrick monastic site and the wealthy Kells, Abbey putative origin-place of the famous illuminated Gospel manuscript, the *Book of Kells* are also located in this area. Contrastingly, the Lough Ramor catchment is located farther from these high-status historical sites and is composed of low elevation lands surrounding the lough, that increase in elevation to the northeast. This area is relatively hilly, compared with the other two selected study areas and contains acid brown earth soils that support pastures and woodlands. The Inny River catchment features a drumlin landscape in the northwest and winding eskers to the south. Its soils are grey-brown podzolic, peats and to a lesser extent acid brown earth and gleysol. The area is dominated by peat bogs in lower elevations and pastures at higher elevations. Pollen studies of the area's bogs suggest a pattern of vegetation evident in many other areas of Ireland (Brown et al. 2005; Caseldine and Hatton 1996; Jessen 1949; Newman et al. 2007). These records illustrate that pastoral and cereal farming activities commenced a substantial decline in woodland cover in these areas from around 500AD. Pollen records indicate a slight decline in farming activity during the middle of the eighth century with a resurgence of oak and hazel trees. Farming activities and decline in tree pollen appear to resume around the middle of the eleventh century.

12.2.1 Methodology

A *ground-truthing* field survey of ringfort sites in the Blackwater Valley study area demonstrated that existing digital archaeological records from the Department of the Environment, Heritage and Local Government archaeological site spatial database, combined with cartographic data from the late nineteenth-century, provide acceptably accurate latitude and longitude coordinates for ringfort sites, and also the physical dimensions of the ringforts (Fig. 12.3). Since these cartographic and archaeological surveys were conducted, most ringforts remain present, but exposure to human activities such as farming has contributed to degradation in the condition of a few of the sites

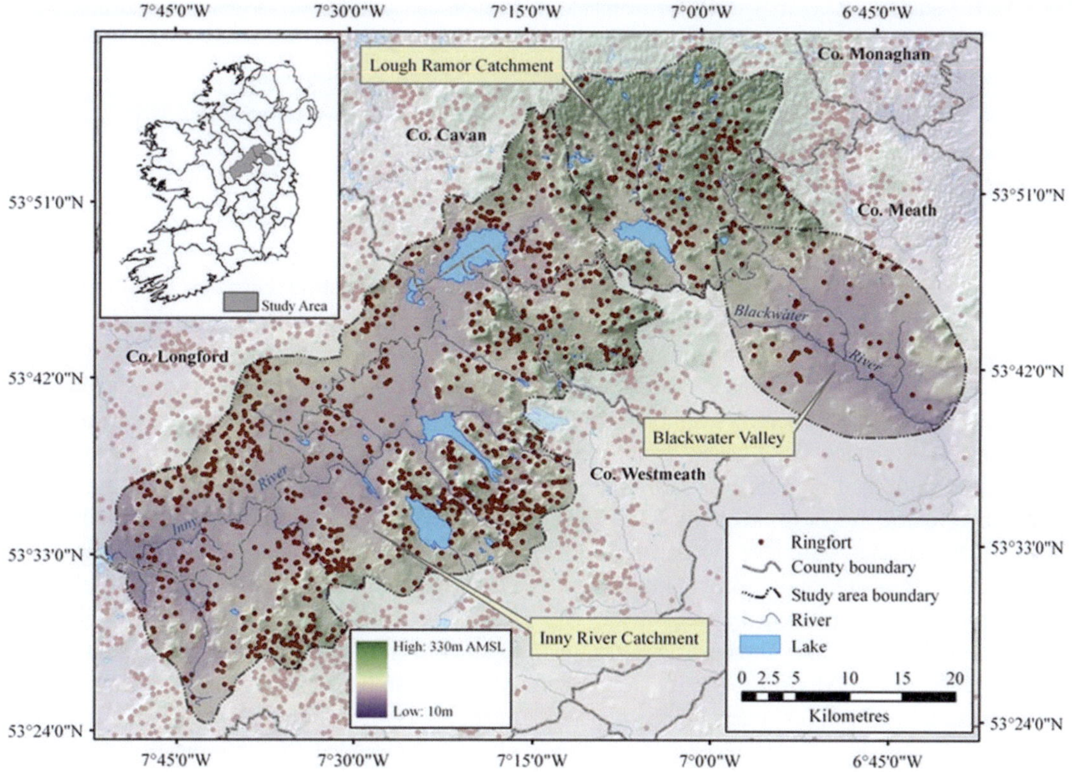

Fig. 12.2 A location map of ringforts in three contiguous study areas. Topographic visualisation is based on a hill-shaded terrain model rendered using Ordnance Survey Ireland 1:50,000 Discovery Series data, 1994

Fig. 12.3 Photographs from fieldwork with photograph **a** showing *univallate* ringfort in the townland of Clonmagaden, County Meath. This ringfort has a maximum overall diameter of 50 m. Photograph **b** is a field survey measurement of the bank heights of the same *univallate* ringfort

(Legg 2006). *Ground-truthing* concluded that pre-existing documentary and cartographic records were—in some cases—more complete, in terms of information readily available than existing ringfort site artefacts. Information on the ringfort sites analysed in the Inny River and Lough Ramor study areas were collected from available documentary and cartographic sources. A full description of the field survey results and methodology is provided by Legg (2006).

Key characteristics of identified ringforts for all three study areas are summarized by the variables detailed in Table 12.1. While additional variables of interest may certainly be specified and considered in

Table 12.1 Variables collected: a listing of morphology, location and distribution variables used to describe the ringforts analysed in this research

Morphology	Location	Distribution
Maximum internal diameter (MID)	Slope	Nearest neighbour (NN)
Maximum overall diameter (MOD)	Elevation	
Number of banks (Banks)		

future work, those considered here were the most frequently present in the available datasets (e.g., morphological data such as maximum internal diameters) or could be calculated based upon locational information (i.e., nearest neighbour data). Morphological variables were taken from measurements in published archaeological inventories and archaeological records kept by the Irish National Monuments Section, Department of Environment, Heritage and Local Government (2005), further supplemented by large-scale 6″ 1:2,500 and 25″ 1:10,560 Ordnance Survey of Ireland (OSI) maps (Moore 1987; O'Donovan 1995). These OSI 6″ and 25″ maps offered precise and accurate scaled-down drawings of each ringfort site. The drawings were suitable for measuring morphological variables unavailable from other sources. Location-based variables were assembled using the tools available in ArcMap GIS. Attributes such as site elevation and slope were determined from a 20-m *Digital Elevation Model* (*DEM*) dataset employing standard GIS overlay functions. Nearest neighbour values were calculated with GIS spatial statistic tools. Mean values for all variables are graphed in Fig. 12.4.

We carried out the grouping process on the remaining sites using a *Two-Step Clustering Process*, a procedure used to process both continuous and categorical variables (SPSS 2004; Francisco 2012). We further interpreted the cluster results by identifying similarities among the five groups in each of the three separate study areas and manually organizing the existing groups into broadly re-occurring typologies. Cluster analysis provides a powerful means to organize large datasets of several hundred or more cases into a small number of groups or clusters. Each group contains cases that are more similar to one another than those in other groups. The number of groups is pre-specified prior to the evaluation. Data is sorted in groups by mathematically evaluating variables associated with each case (individual ringfort) such as their size, number of banks and other available variables. The process is subjective for several reasons (Aldenderfer and Blashfield 1984). For example, the number of groups must be decided upon by the researcher, also, each group as decided by the process must further analysed to interpret its real world meaning. By trial and error, we settled on a *fixed 5* group model (a model with 5 groups) run separately for each of the three study areas. A larger number of groups (more than 5) offered unnecessary detail, while in turn, smaller number of groups provided too little detail. Settling on a *fixed 5* model for each study area offered a reasonable balance between capturing the substantial differences and having too much information (i.e. between having a larger number of groups in which individual cases were differentiated by only minor variations in the assessed variables, and too few groups in which cases that appeared to vary more substantially were aggregated regardless). Groups were calculated using all morphological variables listed in Table 12.1. Although 1312 ringforts were recorded among the study areas, 70% of these sites were ultimately excluded from the analysis due to site damage and the failure to meet our classification criteria (e.g. regarding information completeness). Cluster analysis was carried out in SPSS version 16 and mapping was carried out in ArcGIS 10.0.

12.3 Findings

Our analysis suggests there are seven broadly recurring typologies of ringforts in the overall study area, with one of the typologies, *Trivallate*

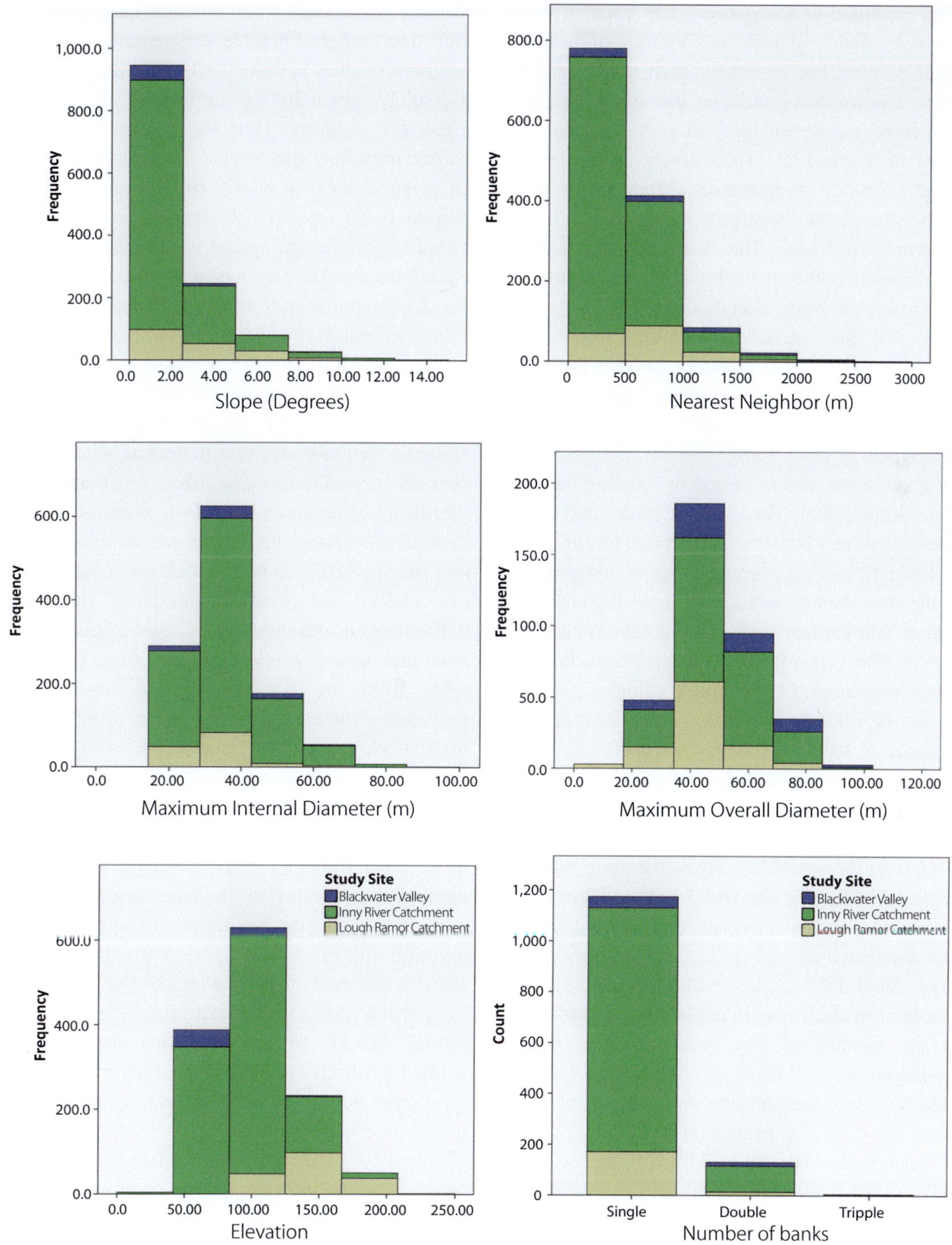

Fig. 12.4 Summary measures graphs for the variables associated with ringfort sites analysed in this study

sites, identified as a separate group external to the *two-step clustering process* (because of the very small number but extremely high status of these sites). The ringfort typologies and number of sites (expressed as percentages) in each typology are listed in Table 12.2. The chosen typologies are based upon our interpretation of the characteristics of the sites within each group as produced by the clustering process. The exercise of defining typologies is somewhat subjective, as with the initial cluster analyses. Specified typologies are sensitive to the number of groups considered appropriate during the clustering process itself, and thereafter the characteristics of each group that are deemed sufficiently meaningful as to be reflected in each typology label. Here, we have been guided by common sense, and informed by existing literature (e.g., Stout 1997). Thus, for example, two of the chief differences between the first and fourth groups in Table 12.2 is the average diameter and elevation of the sites therein, while both share the characteristic of one earthen bank. This is thus reflected in our specified typologies of *Low-lying univallate* and *Large univallate*. Other valid typologies could, of course, be offered, and those offered here are thus designed to facilitate further discussion.

Typifying sites in such a manner allows us to carry out analysis for new patterns associated with these settlements. Irish townlands, for example, are thought to have possible origins dating to the early medieval, with their size and density in a given area thought to reflect physical and socio-economic indicators such as land quality and population densities (Stout 1997; Aalen et al. 2011). To examine whether townland sizes in our study areas associated in any meaningful way with the seven ringfort typologies we identified, an *Analysis of Variance* (*ANOVA*) test was undertaken to evaluate if the mean area of townlands varied by ringfort type (townland sizes were sourced from *H. M. Stationary Office Great Brittan and Cahill and Co.* 1904). The test returned no statistically significant results ($p = 0.70$). This measure suggests that townland sizes do not associate with any of the particular cluster typologies identified by our analysis, but further work is certainly merited on this question.

Ringforts were measured for their proximity to neighbouring ringforts and physical landscape features within each study area, for each typology. Our data suggest notable differences in nearest neighbour values between each typology, often but not solely driven by the number of sites in each respective typology. Thus we see that *Low-status upland univallate* sites are the most proximate, with an average distance of 427 m between each site, despite being considerable less numerous in our sample than other groupings such as the *Low-lying univallate* sites (i.e., 6% versus 19%). By contrast, the *Large univallate sites* are typically 1000 m from their nearest neighbour, despite comprising a similar proportion of our overall sample as the *Low-status upland univallate sites* just discussed (i.e., 3% versus 6%). An *ANOVA* test on these variables demonstrates that these *Large univallate sites* are located farther away from other sites, with significant differences found in contrast to the *Typical univallate* and *Low-status upland univallate sites* ($p<0.05$). In terms of elevation and slope, our *ANOVA* test revealed statistically significant differences in elevation of the *Low-status upland sites* and nearly every other typology ($p<0.05$), aside from the *Large univallate sites*. These *Low-status upland univallate* were found at the highest elevations throughout the study areas. Slope values were found to be similar for all groups with only subtle differences among the groups.

One of the most frequently cited settlement models for Ireland's early medieval period was devised by Stout (1997). This model aims to explain the meaning behind recurring patterns by relating ringfort sites to period textual sources, such as the early Irish law tracts (Kelly 1997b; Fitzpatrick 2009; Aalen et al. 2011). Stout's model, based on a contiguous area in the south-west Irish midlands, identifies five ringfort typologies related to specific features of ringfort morphology, distribution and physical location (Table 12.3). Types vary from perceived *Low-status upland univallete* ringforts, located in upland areas, and progress through to *High-status bivallate* ringforts, located centrally within townland territories. Between the *High* and *Low Status* forts, Stout observed that the most frequently observed type, the *Typical univallate* ringfort, tends towards locations that are suitable for agriculture, without strategic hill-top

Table 12.2 Summary of cluster analysis results. Ringfort typologies based on combined cluster analysis results for the Blackwater Valley, Inny River catchment and Lough Ramor catchment study areas. *Trivallate* sites were separated from the study prior to running clustering procedures. MID refers to maximum internal diameter, MOD to maximum overall diameter, and NN to nearest neighbour Slope values are given in degrees

Variable	Low-lying univallate		Typical univallate		Low-status upland univallate		Large univallate		Very-large univallate multifunctional		High-status bivallate		High-status trivallate		Overall Measures	
	$\bar{x}$	*s*	$\bar{x}$	*s*	$\bar{x}$	*s*	$\bar{x}$	*s*	$\bar{x}$	*s*	$\bar{x}$	*s*	$\bar{x}$	*s*	$\bar{x}$	*s*
MID	28 m	±6.1	33 m	±5.2	32 m	±5.7	39 m	±8.6	52 m	±9.6	36.3 m	±11.8	31 m	±5.35	36 m	±11
MOD	36 m	±9.2	43 m	±6.3	41 m	±6.7	48 m	±9.8	63 m	±10.0	57.2 m	±16.1	68 m	±6.0	48 m	±14
Banks	1	-	1	-	1	-	1	-	1	-	2	-	3		3	
Slope	2.4	±2.2	1.7	±1.5	3.3	±3	1.8	±1.6	2.3	±2.4	2.2	±2.4	1.2	±1.9	2.1	±2.2
Elevation	100 m	±5	109 m	±29	145 m	±16	116 m	±43	115 m	±31	112 m	±30	106 m	±26	111 m	±33
NN	628 m	±447	531 m	±399	427 m	±183	1007 m	±741	470 m	±416	545 m	±428	509 m	±271	568 m	±427
% of 368	19% of 368		35% of 368		6% of 368		3% of 368		16% of 368		20% of 368		1% of 368			

Table 12.3 Hypothetical model of ringfort types as theorized by Stout (1989, 1997)

Groups	Type	Description	% of the total
1	Low-status	Univallate low lying raised living space with an average internal diameter of 39 m	13.5%
2	High-status	Bivallate smaller interiors (*c.* 29 m)	11.5%
3	Typical	Univallate ringfort, small internal diameter (*c.* 31 m) range of elevations	38.0%
4	Large	Univallate multi-functional, large exteriors (**46 m**)	13.5%
5	Low-status upland	Univallate earthworks small internal diameters (*c.* 28 m), typically located in upland locations	23.5%

placement. *Large univallate* ringforts tend to be located on local borders (e.g., townland boundaries) and consequently may have served some military or strategic role. Stout's low-status sites were often sited in lowland and valley bottom locations, and examples can be found scattered throughout Stout's study areas.

The results in this chapter offer some notable parallels to Stout's (1997) findings, but also some contrasts. The data for our study area provide little evidence that ringfort types adhere to consistently ordered sizes and shapes that can be associated with the social standing of their early medieval owners. The data illustrate that ringfort morphology spans a spectrum of sizes, shapes and complexities. Though this observation must be tested for regions outside of our study area, the data suggest a relation between a ringfort's size and the complexity of a society that lacked strong centralised control and was thus more free to develop organically.

Analysis of ringfort sites within our three study areas suggests that differences in the morphologies of the forts may be at least partly associated with areas of greater agricultural potential. *Low-lying univallate* ringforts, for example, featuring the smallest internal and overall diameters of all ringfort groupings, are located mainly in areas of marginal agricultural land quality, indicated by their low lying valley bottom locations. In comparison, the most frequently observed ringforts (i.e. *Typical univallate*), have slightly larger internal and overall diameters than low-lying forts. These are situated in mid-elevation locations and in areas of good agricultural potential, as indicated by soil maps (see: http://gis.teagasc.ie/soils/map.php). High-quality agricultural environments (comprising the majority of the Blackwater Valley study area) can also be seen to support the highest densities of *High-status bivallate* ringforts. By contrast, the Lough Ramor catchment contains poor quality agricultural land and its ringfort sites generally appear as the smallest in our survey. Townland sizes in the study areas averaged 270, 180, and 170 hectares for the Blackwater Valley, Inny River and Lough Ramor Catchments respectively. This itself suggests that townland size may link ringfort type with the local wealth of larger townlands. Located in productive areas, Located in productive areas, larger townlands are often associated with a higher frequency of *High-Status bivallate* ringforts.

Ringfort typologies display recurring patterns throughout our three study areas, often similar to those identified by Stout (1997). We observe differentiation by elevation between different ringfort typologies, with *Low-status upland univallate* sites featuring slightly smaller internal diameters than *Typical univallate* sites and being found at significantly higher elevations than most other typologies. It is possible that many ringforts in these upland areas did not contain permanent dwellings, but instead may have been used seasonally for the summer grazing of cattle, in a practice described by Ó Corráin (2005). By contrast, *High-status bivallate* ringforts can be situated at a range of elevations, but with 95% of sites at elevations between 105 and 119 m (the approximate mean for all ringforts in the three study area cluster analysis was 111 m in elevation). *High-status bivallate* ringforts also exhibit a situational pattern of prominently visible and centrally accessible locations, between hilltop and valley bottom settings, as may perhaps be expected. *Typical univallate* ringforts are located at elevations of 109 m (AMSL ± 29). The mean elevation of these sites agrees with those reported in Stout's (1997) settlement model and indicates that the majority of *Typical univallate* ringforts were located at elevations slightly below the bivallate sites. However, the differences observed in our three study areas are not striking. Consequently, their implications are not fully clear. An *ANOVA* test reveals no statistically significant associations between the slopes and the different ringfort typologies for each of our study areas.

Low-status upland univallate ringforts were found predominantly throughout the Lough Ramor catchment and are associated with poor-quality land. Stout (1997) concludes that such places could have functioned as temporary, seasonal dwellings. There is the possibility, however, that these upland locations may have provided the only available land for a growing population during the main phases of the construction and occupation of ringforts. Even if this were the case, it does not exclude the simultaneous use of some upland sites as temporary, seasonal dwellings. An analysis of the dating of these sites would prove very valuable if, for example, it revealed the ringforts in these areas as being the most recently constructed of all types. Such forts may have been the dwellings for people living in areas with little remaining habitation space.

One exciting pattern did emerge throughout our three study areas (Table 12.2). Across the landscape, we consistently identified *Large univallate* ringfort sites, with sizes of up to 70 m overall. These large early medieval sites may have served multi-functional roles, acting as military enclosures, stockades for seized cattle or for community functions such as meetings and fairs. Empirical evidence gathered by Simms (1986) and Stout (1997) suggests that higher status ringforts were sited closer to monastic settlements than were lower-status ringforts. While monastic sites were not included in our study, future research to determine whether there were associations between ringforts, monastic settlements, and arable farming patterns is warranted. Such patterns may be supported by archaeological evidence, which indicates that monastic sites were engaged in agriculture and milling. According to the *CORINE Land Cover 2000* dataset, 22.7% of land covering the Blackwater Valley area is presently utilized for arable agriculture, compared with only 0.1% of the Lough Ramor Catchment (Byrne 2003). If such differences in agricultural potential have any parallel with the same land in the early medieval period, it appears that monastic settlements could be found more frequently on land with the greatest capacity for arable farming. A working hypothesis, amenable to future testing through geospatial techniques, is that the construction of a greater number of ringforts was precluded in these same areas, with the church gaining control of fertile lands.

12.4 Conclusion

The research presented here acknowledges several uncertainties. In particular, the construction, occupation and abandonment dates of the ringforts in this study are not completely known. Dating ringforts precisely within the early medieval period is difficult. Island-wide, a comparatively small number of ringfort sites are scientifically dated, in comparison to other archaeological sites such as those containing grain drying kilns (Monk and Power 2012), and those that have been dated suggest occupation, construction and abandonment at various different times throughout the period (Stout 1997, 2017; Kerr et al. 2009; Fitzpatrick 2009). In our study, the ringfort sites are analysed as a contemporary group with no chronological distinctions. Despite the comparatively small number of ringforts that have been dated, an examination of those dates that are available for our study area may still prove revealing of trends in the settlement of the region. Such trends may then be linked to the potential forces, such as demographic and climatic changes, that may have informed the decision to inhabit areas of lesser agricultural potential, as seen in our proposed typologies.

It is often suggested that many undocumented sites may have been destroyed before the production of Ordnance Survey maps in the mid-1800s (Reeves-Smyth 1983; Kerr et al. 2009). There are, however, grounds for supposing that ringforts were left largely intact in many areas. Abandoned sites were viewed positively, as they provided free-standing protection for livestock. Superstitions also contributed to their preservation: Irish folklore regards the sites as the locations of fairy forts, and disturbing such grounds incurs a curse (Westropp 2000). Lastly, removing the earthen and stone structures of ringforts would have required considerable manual labour. All these factors have led to the relatively infrequent destruction of ringfort sites, particularly in areas of low-intensity farming. Conversely, areas such as eastern Ireland, subject to particularly intensive agriculture, and from the twelfth-century the greatest Anglo-Norman and English influence, are conspicuous for having a lower frequency of known ringforts, though the numbers remaining as lonely early medieval sentinels on the landscape are still considerable (Aalen et al. 2011; O'Sullivan et al. 2013).

Our research assumes that ringforts were secular, but we recognize that a dynamic relationship between these sites and religious settlements did occur during the early medieval. Towards the end of the period, nucleated dwelling patterns did form around monasteries and ecclesiastical sites. Threats from Viking raids, the introduction of silver for trading, and other factors facilitated the development of proto-towns and an evolution in Irish social behaviour. The likely rapid increase in the construction of ringforts during the early medieval can be viewed in the context of the likely rise in prosperity of this period. The spread of a European Christian network, expansion of monastic settlements, widespread raiding of post-Roman Britain, trade relationships with Celtic Europe and the British Isles, climatic fluctuations between higher precipitation and drier weather, as well as the introduction of new farming practices and introduction of new food sources all contributed in some measure to the proliferation of ringfort sites. Such combined factors likely contributed to both an increase in population and necessitated both the means and the need to build substantial numbers of long-lasting ringfort settlements across the breadth of the Irish landscape.

The morphological, location and distributional patterns associated with ringforts suggest some clear differences between sites and associations with the quality of agricultural land. While the results of our study are limited by the available dating evidence and other uncertainties surrounding some of the data, we hope to have demonstrated the potential of investigating sites using combined GIS and clustering techniques. Ever more sophisticated methods of geographical analysis combined with the sheer number of surviving ringfort sites and the morphological and locational data that can be generated for them offer an immensely rich opportunity to significantly advance our understanding of early medieval human relationships with the physical environment in the western fringe of Europe. A substantial amount of reasonably complete and accessible data on ringforts is available to expand this study and the methods presented to different regions on the island of Ireland. The geographical and historical data gathered and identified in our research of ringfort patterns allows us to examine these sites individually, collectively and in the context of available environmental and societal evidence to better understand the early medieval landscape of the Irish Midlands.

References

Aalen FHA, Whelan K, Stout M (2011) Atlas of the Irish rural landscape. Cork University Press, Cork

Aldenderfer MS, Blashfield RK (1984) Cluster analysis: Sage University paper series on quantitative applications in the social sciences. Sage University Press, London

Barrett G (2002) Flights of discovery: archaeological air survey in Ireland. J Irish Archaeol 11:1–29

Byrne J (2003) Technical report: CORINE land cover 2000. ERA: Maptec Ltd., Dublin

Binchy DA (1954) Secular institutions. In: Dillon M (ed) Early Irish society. Colm O Lochlainn, Cork, pp 52–66

Brown AG, Hatton J, O'Brien CE, Selby KA, Langdon PG, Stuijts I, Caseldine CJ (2005) Vegetation, landscape and human activity in Midland Ireland: mire and lake records from the Lough Kinale-Derragh Lough area, Central Ireland. Vegetation Hist Archaeobotany 14:81–98

Caseldine C, Hatton J (1996) Early land clearance and wooden trackway construction in the third and fourth millennia BC at Corlea, Co. Longford. Proc Royal Irish Acad 96B:11–19

Cody E (2007) A brief look at aspects of ringfort morphology and distribution. In: Manning C (ed) From Ringforts to Fortified Houses: studies on castles and other monuments in honour of David Sweetman. Bray, Co. Wicklow: Wordwell, pp 1–8

Cook ER et al (2015) Old world megadroughts and pluvials during the common era. Sci Adv 1(10):e1500561

Cróinín DÓ (ed) (2005) A new history of Ireland. Oxford University Press, Oxford

Edwards-Charles T (2000) Early Christian Ireland. Cambridge University Press, Cambridge

Fitzpatrick E (2009) Native enclosed settlement and the problem of the Irish 'Ring-fort'. Mediev Archaeol 53:1–37

Francisco D (2012) Poor mental health symptoms among Romanian employees. A two-step cluster analysis. Proc Soc Behav Sci 33:293–297

Geological Survey of Ireland (1962) Rivers and their catchments. Dublin

H. M. Stationary Office Great Brittan and Cahill and Co. (1904) General topographic index: consisting of an alphabetical index to the townlands and towns of Ireland and indices to the parishes, baronies, poor law unions, etc. Dublin

Jessen K (1949) Studies in the late quaternary deposits and flora-history of Ireland. Proc Royal Irish Acade 52B:85–290

Kelly F (1997a) Early Irish farming: a study based mainly on the law-texts of the 7th and 8th centuries AD. Dublin Institute for Advanced Studies, Dublin

Kelly F (1997b) A guide to Early Irish law. Dublin Institute for Advanced Studies, Dublin

Kerr TR, Swindles GT, Plunkett G (2009) Making hay while the sun shines? Socio-economic change, cereal production and climatic deterioration in Early Medieval Ireland. J Archaeol Sci 36:2868–2874

Kiely G, Leahy P, Ludlow F, Stefanini B, Reilly E, Monk M (2010) Extreme weather, climate and natural disasters in Ireland. Environmental Protection Agency, Ireland, Johnstown Castle, Wexford

Legg R (2006) Application of spatial and statistical analysis in identifying the settlement patterns of Irish Ringforts. PhD thesis, Trinity College, Department of Geography, Dublin Ireland

Ludlow F, Stine AR, Leahy P, Murphy E, Mayewski P, Taylor D, Killen J, Baillie M, Hennessy M, Kiely G (2013) Medieval Irish chronicles reveal persistent volcanic forcing of severe winter cold events, 431–1649 CE. Environ Res Lett 8(2):L024035. https://doi.org/10.1088/1748-9326/8/2/024035

Lyttleton J, Monk M (2007) Sites, social change and warfare in the second half of the first millennium AD: a reappraisal. In: Manning C (ed) From Ringforts to fortified houses: studies in castles and other monuments in honour of David Sweetman. Bray, Wordwell, pp 9–20

McCormick F (1995) Cows, ringforts and the origins of Early Christian Ireland. Emania 13:33–37

McDermott F, Mattey D (2001) Centennial-scale Holocene climate variability revealed by a high-resolution speleothem Δ^8 record from SW Ireland. Science 294 (5545):1328–1331

McManus J, Oppo D, Cullen J (1998) A 0.5-million-year record of millennial-scale climate variability, North Atlantic. Science 283(5404):971–974

Meehan RT, Warren WP (1999) The Boyne valley in the ice age. A field guide to some of the valley's most important glacial geological features. Meath County Council and Geological Survey of Ireland, Navan and Dublin

Moore MJ (1987) Archaeological inventory of county Meath. Stationery Office, Dublin

Monk M, Power O (2012) More than a grain of truth emerges from a rash of Corn-Drying Kilns? Archaeol Ireland 26:38–41

Monk M (1998) Early Medieval secular and ecclesiastical settlement in Munster. In: Monk MA, Sheehan J (ed) Early Medieval Munster—archaeology, history and society. Cork University Press, Cork, pp 33–52

Newman C (2005) Re-composing the archaeological landscape of Tara. In: Bhreathnach E (ed) The kingship and landscape of Tara. Four Courts Press, Dublin, pp 361–409

Newman C, O'Connell M, Dillon M, Molloy K (2007) Interpretation of charcoal and pollen data relating to a late Iron Age ritual site in eastern Ireland: a holistic approach. Vegetation Hist Archaeobotany 16:349–365

O'Connor K (1998) The archaeology of Medieval rural settlements in Ireland. Discovery Programs/Royal Irish Academy, Dublin

Ó Corráin D (2005) Aspects of society, c.800. In: O Croinin D (ed) A new history of Ireland, vol 1. Oxford University Press, Oxford, pp 549–608

O'Donovan P (1995) Archaeological inventory of county Cavan. Stationery Office, Dublin

O'Sullivan, A, McCormick F, Kerr TR, Harney L (2013) Early Medieval Ireland, AD 400–1100, The evidence from archaeological excavations, Royal Irish Academy Monographs

Poon J, Eldredge B, Yeung D (2004) Rank size of international financial centres. Int Reg Sci Rev 27 (4):411–430

Papageorgiou I, Baxter M, Cau M (2001) Model-based cluster analysis of artefact compositional data. Archaeometry 43:571–588

Patterson NT (1994) Cattle lords and clansmen: the social structure of Early Ireland. University of Notre Dame Press, London

Reeves-Smyth T (1983) Landscapes in paper: cartographic sources for Irish archaeology. In: Reeves-Smyth T, Hamond F (eds) Landscape archaeology in Ireland. B.A.R, Oxford, pp 119–177

Rice P, Saffer M (1982) Cluster analysis of mixed-level data: pottery provenience as an example. J Archaeol Sci 9:395–409

Simms A (1986) Continuity and change, settlement and society in Medieval Ireland c. 500–1500. In: Nolan W (ed) The shaping of Ireland: the geographical perspective. Mercier Press, Cork, pp 44–65

SPSS Inc (2004) Two step cluster analysis. Technical report, Chicago. http://support.spss.com/tech/stat/Algorithms/12.0/twostepcluster.pdf

Staeheli L, Clarke S (2003) The new politics of citizenship: structuring participation by household, work and identity. Urban Geogr 24:103–126

Stout M (1989) Ringforts in the southwest midlands. Unpublished undergraduate thesis, Trinity College, Dublin

Stout M (1992) Plans from plans: an analysis of the 1:2,500 OS series as a source for ringfort morphology. Proc Royal Irish Acad 97(c):37–53

Stout M (1997) The Irish Ringfort. Four Courts Press, Dublin

Stout M (2017) Early Medieval Ireland, 431–1169. Wordwell, Dublin

Sutton M (1998) Cluster analysis of paleofecal data sets: a test of late prehistoric settlement and subsistence patterns in the northern Coachella Valley. Calif Am Antiq 63(1):86–107

Warner R (2002) Ringfort. In: Connolly SJ (ed) The Oxford companion to Irish history. Oxford University Press, Oxford, pp 512–513

Westropp TJ (2000) Folklore of Clare: a folklore survey of county Clare and county Clare Folk-Tales and Myths. Ennis, Co. Clasp Press, Clare

Part IV

Place, Philology, History

Mapping Power: Using HGIS and Linked Open Data to Study Ancient Greek Garrison Communities

13

Ryan Horne

Abstract

From controlling cities within the Athenian Empire in the fifth-century BCE to maintaining isolated outposts on the border of the Parthian Empire in the second-century CE, the institution of the *phrourarchia* was a critical component of Greek civic and military identity. Despite its longevity and importance to the Greek world, the office has long been overlooked in scholarship, which has largely viewed the office as an isolated regional phenomenon without distinguishing between its local and imperial manifestations. There has also has been no definitive catalog of the institution or its commanders, or any attempt to show the full spatial extent of the institution in the Greek world. Until recent developments in digital gazetteers and Linked Open Data (LOD), identifying and mapping all of the *phrourarchia* in the Greek world was a nearly insurmountable task. However, the advent of the Pleiades project and the Linked Ancient World Data initiative has made such a project feasible. This article illustrates how new advances in HGIS and semantic web technologies has created a robust and expanding academic community and the development of best practices around the concept of sharing geospatial humanities data. After discussing the development of the ancient world LOD ecosystem, this article addresses how these resources were used to identify, locate, and study all of the garrison communities and commanders comprised the institution of the *phrourarchia* until the second-century CE. I discuss the creation of the first map to show all the *phrourarchia*, and how that task reveals that, although the institution was spread across the Greek world, the specific office of the *phrourarchos*, (plural *phrourarchoi*), or garrison commander, was primarily located in Egypt and south east Asia Minor. Mostly known through inscriptions, the presence of *phrourarchoi* signaled a complex interaction between imperial powers and local communities that later Greek and Roman historians often minimized or ignored. *Phrourarchoi* deployed by imperial powers were mostly found in subjugated communities and at the periphery of empire, and were almost unknown in imperial capitals. In contrast, *phrourarchoi* employed by smaller communities could be located anywhere from watchtowers at the edge of a city's territory to fortresses within the very heart of the community. I argue that the varying locations of garrisons and commanders, in addition to the different regulations that governed the office, reveals that imperial *phrourarchoi* were highly specialized individuals who supported the imperial administration of their employers

R. Horne (✉)
History Department, University of Pittsburgh, Pittsburgh, PA, USA
e-mail: rmh74@pitt.edu

C. Travis et al. (eds.), *Historical Geography, GIScience and Textual Analysis*, Historical Geography and Geosciences, https://doi.org/10.1007/978-3-030-37569-0_13

through a blend of intentionally vague civic and military responsibilities. In contrast, smaller communities used *phrourarchoi* who were largely amateurs, and had clear limits on their purely military authority. Following my analysis of the *phrourarchia* and how GIS methodologies aided my investigation, I close the article with a discussion of some of the shortfalls of current GIS approaches (A preliminary digital project based on this investigation is available here: http://awmc.unc.edu/awmc/applications/snagg/).

Keywords
Linked open data (LOD) · *Phrourarchia* · Pleiades project · Linked ancient world data

13.1 Introduction

In 198 BCE, the second Macedonian war entered its second year. The Roman general Titus Quinctius Flamininus and his Greek allies met at Nicaea with their enemy, the Macedonian monarch Philip V, to discuss peace terms for ending the conflict. The talks soon stalled due to demands from Greek representatives that Philip needed to immediately remove his *phrourai* (singular *phroura*, garrisons) from Greece. Philip, convinced by Flamininus that the issue was best decided by the Roman senate, agreed to send representative to Rome. The Greek delegation pleaded so convincingly that the Roman senate immediately demanded to know from Philip's envoys if the monarch would relinquish his *phrourai*. After hearing the envoys reply that they had not received any specific instructions on that point (likely because Philip wished to keep the garrisons), the senate immediately terminated the talks and prepared to continue the war.[1]

Garrisons were not only a locus of raw physical and political power, but were also a nexus of often competing rhetorical, intellectual, and social authority. *Phrourai* and the *phrourarchoi* (garrison commander; singular *phrourarchos*) who commanded them were one of the primary sites of interaction between imperial powers and subject communities in the ancient world.[2] From controlling *poleis* (singular *polis*, city-state) within the Athenian Empire in the fifth-century BCE to maintaining isolated outposts on the border of the Parthian Empire in the second-century CE, *phrourarchoi* played a critical role in Greek civic and military identity. Despite its longevity and importance to the Greek world, the *phrourarchia* (office of the *phrourarchos*) has long been overlooked in scholarship, which has largely viewed it as an isolated regional phenomenon, or has not distinguished between the local and imperial manifestations of the office (Kortenbeutel 1941; Labarre 2004, pp. 237–244; Launey 1987, pp. 633–675). There has also has been no definitive catalog of the institution or its commanders, or any attempt to show the full spatial extent of the institution in the Greek world.

Until recent developments in GIS, digital gazetteers, and Linked Open Data (LOD), identifying, mapping, and studying all *phrourarchoi* in the Greek world was a nearly insurmountable task. However, the advent of projects like Pleiades has made such a task feasible. This chapter, derived from my dissertation *Imperial Power and Local Autonomy in Greek Garrison Communities: The Phrourarchia and the Polis*, illustrates how new advances in HGIS and semantic web technologies created the infrastructure to study the critically important office of the *phrourarchia* (Horne 2015). Through the use of GIS tools and digital gazetteers, I argue that the *phrourarchia* had two primary manifestations. The imperial version of the office blended military powers with ambiguously defined civil authority, and functioned as a method of domination over subject communities. On the other hand, quasi-free and autonomous communities placed strict regulations on their own *phrourarchoi*, who were symbolic defenders of local freedom and autonomy.

[1]*Polyb.* 18.2–18.11.

[2]For some examples see *IG* XII, 1, 900; *Philae* 15; *Thèbes à Syène* 322, 242; Diod. Sic. 14.4.4; 14.53.5; 18.18.5; 18.37.4; 19.86.2; 20.45.2; Plut. *Phoc.* 31.1.

13.1.1 A Gazetteer of Political Power

The creation of a digital gazetteer of garrisons was essential to this study. While much has been said concerning the suitability and application of GIS software to historical research, digital gazetteers have received far less attention (Gregory and Ell 2007; Knowles 2014; Knowles and Hillier 2008; Plewe 2002). Traditionally seen by many scholars as an adjunct product of visual cartography, gazetteers in their most basic form are lists of place names, often derived from or otherwise related to maps. Many print gazetteers serve as indexes, where a sorted list of place names refers to a map grid or other system to locate a place on a map. Some gazetteers also offered additional information and categorization for each entry, such as the location of a place using a standard coordinate system, alternative names, a reference to external scholarly work, or cross-references to other entries (Berman et al. 2016, p. 2; Southall et al. 2011, pp. 127–128). Digital gazetteers build upon this expanded use of their print counterparts. Freed from the constraints of a print medium, digital gazetteers can offer a nearly limitless variety of organizational schemes, additional place data, modeled relationships, and other features. Digital gazetteers are also increasingly pushing the boundaries of what can be considered geographic studies, with some examples almost entirely focusing on relationships between conceptual entities instead of geospatial linkages.[3]

As digital gazetteers focus on place and space, it is worth briefly unpacking these concepts and how they relate to digital systems. Place and space have been the subjects of intense debate and research in the fields of geography and spatial humanities for decades, as they touch upon issues of human perspectives and experiences which are not easily captured by a quantitative approach. Scholars such as Yi-Fu Tuan and Doreen Massey have theorized that place is something created by human experience, or more generally the intersection of relationships between entities at specific times, while space is the medium in which those relationships and experiences happen (Massey 2015, pp. 10, 130; Tuan 1977).

In this view a place is not so much a point on a map as a collection of meanings associated with the same conceptual entity. There can be many different names, definitions, hierarchies, and conceptions of a place entity; the contentious nature of the concept of "western civilization", including its constituent countries, its role in global affairs, and even its very existence is a perfect example. Even if there can be no agreement on any aspect of "western civilization", it can still be agreed that it is a *conceptual* place which has attested linkages to other places, some of which have a geospatial footprint and others which may only be mental constructions. Despite the potential qualitative "fuzziness" of place, there has been some preliminary work to more fully represent the conceptual difficulties of place within a GIS environment, although these efforts are largely experimental (Giordano and Cole 2018, pp. 671–674).

Due to the limitations of traditional GIS approaches to express the multiplicity of meanings surrounding place, the gazetteer model described above was adopted for this study. As archaeological evidence is largely insufficient to associate extant physical remains with specific garrisons, the data for this project is largely based on textual attestations. Given the expansive variety ancient Greek technical terms for different types of fortifications and commanders, this investigation is concerned only with a limited set of vocabulary related to the verb φρουρέω (*phroureō*), to watch, which can be associated with specific empires and poleis. The term *phroura* (plural *phrourai*) generally referred to a garrison in an abstract sense, while *phrourion* (plural *phrourioi*) was mostly used to describe a physical fortress or watchtower. Finally *phrouros* (plural *phrouroi*) was most often used to describe the men who guarded a *phroura* or *phrourion*. These terms all had a strong association with observation within a military setting (*LSJ* s.v. φρουρά, φρουραρχέω, φρουρέω, φρουρός; Nankov 2009, p. 4).

[3]See *PeriodO*, a self-described "…gazetteer of scholarly definitions of historical, art-historical, and archaeological periods" at http://perio.do/en/.

Using this limited vocabulary, I created a list of attestations of specific terms and their references. These attestations were then disambiguated and associated with individual places and commanders and placed into a relational database. This process was largely organized on a semantic model, where information about a place is not based on its precise location on a map, but instead on relationships, in this case expressed through text. Different garrisons could be cataloged in the gazetteer based on their connection to other entities.

This project focused on semantic relationships as they are a critical component of many digital gazetteers. Instead of tying entries to a specific coordinate system, a gazetteer based on semantic relationships associates different entities together, often using a fixed vocabulary. For instance, the *Getty Art & Architecture Thesaurus* (AAT) offers an extensive vocabulary of possible connections, including political, conceptual, and spatial relationships.[4] Using such a vocabulary, a digital gazetteer can express relationships between entities that are machine actionable and understandable to other researchers and digital projects, which closely mirrors LOD systems (Bizer et al. 2009, p. 2).

13.1.2 Gazetteers, LOD, and GIS

Many humanities data sets are derived from textual sources which offer little geospatial data, which severely complicates the use of GIS approaches (Berman et al. 2016, p. 4). Fortunately, advances in GIS science and LOD systems have continued to improve the process of reconciling gazetteers with geospatial information. The main goals of LOD were outlined by Tim Berners-Lee, and include the creation of data that is available on the web at an unchanging URI, is machine readable, uses non-proprietary data formats, released under open standards, and has links to content from other data providers (Berners-Lee 2007).

Underlying most LOD implementations is the Resource Description Framework (RDF). RDF models statements about resources in a *triple* format that follows the pattern of *subject–predicate–object*. For example, I could say that Nikanor (*subject*) was *phrourarchos* at (*predicate*) Athens (*object*). Much like the semantic relationships described above for gazetteers, the different components of a RDF triple can be drawn from a fixed vocabulary; indeed there are an increasing number of standards and ontologies for describing LOD from different expert communities.[5]

Spatial studies and LOD are two separate areas of research, each with their own standards and communities. However, there is an increasing amount of spatially enriched data in the LOD cloud, and some attempts by different groups to standardize the expression of geospatial data through RDF technologies (Iwaniak et al. 2016, pp. 323–324). A robust digital gazetteer ecosystem has developed around the use of common data formats to publish and share open-licensed geospatial data, like *Pleiades*, *GeoNames*, and the forthcoming *World History Gazetteer* (WHG).[6] As many of these projects and standards are based on common file formats that can be read and manipulated by GIS software, LOD design principles are increasingly being adopted by spatial studies.

For this project, alignment with *Pleiades* was essential to its success. *Pleiades* is a collection of authoritative information concerning over 36,000 places, names, locations, and geospatial data in the ancient Mediterranean world (Fig. 13.1).[7] Currently maintained and edited by volunteers, *Pleiades* creates stable URIs for each place in the data set, and provides a user interface, API, and direct download that conforms LOD principles. Due to its open nature, *Pleiades* has become a cornerstone of the Linked Ancient World Data

[4] *Getty Art and Architecture Thesaurus* http://www.getty.edu/research/tools/vocabularies/aat/index.html.

[5] See https://github.com/lawdi/LAWD, http://nomisma.org/ontology for some examples in the field of ancient studies.

[6] https://pleiades.stoa.org/; http://www.geonames.org/; http://commons.pelagios.org/; http://whgazetteer.org/.

[7] *Pleiades* https://pleiades.stoa.org/.

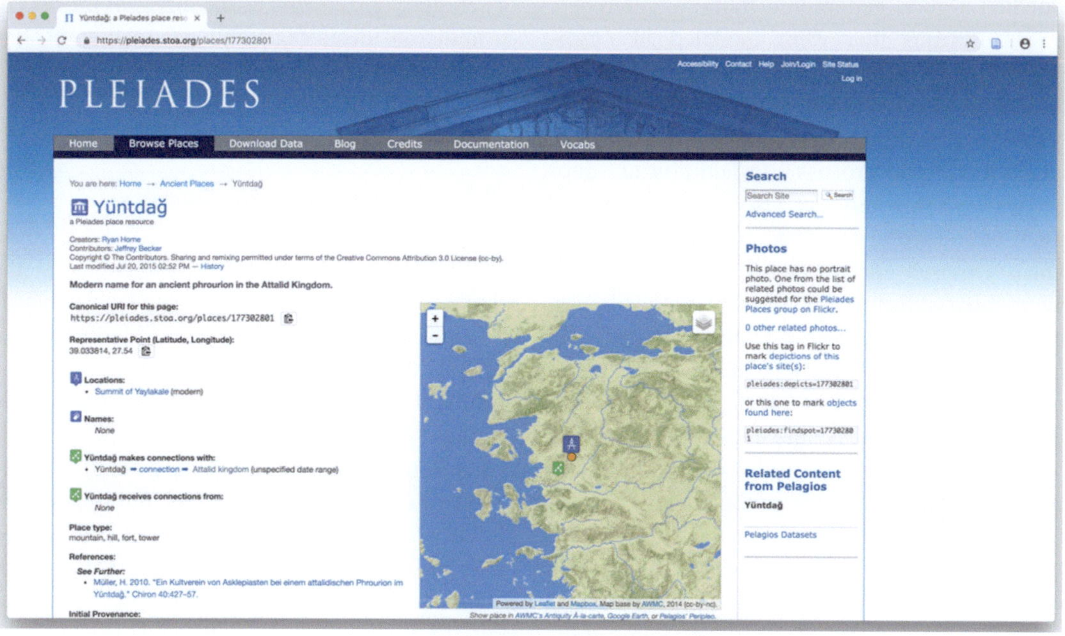

Fig. 13.1 *Pleiades* place page for Yüntdag. (https://pleiades.stoa.org/place/177302801)

community, with its URIs used in numerous projects (Fig. 13.2).

Given its coverage, it was a trivial matter to associate many of my gazetteer entries with *Pleiades* IDs. For places that were not in the Pleiades data set, the process of using LOD resources more involved. For some places there was a rich trail of excavation reports which

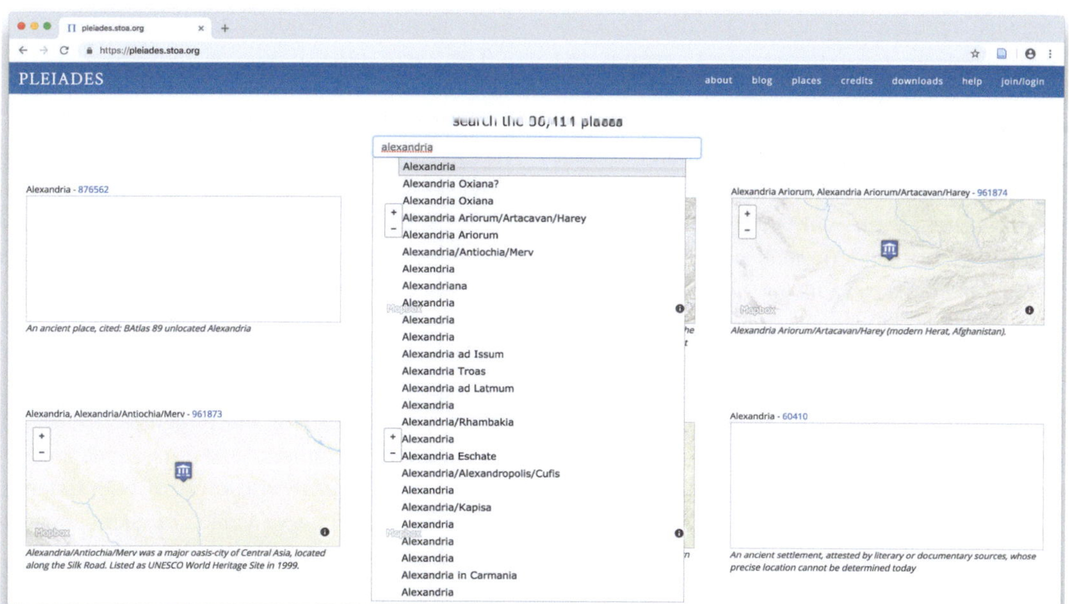

Fig. 13.2 *Pleiades* search interface

provided a good geographic frame of reference, or extant ruins that were modeled in *OpenStreetMap*, *GeoNames*, or *Wikipedia*; this information was used to create new place resources in *Pleiades* which were submitted through the standard editorial process.[8] In addition to editorial review, this process also allowed for the micro-publication of that place information which exposed the place to the larger LOD community. After the process of reconciliation and disambiguation, the garrisons gazetteer/database inherited all of the spatial information present in *Pleiades* and other geospatial providers. As the database is based on unchanging URIs and well-documented APIs, a simple python script can automatically refresh the database, gazetteer, and maps with current data in a manner of seconds instead of the hours it would take to do so by hand.

Despite its advantages, there are some difficulties with using LOD resources. The vast majority of places in *Pleiades* are centers of geometry generated automatically by GIS software. Where possible these representative points are associated with extant remains by the *Pleiades* editorial board; where this proves impossible points are derived from excavation reports or modern locations, which can give a somewhat modern bias to the results. In cases where there are no extant ancient remains the location is even less tethered to a precise location. As a result, it is not currently advisable to use the data provided by *Pleiades* and similar gazetteers to perform spatial analysis at a micro or local level. Even with these difficulties, the process of creating any such gazetteer or database that is linked to LOD geospatial resources is itself a product of intense scholarship. As Stephen Ramsay notes, the act of creating a database is "… often as vital to the experiential meaning of the scholarly endeavor as the use of the final product" (Ramsay 2004, p. 178). Ramsay's assessment is especially pertinent to the creation of scholarly gazetteers, as the different process of compilation, disambiguation, reconciliation, and publication are all significant scholarly endeavors in their own right.

13.2 Examining the Results

This investigation, given the limitations of the source material, focuses on macro-level analysis of historical trends in the ancient Mediterranean along with the creation of maps that provide a reasonably precise, yet still impressionistic, view of ancient geospatial data. By doing so, it is possible to discuss some of the general features of the *phrourarchia*, its spread throughout the Greek world, and its impact on later authors and historians in the Roman Empire. Visualizing the footprint of *phrourarchoi* and *phrourai* (Fig. 13.3) shows garrisons throughout the Greek areas of the Mediterranean, with literary sources placing them in areas that historically had minimal, if nonexistent, Greek presence.[9] This is largely an artifact of later authors under the Roman empire who used Greek terminology to describe Roman institutions; removing these later sources centers the *phrourarchoi* and *phrourai* in the eastern Mediterranean, but still covering nearly all areas of Greek society (Fig. 13.4).

The most intense use of *phrourarchoi* and *phrourai* was in the Hellenistic period ($\sim$323 BCE to $\sim$31 BCE); the famous "Successors" of Alexander the Great used *phrourarchoi* to help secure empires based on the ideology of "spear-won land", or territory which was the personal property of the monarch due to conquest (Billows 1990, p. 135; Holt 1999, p. 125; Mehl 1980, pp. 173–212).

The Successors did not just use *phrourarchoi* to police their frontiers; *phrourarchoi* are known from the very heart of the Ptolemaic, Seleucid, and Antigonid empires, and quasi-independent poleis used *phrourarchoi* to guard both external fortresses and the very center of their communities (Figs. 13.5 and 13.6).

[8]In the interests of full disclosure, I currently serve as an editor for *Pleiades*; however, all of my additions are subject to the same peer-review process as all other contributors.

[9]Arr. Anab. 5.24.8; 6.17.1; Plut. Vit. Alex. 57.3.

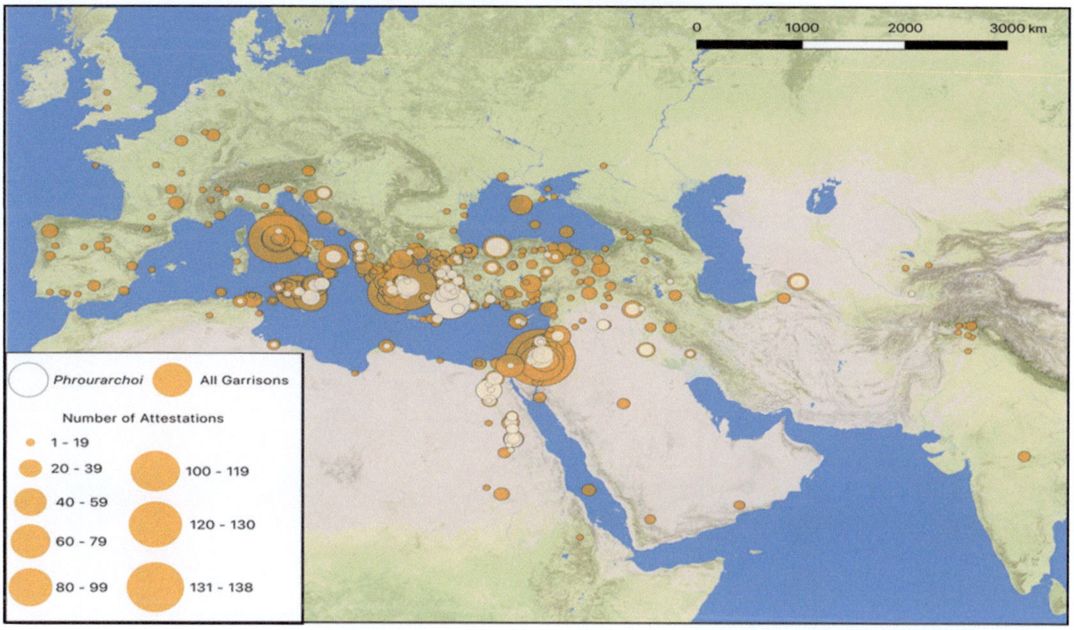

Fig. 13.3 All *phrourarchoi* and all garrisons

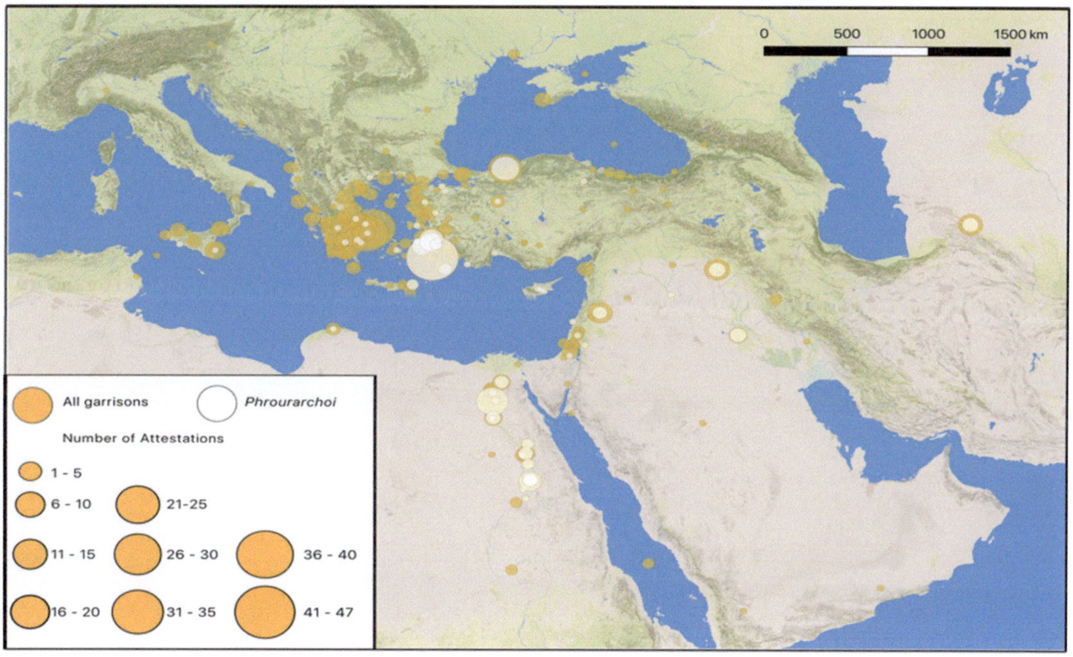

Fig. 13.4 *Phrourarchoi* and garrisons from Hellenistic sources

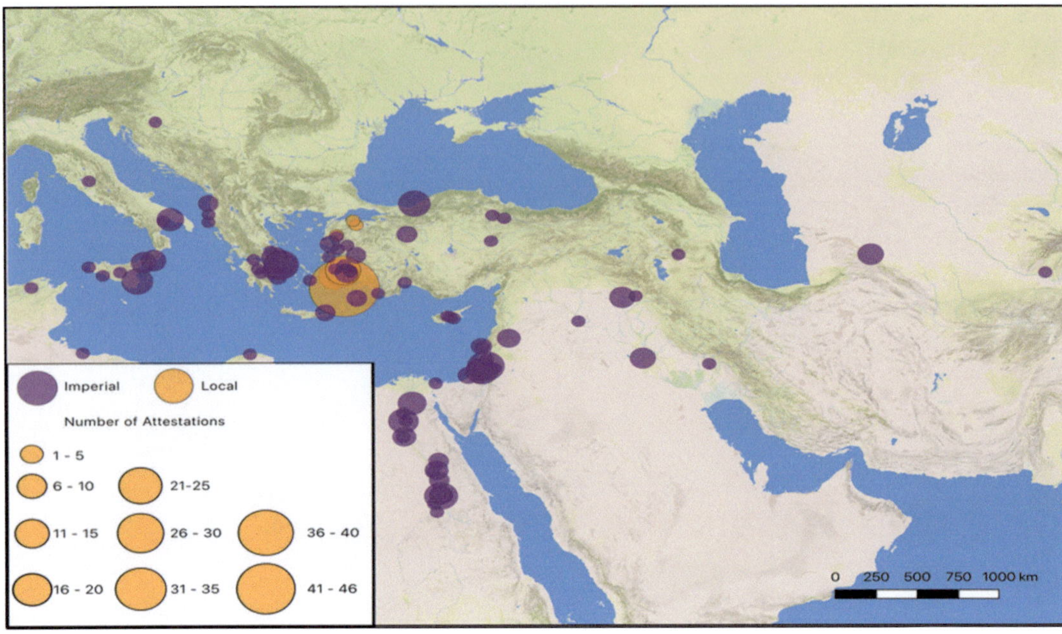

Fig. 13.5 Hellenistic imperial and local *phrourarchoi*

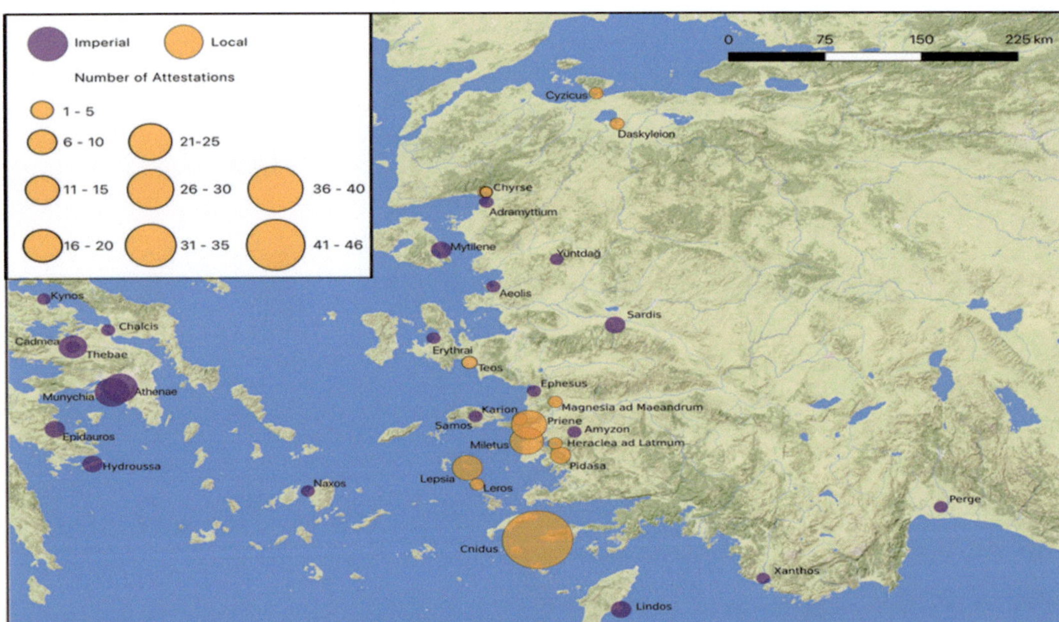

Fig. 13.6 Imperial and local *phrourarchoi* on the coast of Asia Minor

Although quantifying the evidence is difficult due to the gaps in our knowledge and the uneven survival of sources, some interesting patterns are present in the data. This investigation found 2,895 uses of *phrourarchoi*, *phrourai*, *phrouroi*, and their variants. Of these, 2,434 (~84%) could be associated with a mappable location; the remainder were primarily used rhetorically. Of the locatable garrisons, only 11 different commanding offices were attested more than four times, while the vast majority of attested garrisons do not mention a commanding officer (Tables 13.1 and 13.2).

Even if their commanders are unknown, it is obvious that garrison played an important part in Greek civic and military life. Despite this ubiquity, the cost of garrisons and *phrourarchoi* was high. A Third century BCE inscription from Teos helps to place this in perspective:

> …the *phrourarchos* will have no less than twenty of the citizens as *phrouroi* and three dogs; the polis, purchasing the dogs, will hand them over to the *phrourarchos*; the *phrourarchos* will maintain the dogs… The pay shall be given to him [the *phrourarchos*], according to the laws, every four months by the *tamiai* (treasurers) whenever he sets out for the chorion (land); the pay for the *phrourarchos* shall be four Alexander *drachmas* [per day], and for each of the *phrouroi* one Alexander *drachma* [per day].[10]

The cost to Teos came to at least 8,760 *drachmas* a year to maintain a single *phrourion* and its small complement of 21 men. This total does not account for the cost of the fortification itself, provisions, or maintenance; to put such costs in perspective, the construction of a single tower in the Hellenistic period could cost more than 200,000 drachmas (Chaniotis 2005, p. 116). Quantifying the economic impact on Teos of maintaining the *phrourion* is difficult, but it has been argued that the expense was typical (Fischer-Bovet 2014, pp. 73–74; Robert and Robert 1976, p. 216). Although the economy of Teos remains a mystery, Miletus is better understood, and the *polis* is known to have had at least four active *phrourarchoi* in the Hellenistic period.[11] Using Teos as a guide, the cost for maintaining four *phrourai*, each with a complement of no more than 21 men, would require an outlay of 52,560 *drachmas*, or nearly six talents a year.

Table 13.1 Word attestations

Type	Count
phroura	1053
phrourion	824
unspecified	280
phrouros	136
phroureo (verb)	101
phrouroi	38
phrourion (?)	9
polis	7
phroura (?)	5

Table 13.2 Number of commander attestations

Commander	Count
unspecified	1908
phrourarchos	299
strategos	45
hegemon	44
harmost	26
archon	23
archiphrouros	16
polemarchos	7
harmostes; dekadarchiai	7
epi	5
epi phroura	5
phrourarchos (?)	4

[10]Robert and Robert 1976, 153–235. ll. 18–21; 27–31: "διδόναι | δὲ αὐτῶι τὸμ μισθὸν τὸν [ἐκ τῶν νόμων ἑκάστου τε]τραμήνου τοὺς ταμ[ίας ἐ]|πάναγκον ὅταν πορεύηται [εἰ]ς τὸ [χωρίον· μ]ισθὸν δὲ εἶναι τῶ[ι μὲν] | φρουράρχωι τεσσέρας δραχμ[ὰς] ἀ[λεξ]ανδρε[ίας,] τῶν δὲ φρουρῶ[ν] | ἑκάστωι δραχμὴν ἀλε[ξ]ανδρ[είαν μίαν·… φρ[ου] | ροὺς δ' ἔχειν τὸ[μ] φρο[ύραρχον μ]ὴ ἐλάττους ἢ εἴκοσι τῶμ πολιτῶν καὶ | κυνὰς τρεῖς· τοὺς δὲ κυνὰς [π]ριαμένην παραδοῦναι τῶι φρουράρχω[ι] | τὴμ πόλιν· τρέφειν [δὲ] τοὺς κυνὰς τὸμ [φ]ρο ύραρχον…".

[11]*Staatsverträge* III 539I = *I. Mylasa* II T51.A; *SIG*3 588; *SIG*3 633 = *SEG* 34.1173 = *SEG* 37.984; Manganaro 1963 #18; Sakkelion 1890 col. 221 = Manganaro 1963, #22B; Sakkelion 1862 col. 265-266 = Brent and Gardner 1886, 144 = *SEG* 18, 388 = Manganaro 1963, #21B.

The coinage of Miletus helps to place such costs in perspective. Over two production periods, c. 225-195 BCE and c. 175-86 BCE, the Milesian mint made the equivalent of 1,190,000 drachmas (Ashton and Kinns 2003, pp. 16–26; Kinns 1998, pp. 175–183). Although production was certainly not uniform over this period, seeming to have ceased altogether for two decades, this averages to the equivalent production of ~8,561 drachmas per year, an amount that is even lower than the pay of a single garrison of 21 soldiers and one *phrourarchos*. Even though minting coinage is not the entirety of all economic activity or resources, the cost of *phrourarchoi* to an autonomous *polis* was staggering.

For imperial powers the cost of *phrourarchoi* was also burdensome. It is difficult to arrive at an exact estimate for the tax revenues of Hellenistic empires; modern scholars have estimated between 84,000,000 and 114,000,000 *drachmas* of total revenues for the Seleucid Kingdom and a comparable number for Ptolemaic Egypt. Using Teos as a rough guide, some scholars have estimated a total cost for all of the garrison forces in the Ptolemaic empire at 7,200,000 *drachma* a year, equivalent to between ~6.3 and ~8.6% of total revenues (Fischer-Bovet 2014, pp. 67–74). The outlay was hardly less for other Hellenistic kingdoms. Strombichos, who was phrourarchos in Arcadian Orchomenus under Polyperchon, commanded at least 2,000 mercenaries, who would require a minimum pay of 731,460 *drachmas* a year. The maintenance of just ten *phrourai* on this scale would equal the entire estimated cost of all Ptolemaic garrisons. Indeed, the high costs of garrisons were so well known that it was the basis of a proverbial choice: "To get rich or to garrison".[12]

Despite the near ubiquity of *phrourai* and the prohibitive expense of maintaining *phrourarchoi* and garrisons, local communities and imperial powers gained little in military effectiveness in return. *Phrourarchoi* who successfully withstood assault are exceedingly rare, even though many of them refused to surrender when faced with enemy forces.[13] This is hardly surprising given the scale of Hellenistic armies. Even relatively small monarchies like the Attalid Kingdom could field an army up to 12,000 strong, while larger empires, such as the Seleucids and the Ptolemies, could generally field armies of ~30,000, with up to 60,000 in emergencies (Bar-Kochva 1976, pp. 7–8; Ma 2013, pp. 59–60; Roth 2007, p. 379). Although it has been argued that one of the primary duties of a *phrourarchos* was to protect the territorial possessions of a *polis* against external threats, against these numbers a *phrourion* like that established by Teos would be little more than an annoyance (Dmitriev 2005, p. 19; Labarre 2004, pp. 221–222).

Given the ubiquity of *phrourarchoi*, the expense of the office, and their limited military efficacy, I contend that *phrourarchoi* were more concerned with internal security than external enemies, and the different values and assumptions of power systems in the Hellenistic world shaped and radically altered the institution of the *phrourarchia* to fit particular needs. When *phrourarchoi* were employed by an imperial power, they maintained that power's control, often to the detriment of the subject community. In contrast, garrisons maintained by independent poleis served as physical and symbolic security for freedom and independence, and the local *phrourarchoi* who commanded them were expected to adhere strictly to the laws of the community.

13.2.1 Imperial *Phrourarchoi*: Ambiguity in Power and Space

Imperial *phrourarchoi* were instrumental components of the delicate balancing act between Hellenistic monarchs and subject communities. Even though some scholars view Hellenistic governments as little more than tools of imperial exploitation, kings still had to take into account

[12]*Suda* s.v. Φρουρεῖν ἢ πλουτεῖν: "Φρουρεῖν ἢ πλουτεῖν…".

[13]*SEG* 4.671 ll.1–6; Diod. Sic. 12.65.9; Polyaenus, *Strat.* 3.9.48, Arr. *Anab.* 4.16.5, Diod. Sic. 20.45.

the internal politics and desires of supposedly subject cities in their kingdoms, lest they face internal revolt and the loss of revenue and communications (Green 1990, p. xv). This concern led the monarchies to exercise a limited hegemony that required negotiation and appeasement as much as raw might (Lévêque 1999, p. 361; Ma 1999, pp. 179–242; Shipley 2000, p. 59). Hellenistic monarchies turned to *phrourarchoi* as a method to control and dominate their spear-won land, suppress local dissent, and even to soften the image of their power.

Many imperial *phrourarchoi* in the Hellenistic period were mercenaries or drawn from the inner circle of the monarch's *philoi* (friends) and had little connection with their assignments (Chaniotis 2002, p. 100; Parke 1933, p. 1.)[14] As a rule, such men were motivated more by profit than a sense of responsibility to a local community or *polis*. Thus, a symbiotic relationship developed in which garrisons and commanders increasingly looked to Hellenistic monarchs to provide and defend their social and economic status, while those same monarchs depended on *phrourarchoi* to protect the integrity of their kingdoms. Competition for quality mercenaries was fierce; the Ptolemies even patronized poets who wrote what were effectively advertisements for the benefits of mercenary service in Egypt.[15] *Phrourarchoi* were not necessarily tied to a single location, and they could be called on to serve in the navy, hold multiple *phrourarchiai*, or to move between *poleis* as needed.[16] Isolated fortresses could also fall under the administration of a *phrourarchos* and were critical components of the system of observation, control, and exploitation practiced by imperial powers. There are no surviving regulations that outline any limits to the authority of a *phrourarchos* over a subject community. The mercenary *phrourarchoi* of the Successor kingdoms were not bound by local laws or any clear constitutional limits to their authority and indeed served as a break on local autonomy.[17] Antipater's treatment of Athens in 322 BCE is typical; the polis was compelled by him

> …to accept Menyllus as *phrourarchos* and [accept] a *phroura*, its purpose being to keep anyone from making revolutionary changes.[18]

The Athenian experience vividly shows that an imperial *phrourarchos* was capable of, and almost certainly expected to, interfere with local politics for the benefit of his employer. Beyond high level politics, *phrourarchoi* could also be expected to dispense royal justice for even mundane offenses. Hellenistic Seleucid and Egyptian *phrourarchoi* judged property disputes, intervened in the business dealings of wine sellers, and even adjudicated assault cases between brothers.[19] Despite such involvement, imperial *phrourarchoi* never lost their military character, served for the express benefit of their employers, and were not subject to the laws of their subject communities.

13.2.2 Local *Phrourarchoi*: Restricted Space and Power

The amateur *phrourarchoi* in the quasi-independent communities of Asia Minor dramatically differed from their imperial counterparts. In the multi-polar Hellenistic world, small poleis were unable to compete economically or militarily with the major imperial polities that surrounded them. Even so, strong fortifications and garrisons, devoid of external influence, were viewed as a fundamental requirement of a free and autonomous community.[20] Unlike the *phrourarchoi* employed by Hellenistic monarchies, *phrourarchoi* from these communities

[14] Plut. Eum. 3.7; AbhBerlin (1937.6) 3–63; Philae 13; Thèbes à Syène 242, 243, 318, 320; IC III iv 14; OGIS 111.

[15] Theoc. *Id.* 14.

[16] *SEG* 31.1348; See also *Thèbes à Syène* 320 where in 116 BCE a *phrourarchos* named Nestor commanded the garrisons of Syene, Elephantine and Philae.

[17] *SEG* 31.5.

[18] Diod. Sic. 18.18.5: "…φρούραρχον δὲ Μένυλλον καὶ φρουρὰν ἠναγκάσθησαν δέξασθαι τὴν οὐκ ἐπιτρέψουσαν οὐδενὶ νεωτερίζειν…".

[19] *IG* XII, 6 1:11 = *AM* 1919, 25–29 #13 = *SEG* 1 366; *P. Diosk.* 8; *P. Diosk.* 7.

[20] Polyb., 18.11.5–10, discussed above, is a striking example of this.

were amateurs who were largely drawn from the local population. These offices were highly regulated with clear constitutional limits on their authority.

After choosing a *phrourarchos*, most *poleis* took great pains to ensure that he remained in his assignment. In inscriptions celebrating past *phrourarhcoi*, the *polis* of Priene stressed that these officers remained in the fortifications for the duration of their assignments according to the law, and inscriptions from Milesian emphasized the importance of remaining physically at an assignment.[21] Penalties for failing to remain at a fortress, or conversely holding it too long, were harsh, including complete property confiscation, exile, and death.[22]

There is no surviving evidence to suggest that *phrourarchoi* from independent *poleis* exercised any form of civic power in the *polis* itself. Even after Roman expansion in the east, the local *phrourarchiai* of Asia Minor retained a consistent form: they were established and selected by the demos of the community, only exercised military authority, and remained subordinate to the laws of the *polis*. The location of these *poleis* on the borderlands of the Hellenistic world assured the continued survival of the institution, even as a symbolic reminder that these communities were free from foreign occupation.

13.2.3 A Question of Location

As we have explored the cost of garrisons and the difference between imperial and local commanders, the *phrourarchoi* on the southwestern coast of Anatolia in the regions of Caria and Ionia is worth examining further. This was a highly militarized and hotly contested area between the Ptolemaic, Seleucid, and Attalid kingdoms. After the treaty of Apamea in 188 BCE the southern part of the region was nominally under the Rhode's sphere of influence, although many of the communities were in effect autonomous. The north of the region was placed under control of the Attalids; Rome annexed this after the dissolution of the Attalid kingdom in 133 BCE and likely incorporated the south in 85 BCE following the wars against Mithridates (Allen 1983, pp. 38–39; LaBuff 2016, pp. 42–43; Ma 1999, pp. 42–45). Given such volatility one may expect to find a number of imperial *phrourarchoi* securing the interests of imperial powers, but instead the opposite appears to be true. This could be an artifact of the surviving evidence, but even literary sources make scant mention of imperial *phrourarchoi* in the region.

There is a possible explanation for this phenomenon. Trade and profit, along with secure networks of communication, were critical to the administration of spear-won land. Securing reliable and predictable transportation between the often far-flung reaches of Hellenistic empires was a critical and difficult task, one that was exacerbated by the myriad difficulties of pre-modern extended travel. Given this background, there have been some attempts to model transportation times and costs in the ancient Mediterranean, notably the *ORBIS* project at Stanford.[23] By combining literary evidence with elevation, wind, ocean currents, and other data, *ORBIS* attempts to model a variety of different routes between ancient places, modeling different seasons, transportation types, and costs, albeit on an abstract and somewhat impressionistic level. Using the *ORBIS* application and data it is possible to model the least cost, fastest, or shortest paths between a select number of ancient places. Using route and cost information from *ORBIS*, combined with the pgRouting library and place data from the *Pleiades* project, this investigation created a least-cost map of connections between the constituent places in the Attalid and Seleucid kingdoms, who were major powers in Caria.[24]

When visualized, the results are somewhat surprising. Shown at the height of their

[21] *I. Priene* 21; *I. Priene* 22; *SEG* 19.678 = *SEG* 37.984 & 987.

[22] Robert and Robert (1976) ll. 22–27.

[23] *ORBIS*, http://orbis.stanford.edu/.

[24] Rhodes and the Ptolemies were left out of this analysis for the simple reason that the only connections between the region and the rest of their respective polities were direct sea routs with little to no connection to other paths.

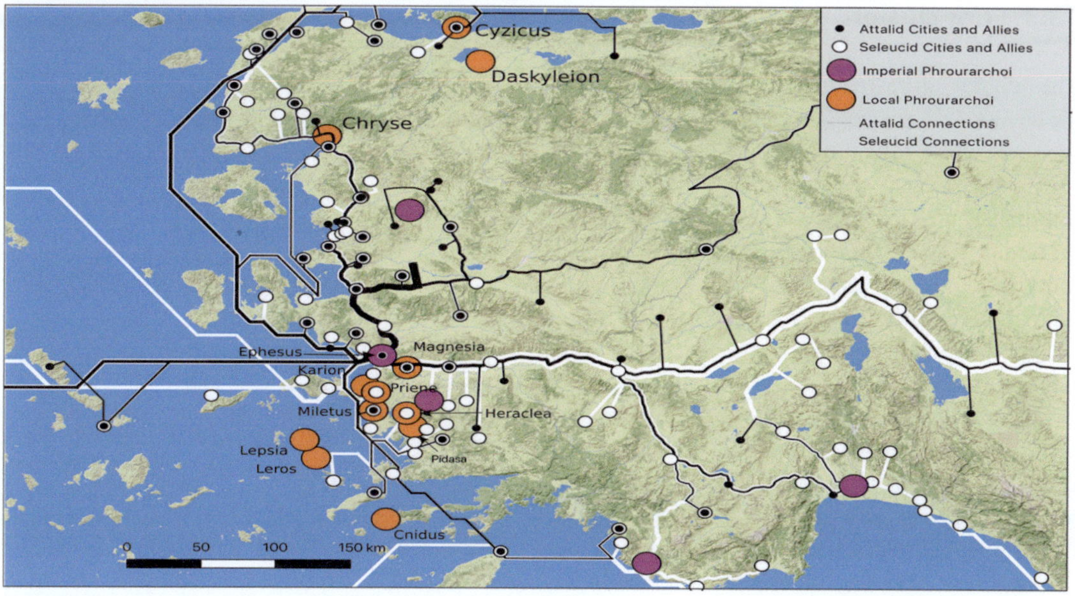

Fig. 13.7 *ORBIS* and *Pleiades* Least Cost Option Path between the constituent places in the Attalid and Seleucid kingdoms

respective kingdoms, the least-cost paths of each known affiliated or allied city to the capitals of the kingdoms are shown. Almost none of the local *phrourarchoi* are known to operate along a major branch of these paths; instead they are located mostly on "dead ends" that are not connected to any other communities. The only exception is Magnesia, where the position of the *phrourarchos* is attested in the third-century BCE. Little is known about the office, but the history of Magnesia is better understood. The city was a Seleucid possession from 281 BCE, fell under Attalid control in 220 s BCE, but was reconquered by the Seleucids no later than 201/After 188 BCE it seems that the Romans granted the city its autonomy, after which it remained independent until its eventual annexation by Rome (Allen 1983, pp. 98–99; Grainger 1997, s.v. Magnesia-on-the-Maiandros).

The relative ease and frequency of Magnesia changing hands between competing powers is yet another example of the relative ease in which local garrisons could be overcome by external forces. Combined with the colossal expense of garrisoning, and with the key port of Ephesus already garrisoned and protecting the same route, it may have been cost-prohibitive to replace a local manifestation of the office with an imperial one. It may have made more fiscal sense to let Magnesia retain its own *phrourarchoi* for internal security and low-intensity conflicts while concentrating on the far more important city of Ephesus to secure communications across the Aegean.

This is, of course, a highly speculative analysis supported by comparative literary evidence, abstract models, and basic quantification. Due to the scant survival of ancient evidence, any such study on the networks of communication and interaction will be by necessity abstract; however, the convergence of new GIS systems, digital gazetteers, and the continued embrace of LOD publication by scholars is creating an exciting new digital ecosystem for exploring spatial humanities. No longer the sole domain of specialists in geography or geospatial humanities, new tools and methodologies are increasing access and usability for non-technical specialists, presenting exciting new collaborative opportunities, and pushing traditional disciplinary boundaries (Fig. 13.7).

13.3 Conclusion

The digital infrastructure surrounding digital gazetteers, GIS science, and LOD has matured to the stage where different aggregations of data can now be leveraged to explore specific research questions and create specialist projects without requiring extensive expertise or training in GIS methodologies or semantic web technologies. Although significant care and caution still need to be exercised to ensure the accuracy and integrity of data alignment, new tools produced by the LOD GIS community are making this process accessible to a far wider audience. As a result, a rapidly increasing number of researchers and students are now able to create gazetteers linked with geospatial resources, thereby creating exciting new avenues for exploration, research, publication, and pedagogy.

References

Allen RE (1983) The Attalid Kingdom: a constitutional history. Clarendon Press, Oxford

Ashton R, Kinns P (2003) Opuscula Anatolica II. Numis Chron 163:1–48

Bar-Kochva B (1976) The Seleucid army: organization and tactics in the Great Campaigns. Cambridge University Press

Berman ML, Mostern R, Southall H (2016) Introduction. In: Berman ML, Mostern R, Southall H (eds) Placing names. Enriching and integrating gazetteers. Indiana University Press, Bloomington, pp 1–11

Berners-Lee T (2007) Linked Data [WWW Document]. W3 Des. Issues. http://www.w3.org/DesignIssues/LinkedData.html

Billows RA (1990) Antigonos the one-eyed and the creation of the Hellenistic state. University of California Press, Berkeley

Bizer C, Heath T, Berners-Lee T (2009) Linked data—the story so far. Int J Semantic Web Inf Syst 5:1–22

Chaniotis A (2005) War in the Hellenistic world: a social and cultural history. Blackwell Pub., Malden

Chaniotis A (2002) Foreign soldiers—native girls? Constructing and crossing boundaries in Hellenistic cities with foreign garrisons. In: Chaniotis A, Ducrey P (eds) Army and power in the ancient world. Steiner, Stuttgart, pp 99–113

Dmitriev S (2005) City government in Hellenistic and Roman Asia minor. Oxford University Press, Oxford

Fischer-Bovet C (2014) Army and society in Ptolemaic Egypt. Cambridge University Press, Cambridge

Giordano A, Cole T (2018) The limits of GIS: towards a GIS of place. Trans GIS 22:664–676. https://doi.org/10.1111/tgis.12342

Grainger JD (1997) A Seleukid prosopography and gazetteer. Brill, Leiden, New York

Green P (1990) Alexander to Actium: the historical evolution of the Hellenistic age. University of California Press, Berkeley

Gregory I, Ell PS (2007) Historical GIS: technologies, methodologies, and scholarship. Cambridge University Press, Cambridge, New York

Holt FL (1999) Thundering Zeus: the making of Hellenistic Bactria. University of California Press, Berkeley

Horne R (2015) Imperial power and local autonomy in Greek Garrison communities: the phrourarchia and the polis (PhD Dissertation). University of North Carolina, Chapel Hill

Iwaniak A, Kaczmarek I, Strzelecki M, Lukowicz J, Jankowski P (2016) Enriching and improving the quality of linked data with GIS. Open Geosci 8:323–336

Kinns P (1998) CH 8, 474: Milesian silver coinage in the second century BC. In: Ashton R, Hurter S (eds) Studies in Greek numismatics in memory of Martin Jessop Price. Spink, London, pp 175–196

Knowles A (2014) The contested nature of historical GIS. Int J Geogr Inf Sci 28:206–211

Knowles AK, Hillier A (eds) (2008) Placing history: how maps, spatial data, and GIS are changing historical scholarship. ESRI Press

Kortenbeutel H (1941) Phrourarchos. RE

Labarre G (2004) Phrourarques et phrouroi des cités grecques d'Asie Mineure à l'époque hellénistique. In: Couvenhes JC, Fernoux H-L (eds) Les Cités grecques et la guerre en Asie Mineure à l'époque hellénistique. Actes de la journée d'études de Lyon, 10 octobre 2003. Presses universitaires François-Rabelais, Tours, pp 221–248

LaBuff J (2016) Polis expansion and elite power in Hellenistic Karia. Lexington Books, Lanham

Launey M (1987) Recherches sur les armées hellénistiques. De Boccard, Paris

Lévêque P (1999) La guerre à l'époque hellénistique, in: Problèmes de La Guerre En Grèce Ancienne. École des hautes études en sciences sociales: Seuil, Paris

Ma J (1999) Antiochus III and the cities of Western Asia minor. Oxford University Press, London, New York

Ma J (2013) The Attalids. A military history. In: Attalid Asia minor: money, international relations, and the state. Oxford, pp 49–82

Massey D (2015) For space. Sage, Los Angeles

Mehl A (1980) Doriktetos chora. Kritische Bemerkungen zum 'Speererwerb' in Politik und Völkerrecht der hellenistischen Epoche. Anc Soc 173–212

Nankov E (2009) Phrouria Lokrika: aspects of military presence in Hellenistic Opountian Lokris (PhD Dissertation). Cornell

Parke HW (1933) Greek mercenary soldiers, from the earliest times to the Battle of Ipsus. The Clarendon Press, Oxford

Plewe B (2002) The nature of uncertainty in historical geographic information. Trans GIS 6:431–456

Ramsay S (2004) Databases. In: Schreibman S, Siemens R, Unsworth J (eds) A companion to digital humanities. Wiley Online Library, Malden, pp 177–197

Robert J, Robert L (1976) Une inscription grecque de Téos en Ionie. L'union de Téos et de Kyrbissos. J Sav 153–235

Roth J (2007) War. In: Sabin PAG, van Wees H, Whitby M (eds) The Cambridge history of Greek and Roman Warfare. Cambridge University Press, Cambridge, New York, pp 368–398

Shipley G (2000) The Greek world after Alexander, 323-30 B.C. Routledge, London, New York

Southall H, Mostern R, Berman ML (2011) On historical gazetteers. Int J Humanit Arts Comput 5:127–145

Tuan Y-F (1977) Space and place: the perspective of experience. University of Minnesota Press, Minneapolis

The Preservation of Paradox: Bismarck Towers as National Metaphor and Local Reality

14

Gordon Cromley and Chris Post

Abstract

Solving geographic questions is a complex task. To address key geographic issues, the discipline has developed a wide variety of research techniques. Geographic Information Systems (GIS) have, as a tool and sub-discipline, become ubiquitous for scholars looking to address issues tied to patterns on the surface of the earth. A key trend within GIS has been to incorporate alternative frameworks for software and research design. Narrative tools have been used in a variety of ways with respect to such systems while alternative types of data beyond traditional GIS archetypes have been explored. In the past several years, humanities-based scholarship has made key contributions to reframing what GIS can be. This chapter follows a humanities GIS framework while incorporating standard GIS tools to allow us to rethink geographical and historical processes. Commemorative research represents the complexity inherent in a geographic approach. While memorials are the realization of a myriad of local drivers over time, memorials also have contexts grounded in larger geographic realities. The network of memorials to Otto von Bismarck, first minister-president of the Kingdom of Prussia and later the first chancellor of the German Empire, represents the tension between place and wider geographic discourses. Approximately 240 Bismarck Tower memorials were built between 1867 and 1935. While each tower is grounded within a specific local narrative, they were meant to provide a link to a larger discourse on national identity within the community. This chapter uses *kernel density estimation* (KDE) as a *geo-visualization* and *Exploratory Spatial Data Analysis* (ESDA) technique to uncover national patterns of the tower network that can guide our thinking of specific memorials and how we might better understand their development.

Keywords

Commemoration · Nationalism · HGIS · Memorialization · German empire

14.1 Introduction

Commencing in 1867 with construction growing after Otto von Bismarck's death in 1898, the network of commemorative structures known colloquially as ‘Bismarck Towers’ remains an enduring feature of the landscape of the ‘Second German Empire,’ known idiomatically as the

G. Cromley (✉) · C. Post
Department of Geography, Kent State University at Stark, North Canton, OH, USA
e-mail: gcromle2@kent.edu

C. Post
e-mail: cpost2@kent.edu

C. Travis et al. (eds.), *Historical Geography, GIScience and Textual Analysis*, Historical Geography and Geosciences, https://doi.org/10.1007/978-3-030-37569-0_14

German Empire which existed from 1871–1919 (indeed, the notion of a first German Empire and a second German Empire was developed by the Nazis to frame their regime as a third Empire.) This network of memorials is perhaps as little studied as their namesake, Otto von Bismarck, has been exhaustively studied. Yet, the towers represent an interesting case study at the intersection of nationalism, commemoration and spatial analysis. This study uses the framework of landscape as discourse materialized, together with a GIS-based spatial approach to analyze: (1) the towers' presence on the landscape as a product of the development of German nationalism during the Second Empire, and (2) the relationship between the spatial structure of the memorial network created by this memory work. It approaches cartographic representation through Geographic Information Systems as a text which can be interpreted as a form of visual and spatial hermeneutics to provide a contextual underpinning of specific sites of memorialization. We use this combined approach and then analyze the presence and repurposing of multiple monuments in Dresden.

14.1.1 Linking Commemoration and Nationalism

Nationalism is one of the most debated ideas in the academy. What many citizens and politicians take for granted, many scholars suggest is a myth (Giddens 1987; Smith 1999), an invented tradition (Bowden 1992; Hobsbawm and Ranger 1983) or a local metaphor (Confino 1993). Those individual and long-standing institutions seeking control of a national narrative use the landscape to forge their sense of it (Harvey 1979; Nora 1989; Osborne 2001; Savage 2011; Zelinsky 1988). One of the most influential works on constructing national identity is Anderson's (2006) book *Imagined Communities*. Key to his argument is the creation of process typologies in which various historical narratives are binned into specific processes. One of Anderson's main points is that nationalism is not formed through a 'pure' process, but rather incorporates several processes at different stages of development. As these various processes develop along a variety of national-scale trajectories, the local expression of these imaginaries varies from location to location. In effect, the unitary nation-state must always be on guard for the emergence of sub-regional identities. Yet in many instances it is then the 'myth' of the nation that acts to counterbalance these separatist tendencies. In the case of Germany, nationalism was consciously constructed through a variety of institutions following the structures articulated by Anderson (2006): Religion, political conservatism, and various strands of liberalism and dynastic and state identity and collective memory (Blanke 1974; Clark 1996; Johnson 2011) all engaged with local articulations of an historical national German identity. These nationalisms were often realized at a local level as monuments to various historical elements of the German nation that had particularly local salience (Mueller 2011). The narrative of the German response to the Napoleonic War, illustrates how any official attempt at constructing a unitary national narrative is fraught with alternative voices (Clark 1996).

14.1.2 Constructing a Nationalist Commemorative Landscape

As a product of memory work (Till 2005), memorials serve to bond communities to specific historical events. National memorials serve to project a specific discourse of national identity (Savage 2011). The network of Bismarck towers created during the period from 1867 to 1935 represents an opportunity to understand the spatial relationships between the commemoration of an important individual in the formation of the German Empire, Otto von Bismarck (1815–1898), and the changes in the structure of that network with respect to changes in the boundaries of Germany in 1914 and present-day Germany. Additionally, this study will use data that categorizes the organizations which initiated the memorialization process of constructing each individual Bismarck tower, in order to further understand the interplay between national identity and the memorialization process. Many studies on the geographies of Western nationalist commemoration speak to this use of landscape to establish power. In fact, it can be concluded that nationalism—the source of politics surrounding commemoration—provides the crux of memorial studies in geography over the past three decades. Studies focusing on this intellectual intersection take many different forms, from toponomy (Azaryahu 1996; Post and Alderman 2014; Drozdzewski 2014) to war memorials (Azaryahu 2016; Gillen 2014; Johnson 1994, 1995; Mayo 1988; Waterton and Dittmer 2014; Winberry 1982, 1983), and from the commemoration of civil rights politics (Dwyer and Alderman 2008) to the negotiation of identity in post-conflict states (Johnson 1996, 2011; Kruk 2008; McDowell and Braniff 2014; Osborne 2001; Till 2005; Zelinsky 1988). This work sits within the specific subfield that analyzes commemoration as an invocation of nationalism in a fragmented post-conflict society. A common conclusion of these studies resides in the landscape's role in the formation of a national ideal through its ability to concretize a powerful discourse through permanence and accessibility. According to Robin Wagner-Pacifici and Barry Schwartz, 'Memorial devices are not self-created; they are conceived and built by those who wish to bring to consciousness the events and people that others are inclined to forget' (1991, 382). Sarah McDowell and Máire Braniff agree, claiming, "Commemoration and memorialization may offer one such *modus operandi* which allows on-going fealty to the 'cause,' often predicated on assumptions of victimhood, while also serving in its material manifestations to mark out the territorialization upon which political support generally depends in conflictual and 'ethnocratic' societies" (2014, 5; see also Yiftachel and Ghanem 2004). Nuala C. Johnson links this activity particularly to the public commemorative landscape as it 'attempts to solicit public participation in the politics of the day' (1994, 79). Thus, the commemorative landscape, simultaneously both political and cultural, plays a crucial role in this discourse of power, state, culture and nation.

14.2 Bismarck, Nationalism and German Unification in the Nineteenth Century

From 1871–1918, Germany, as a nation-state, was constituted as the German Empire, a federation of associated kingdoms and sovereign territories. It was ruled by the King of Prussia as the German Emperor with the Kingdom of Prussia as the dominant political and territorial entity within it. Despite its relatively short existence, about 47 years, the German Empire as a political, cultural and scientific entity has had a tremendous impact on the history of the twentieth century. At the time, the empire was seen as the culmination of a century-long attempt at uniting the patchwork quilt of central European principalities and kingdoms into a cohesive nation-state. It was an amalgam of various dynasties and territorial entities, many of whom were mutually distrustful of one another and did not share an institutional legacy. At the conclusion of the Franco-Prussian War, the newly formed German Empire appeared to be held together by nothing more than the strength of the Prussian Army and the political acumen of its chancellor Otto von Bismarck. Yet, despite this, the Franco-Prussian War revolutionized the power balance of Europe and gave rise to this new territorial entity that dominated Europe until the end of 1945 and continues to impact the region today (Howard 2005). To contemporary observers, Prussia's rise to the core of the German Empire was miraculous. It had few 'natural' resources and has been termed a triumph of man over nature. However, just as quickly as Prussia emerged as a major territorial entity, it disappeared. On 25 February, 1947, the Allied Military Government occupying Germany chose to dissolve Prussia out of a belief that its continued existence would perpetuate the violent militarism that had characterized Germany in the first half of the twentieth century (Taylor 1949; Clark 2008). The constant evolution of the territorial boundaries of Prussia-Germany provides a clear example of the warning to beware of the 'territorial trap' by which nation-states are conventionally understood (Agnew 1994). The continuing reinvention of the German nation-state has led to competing forms of what constitutes the German nation (Herb 2004). A key aspect in constructing this seemingly improbable national identity has been the establishment of a memorial landscape that mythologizes the nation (Osborne 2001) Starting in 1867 and ending in the early 1930s, approximately 240 structures of varying design—Bismarck towers —were constructed or appropriated to honor the memory of Otto von Bismarck, first Chancellor of the German Empire and a key architect in its creation and organization

14.2.1 Bismarck's Kulturkampf

Bismarck sought to create a country that was not only superficially unified as a single geographic territory forged through conquest—the Danish-Prussian War of 1864, the Austro-Prussian War of 1866 and the Franco-Prussian War 1871—but also unified more deeply by a common cultural identity. He attempted this latter form of unification by legislating a fourth war of unification—the 'Cultural War' or 'Kulturkampf.' The three wars of territorial expansion had brought together an amalgam of cultural identities, and the individual citizens of the newly integrated territories did not necessarily share the same 'German reality' as Bismarck and the new German leadership had intended. The new German Empire with Prussia as the core was a trans-national state with multiple cultural groups even among the Germanic-speaking peoples, and it was the conception of Prussia as the cultural model that led to an internal debate over how to strengthen the nation (Nelson 2010). The three wars of territorial expansion had wrought violence against the body of the male citizenry, whereas the Kulturkampf now promoted violence against the minds all of its citizens. The Bismarckian Kulturkampf was based on eradicating the cultural consciousness that predated German unification, in order to bind the new territorial acquisitions more closely to Prussia as the keystone of the empire (Retallack 1997). The cultural connections most targeted were those of the urbanized, and often Catholic, Germany located in the south and an ethnically polyglot German East. Unlike old Prussia, which was sparsely populated and heavily reliant on agriculture, the rest of Germany was urban, industrializing, and had a more self-realized cultural heritage evidenced on the landscape (Sutherland 2011). Kulturkampf represented a uniquely Prussian expression of the traditional antipodal nature of Germany—the protestant/catholic dichotomy. Although other dichotomies existed within the German nation, this particular expression directly pitted the ideologies of Prussian nationalism against a more 'normative' south-German nationalism. The strains were partly the result of continuing resentment of Prussia's reaction during the years of 1848–49, but it also represented older, more entrenched legacies. The concept of an internal colonization of an imperial nationalism was utilized, 'by virtue of their confessional difference, German Catholics were imagined as Germany's internal foreigners and increasingly marginalized from German society and politics as enemies of the new Reich' (Landry 2005, 285).

Two important laws imposed during the Kulturkampf period were the Falk and Congregation laws. The Falk laws in 1872 turned over training and nomination of clergy to the state. The Congregation laws in 1875 stopped state subsidies of the Catholic Church, abolished monasteries, and took away constitutional protection for the Catholic Church within the empire. While some scholars see this as a conflict between Germans, it is more correctly a conflict of Prussians against south-German Catholics and especially Polish Catholics. The austere and humble conceptualization of the Prussian national sentiment contrasted against the more cosmopolitan German South. Prussia's distinctiveness within the more collective German whole was represented by the Ut Omnes Unum journal that emerged from the Wilhemine Kulturkampf (Landry 2005). The general ethos behind this journal was that, 'despite the political unification of Germany, the author claimed that the German people remained internally divided' (Landry 2005, 289). After 1878, the targeting of Catholics in the Kulturkampf receded.

Instead, Bismarck allied with the powerful Catholic voting bloc in the south to marginalize other minorities and separate himself from the liberal Germans who supported his anti-Catholic policies (Clark 2008). Ethnic minorities in the empire such as Poles in Western Prussia became new targets of the Kulturkampf (Blanke 1974). Because Germany extended far into Poland after it was partitioned, much of Poland was constructed in Germany as 'colonial spaces' before, during and after the First World War (Brunhes and Vallaux 1918; Kopp 2012). The creation of the German state was viewed with concern even by Prussians particularly in the East; these citizens who had been viewed as Prussian were now afraid that they might not qualify as German in the new empire (Clark 2008). The Kulturkampf did not just impact those individuals whose 'Germanness' was questioned. States that possessed strong independent identities separate from the overarching German identity also felt the strain of this period. Saxony was such a state and it resisted the Kulturkampf. Saxony through the eighteenth and early nineteenth centuries had remained an ardent foe of Prussian expansion. The legitimacy of its constitutional monarchy was based on its elevation to that status by Napoleon during the Napoleonic Wars. From their perspective, Prussia represented a normative threat to the culture of Saxony as well as the rest of Germany (Retallack 1997). Throughout the German Empire, an internal struggle for culture continued to decide where the cultural center of the new nation would be (Steinberg 2012).

The Kulturkampf also served as an attempt to destroy the trans-national identities of those spaces that existed outside of the 'German' ethnic core (Frieberg 2010), to create a more cohesive sense of 'Germanness.' Prussia's eastern territories had long been an embarrassment to the more progressive elements within Prussia (Nelson 2010). This embarrassment did not come from the region's ethnicity, but from its delineation as a cultural periphery to the new German Empire (Wilson 2007). An east-Elban identity (east of the Elbe River) anchored the core of Prussian power. The perceived backwardness and constrained social structures of the eastern provinces of the German Empire were of great concern to administrators in Berlin (Wilson 2007). As a matter of policy, these eastern territories, while central to the conceptualization of the Prussian state, were treated no differently than the imperial territory of Alsace-Lorraine (Nelson 2010). Throughout the period of empire, Prussia's eastern territories were subject to a slow drain of 'ethnic' Germans to the western parts of the empire (Nelson 2010; Wilson 2007), and these territories were not overly Prussian from the beginning. The atrophy of these national spaces was seen as an ideological problem for the state. Indeed, the landscapes of the east were promoted in an attempt to create a Prussian landscape that could compete with the power of the more quintessentially German landscapes in the south and west (Wilson 2007).

After the end of the Kulturkampf, there was a ripple effect of many nationalistic movements being formed such as the Eastern Marches society which promoted the resettlement of German territory by Germans. Minority populations, particularly the Poles, were the target of attempts to remove them from the territory. As there was not a large enough population of 'true' Prussians, the administrators in the Eastern lands were often from other parts of the German Empire (Wilson 2007). Because these administrators were not tied into the local political infrastructure, they could appeal directly to the national government. The solution reached by intellectuals such as Max Sering was to re-colonize the Eastern provinces through a comprehensive policy of land reform (Nelson 2010).

With the advent of unification, the Junker class also began to be seen as a reactionary element that was antithetical to the Prussian nation. Sering characterized this form of nationalism as an 'internal imperialism' that sought internal colonization to ensure of the health of the nation (Nelson 2010). Because of Prussia's internal identity crisis the state came to represent a harsh chauvinistic regime that sought to undermine the regional nature of the nation. Even after all of these attempts at expanding a German nationalism, the institutional control of the empire was never totally centralized. Much like the earlier title of King in Prussia, Kaiser William I was not allowed to adopt the title Emperor of Germany, he was instead styled German Emperor (Mueller 2011). This again represented the limits of Prussian militarism—Prussia was not to be synonymous with Germany.

14.3 The Bismarck Tower

Throughout this period of change and restructuring the one common element was the power and governing hand of Bismarck. Bismarck bestrode the era and governed the new nation so thoroughly that it could not successfully exist without him. Steinberg (2012) discusses Bismarck's 'sovereign self,' as a unique characteristic expressing absolute control, an authority that emanated from his personal charisma. Additionally, the fickle nature of Bismarck's policy and his willingness to work with any political group to further his own ends made Bismarck a highly ambiguous, even contradictory character. Yet, with the removal of Bismarck from power in 1891 and his death in 1898, a cult around Bismarck emerged as a personification of the edifice of the German nation-state. Indeed, during that era, Bismarck became, for a specific segment of the population, the personification of national ideals (Pflanze 1955) that was surprisingly malleable to the variety of groups and often acting as a unifying feature in national discourse (Gerwarth and Riall 2009).

At their core, the purpose of Bismarck towers was to impose the authority of this new German state on the landscape through the materialization of the 'cult of Bismarck.' These structures embedded the discourse of the state as materialized in the ordinary landscape of the state. Bismarck was an essential presence for controlling potential internal conflicts made possible by unification. Bismarck towers were primarily funded by private civic groups rather than the state.

They represented the people's desire to keep Bismarck in their memory and to ensure that his 'presence' was felt upon the landscape as a sign for the still young German state. While the later development of German history linked the Bismarck Myth to an articulation of a specific strand of German conservatism (Frankel 2003), even in a post-totalitarian environment, the Bismarck persona continues to oscillate between various interpretations of Bismarck himself (Urbach 1998). Bismarck towers were constructed throughout the German Empire and its colonies (one was even constructed in Chile).

Overall a total of 240 Bismarck towers were constructed on four different continents (Bielefeld and Büllesbach 2014). Of the original 240, a total of 173 remain. In Germany, most of these towers are still in existence. What has changed since the last Bismarck tower was completed in 1935 has been the political boundaries (externally and internally) of Germany. Bismarck towers are still located in six European countries: Germany, France, Austria, the Czech Republic, Poland and Russia—mostly on territory that was once part of the German Empire. In some instances, these towers have been appropriated for other uses, such as in Denmark which turned its monument into a peace memorial. The first Bismarck tower was built in 1867 in Silesia.

Figure 14.1 shows the cumulative percentage of completed Bismarck towers by year. The vast majority of these towers were built in the sixteen-year period immediately after the death of Bismarck in 1898 up to the outbreak of World War I. A more telling aspect of their construction was that many of the Bismarck towers began construction on the anniversary of important national dates. While using birth and death days of Bismarck to construct these towers is expected (Foote 2003) using the anniversary of important battles in the nineteenth century shows how varying elements of force used to construct the nation were tied to the public's psyche in the construction of these towers (see Fig. 14.2). Construction of the largest number of towers was started on Bismarck's birthday. The next largest number, commenced on the anniversary of the Battle of Sedan or the eve of this battle. The battle of Sedan was the engagement that led to the capture of Emperor Napoleon III and a significant portion of the French Army in 1870. A smaller number were started on the anniversary of the Battle of Leipzig. The Battle of Leipzig was the decisive defeat for Napoleon in 1813 that broke his control over the German states. A fewer number were initiated on the anniversary of the Battle of Noisseville which also occurred during the Franco-Prussian War. There were also holidays such as the Summer Solstice and Johannisnacht that were chosen as foundational days for the construction of Bismarck towers. Through this construction process, Bismarck was being linked by the German people to the mythology of their struggle for unification.

14.4 Navigating Germany's Other Bismarck Archipelago: An ESDA Approach

Memorialization as a field of inquiry has a long tradition within geography (Dwyer and Alderman 2008). However, while the study of individual monuments has a strong attestation within the literature (Crampton 2001; Harvey 1979) and groups of monuments have been used to understand aspects of how memorials are reinterpreted through political change (Forest et al. 2004), there are very few studies (Drozdzewski et al. 2016a, b; Kruk 2008) that engage with the analysis of a commemorative landscape with a comprehensive spatial understanding of the entire distribution of a given memorial landscape. This dataset offers a unique opportunity to test study the relationships between the 'national' and the local in a commemorative landscape. Following Dwyer and Alderman (2008), the Bismarck tower landscape can be thought of as an arena for state discourse on the expression and identity of the recently formed German nation in the personification of Otto von Bismarck. Building on Flint (2002), this study utilizes basic spatial analysis techniques to provide context for understanding historical processes.

Bismarck towers represent a material manifestation of the nation concept operationalized at a local level. The regional nature of the Bismarck tower memorial landscape can be displayed and related to this wider commemorative relationship between Bismarck and the nation-state. A database consisting of 235 Bismarck towers was constructed using GIS. Five towers, constructed outside of Europe, were not included within this dataset. The website www.bismarcktuerme.de was used as a data source for the creation of this database along with Siegelinde Seele's *Lexicon der Bismarckdenkmäler* (2005). Bielefeld and Büllesbach's book (2014) on Bismarck towers was then used to determine which pertinent attribute data to include in a digital gazetteer of the towers. Pertinent data was divided into two data types: temporal state data and memorial attribute data. Data involving the temporal stages of the towers include the date of both the beginning of construction and the dedication of the tower (if available). If the temporal resolution was not available for the day of the month, then the month or year was applied. If neither scale was available then the date was coded as 'UNK' for the unknown date. The construction of each tower was assigned to one of four 'eras': (1) before Bismarck's death in 1898, (2) before the Student Association competition in 1899, (3) before the outbreak of World War I in 1914, and (4) after the outbreak of World War I in 1914. The era when individual towers were destroyed were: (1) pre-World War I (1914), (2) the Weimar era (1919–1932), (3) the National Socialist Era before World War II (1933–1939), (4) World War II (1939–1945), (5) the period of East and West Germany (1949–1989), and (6) the post-unification era (after 1989).

The attributes of the physical aspects of a tower can be broken into two types: conceptual and material. The conceptual component codes two distinct typologies. The first typology divides the civic groups that initiated fundraising and construction of these monuments into four categories: (1) Bismarck honoring groups, (2) student organizations, (3) national associations and (4) nature/beautification associations. This typology is again taken from Bielefeld and Büllesbach (2014). The second typology relates to the underlying design upon which the towers were constructed. Forty-eight of the Bismarck towers were constructed using the

'Götterdammerung' design created by the architect Johan Kreis. This design had a square base in which the four corners were enclosed by column capped by a superstructure for a firepit. The Götterdammerung towers came in different sizes and proportions. There were also towers constructed of both wood and metal construction that were built as lookout towers. The final category is a catch-all category that collects all towers of primarily stone and brick construction. The height of each individual tower was also recorded. The functional characteristics were coded according to the twin function of most towers: as lookout towers and fire columns. It was intended that these towers act as participatory monuments where people could both climb the tower to survey the surrounding countryside and could also light a fire column on the top of the towers to act as beacons for the surrounding region. It was intended that these towers would act as a network uniting the country. Construction of the first Bismarck tower was completed in 1867, in Silesia, a province of the Kingdom of Prussia that is currently part of Poland. Unlike later towers, only one of the early structures had the ability to produce a visible flame at its top.

At the time of Bismarck's death, the number of Bismarck towers that had been built was still relatively small. It is only after the German Student Organization's contest for a standard Bismarck tower design that the rate of Bismarck towers being constructed and completed increases. The first decade of the twentieth century is the period during which most of Bismarck towers were constructed (see Fig. 14.1). The politics behind the construction of these towers also plays an important role in understanding how building these towers reinforced the larger discussion of constructing a national identity during this period. Although the latter half of his career was spent as the German chancellor, Bismarck derived his position from his office as first minister of Prussia. Therefore, while Bismarck directed the government of the empire, his authority was still granted in the institutions of the Prussian state instead of a truly 'German' political institution. Likewise, German nationalism was rooted in a variety of various forms such as a focus on an authentic natural landscape (Clark 1996, 2008) a conservative and ethnically based nationalism (Frankel 2003; Retallack 2017), a nationalism based on a new urban and middle-class population that sought to join the larger discussion of national identity (Russell 2000). Additionally, Bismarck towers also served to reinforce the specific Bismarck cult as well. Kernel density surfaces were created to show the spatial aspects of these forms. *Kernel density estimation* (*KDE*) is a tool to estimate the likelihood of an observation occurring in a certain location given a pattern of existing observations—it helps one to make spatial generalizations of the pattern. Analyzing the spatial pattern of towers associated with each of the four different types of civic groups that built them can help shed light on the cultural and national forces behind the commemoration of Bismarck, and the areas where these forces have a higher likelihood of occurring.

A *KDE* surface of towers associated with nature organizations and student organizations (Fig. 14.3) has a pattern in which the peak likelihood is located to the west within the German Empire and has a much lower likelihood in the eastern territories. These two civic groups built their towers more in the core of modern Germany and spread north and south. The Bismarck and national organizations' *KDE* is centered again in the middle of modern Germany but spreads out and follows the outline of the Kingdom of Prussia much more closely to the east (Fig. 14.4).

The distribution of all Bismarck towers throughout Europe vaguely follows the outline of the German Empire (Fig. 14.5). The areas that have the highest density are in the urbanized Rhine River valley, and in the more central part of Germany, in the states of Thuringia and Saxony. It is these regions that contained Weimar and other cities that through the eighteenth and nineteenth centuries proved to be the cultural core of Germany (Clark 2008).

This distribution of the memorials in one sense is surprising given the number that lie outside the traditional core of Prussia prior to 1815. During the nineteenth century, Prussia grew in size from roughly 276,000 square

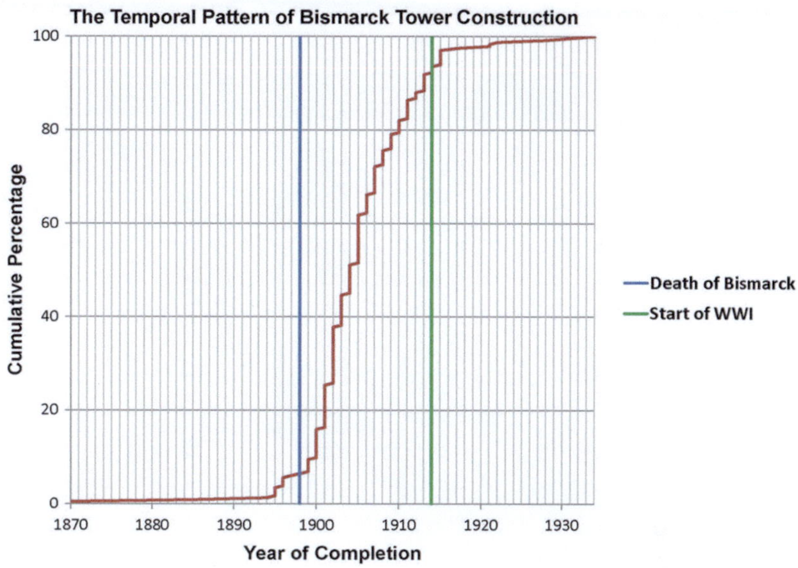

Fig. 14.1 Graph dating the construction of Bismarck towers

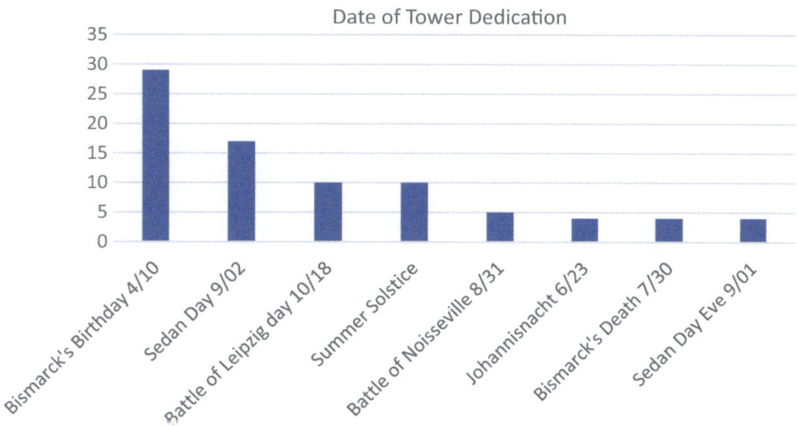

Fig. 14.2 Important dates and numbers of dedication of Bismarck towers (adapted from Bielefeld and Büllesbach 2014)

kilometers to 350,000 square kilometers in only about 50 years. During the Congress of Vienna, Prussia was given a swath of territories along the Rhineland and roughly 40% of Saxony (Clark 2008). The core of the Prussian territories can be thought of as the territorial extent of Prussia at the death of Frederick II in 1786 (Fig. 14.5). Yet, of the 141 Bismarck towers built after 1867, only 47 were built on territory that had been under the control of Prussia since 1786. Construction of Bismarck towers is more closely aligned with the population distribution during the 2nd Empire. With the Prussian Rhenish provinces forming one zone of intensive construction and the region around Saxony forming a second. In terms of density, Saxony has the highest rate of Bismarck tower construction in any of the major German states of the Second Empire. Construction of towers was at a rate of over 2½ per 2,000 km^2, while the national average was under 1 tower per 2,000 km^2 and the density of towers in the Rhineland was slightly over 1.5 towers per 2,000 km^2.

Although the state of Prussia has by far the largest numbers of Bismarck towers of any of the constituent states within the Empire, this map clearly shows that the core of conservative support, particularly the core of the Prussian

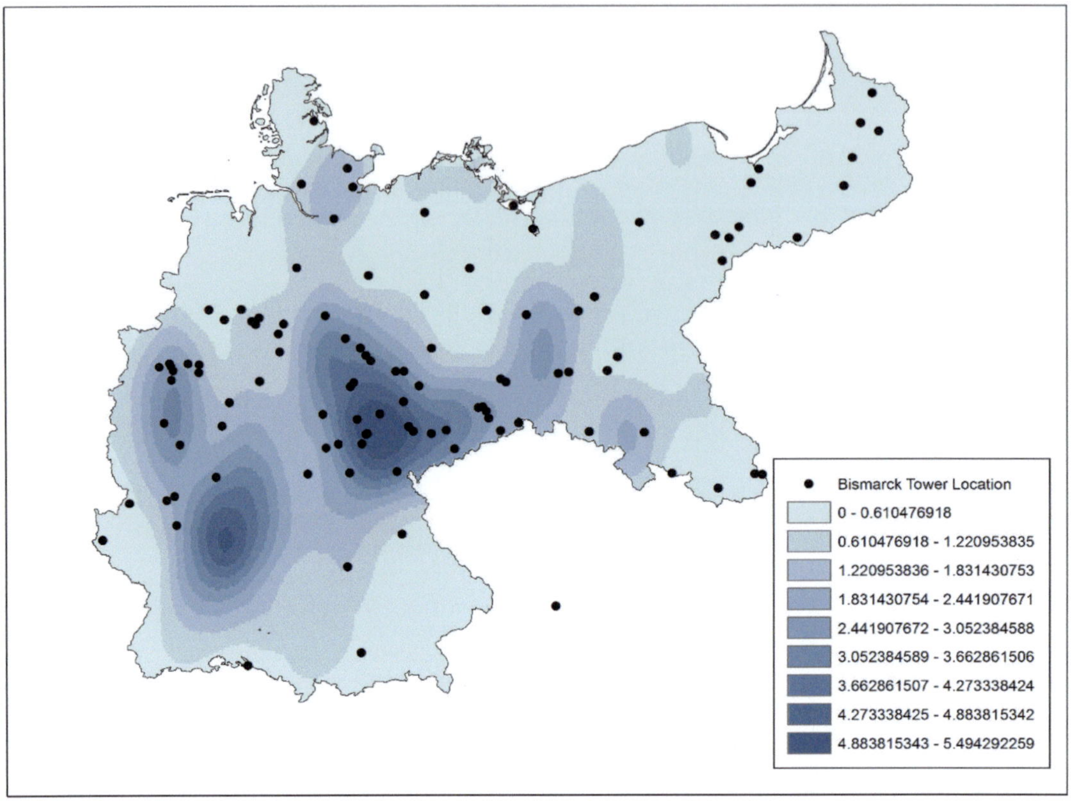

Fig. 14.3 *KDE* surface associated with Nature and Student Organizations

aristocracy in the eastern third of the country, did not have high densities of Bismarck towers. Another region that did not have high densities of towers was the Catholic South. This is not surprising, though, because Bismarck was a nemesis for this region during the Kulturkampf.

The distribution of Bismarck towers then represents a reality that seems somewhat counterintuitive to what one might expect. Because these structures were funded at a municipal level, one would expect that the greatest concentration of these structures would be in the Prussian heartland rather than in areas that were often quite hostile to the type of nation that Bismarck was trying to forge through his series of conflicts. Yet, the history of the Kulturkampf suggests that this distribution is expected. While the core of the Prussian state might have been east-Elbian, the periphery of Prussia east of the Oder River was marginal with respect to the new Germany created by Bismarck and was subject to the harshest treatment during the Kulturkampf.

Consequently, it is not surprising that East Prussia, like the Catholic South, did not construct many towers. Alternatively, the areas that were least affected by the Kulturkampf and the regions that had always been active in expressing a German ethnic nationalism were more likely to have constructed a tower. For example, the Kulturkampf did not have a major impact in Saxony, and although Saxony was a combatant against Prussia during the Austro-Prussian war, it did not have significant non-German populations and was a strong supporter of German nationalism.

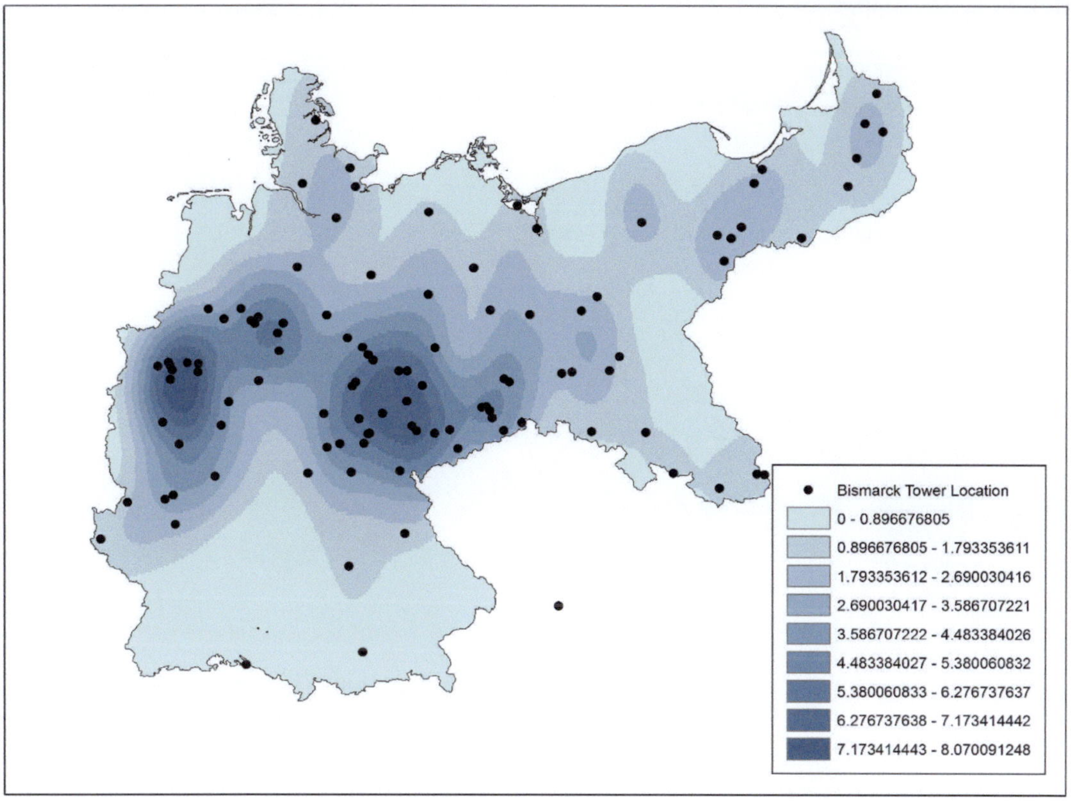

Fig. 14.4 *KDE* surface of Bismarck towers associated with Nationalist and Bismarck organizations

14.5 The Dresden Markers as the Discourse Materialized of a Memorial Movement

Dresden stands as a significant city in the German Empire of Bismarck's time. The two different *kernel density analyses* that inform our understanding of the spatial distribution of Bismarck tower's as a commemorative landscape highlight an interesting visualization of several aspects of the memorials. First, there is a definite pattern of the tower's representation on the landscape mimicking the politics of the Bismarckian political era, namely the Kulturkampf. The greatest density of towers does not extend into the predominantly Catholic areas of the country: namely South Germany and the predominantly Catholic Polish and Silesian provinces of Prussia in east- Elbian Prussia. Instead, the densest concentrations of Bismarck towers are in the Prussian Rhine provinces and the so-called Thuringian principalities and surprisingly in the Kingdom of Saxony. Additionally, Saxony, and specifically the capital of Dresden, straddles the Elbe River making it a territory that warrants greater consideration. Standing almost due south of the old city center, a stone tower stands as a relic of a different Germany. Constructed on the site of a Napoleonic Battlefield, this site—now surrounded by apartment complexes and just south of the Dresden Technical University—represents layers of German national identity and iconography. The tower itself, one of three Bismarck towers in the city of Dresden, is known as the Dresden-Raecknitz Bismarck Tower and has undergone a complex series of evolutions and even name changes. Built after the death of Otto von Bismarck, the tower was constructed in the 'Götterdamerung' style. This style, of which 47

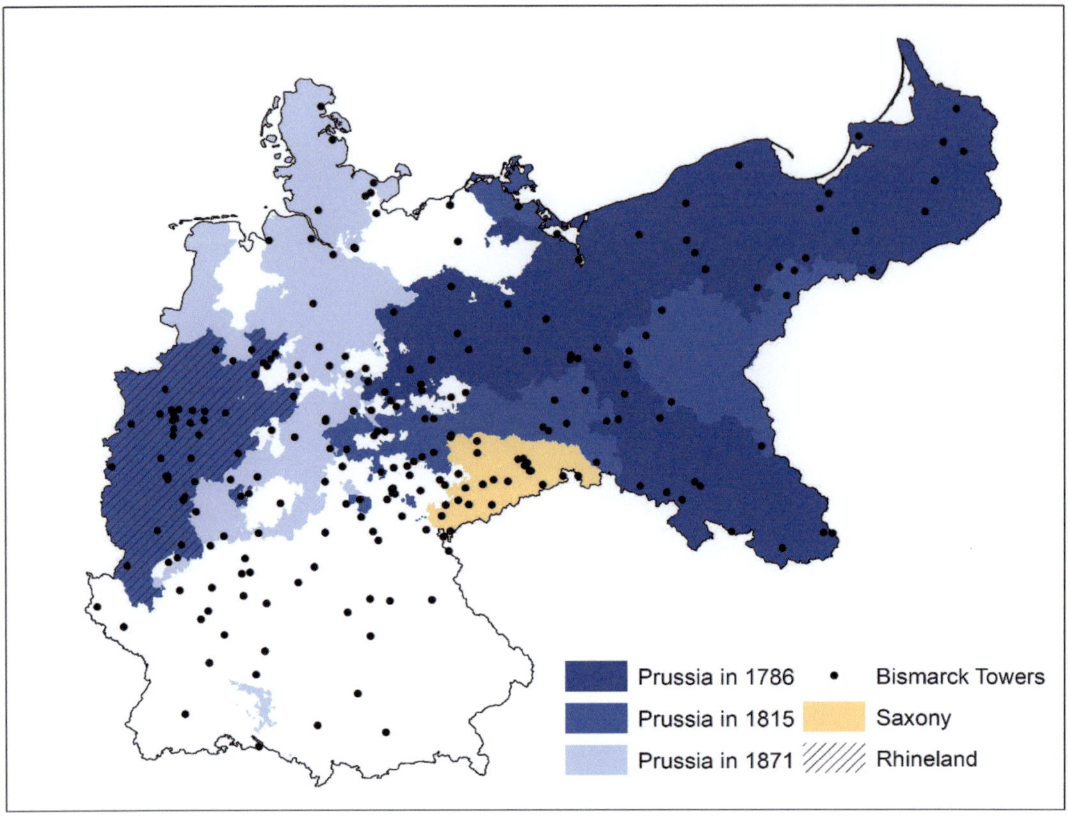

Fig. 14.5 The territorial evolution of Prussia and the distribution of Bismarck towers

towers of varying size and proportions were constructed, was the result of a national contest run nationally by the Studentenschaft, a national organization of German students to create a uniform design for a memorial of Bismarck, a former member. The winning design was submitted by the architect Wilhelm Krebs who not only designed Bismarck towers in other styles but also was a leading architect into the National Socialist period.

The initial construction of the tower included a flame pit at the top which could be lit and used for a variety of celebrations commemorating important dates for the nation. Later, stairs leading to an observation deck were constructed bringing the tower in line with the majority of other Bismarck towers. One can imagine the feeling that a visitor to the site would feel climbing from the base of the tower grounded in the history of German unification in the form of the Battle of Dresden battlefield all around the monument, and then slowly rising up to look out upon the surrounding countryside to view at the present the landscape of the empire that had been made by Bismarck. Figure 14.6 illustrates the modern experience of entering the tower and climbing up the stairs to the observation deck. Much like Wylie's (2003) discussion of ascending the Glastonbury Tor, the symbolism underlines the Hegelian idealism of the site's role in merging Saxony up into the ideal of the German nation-state.

Later, during the National Socialist period, the tower became the focus of various activities of the Third Reich and was even the location of a large book burning in Dresden. After World War II, the Raecknitz tower along with all of Saxony fell behind the proverbial Iron Curtain. During this period, the tower fell into disrepair and was renamed the Peace Tower in honor of

Fig. 14.6 Ascending the Raecknitz Bismarck tower (photo by authors)

World Peace Day. It came to embody the antipathy of the Communist Party's ideology towards the Prusso-German legacy that the tower originally represented. Numerous attempts were enacted to raise the funds to destroy the site. However, unlike other towers in the successor states to Bismarck's German Empire, this tower endured. With the *Wende* (*Turnaround* -signaling the shift from a socialist East Germany to a reunited democratic Germany) at the end of the twentieth century and the creation of a new Germany, numerous towers including the Raecknitz tower slowly began to undergo a process of repair and renewal. Still, it cannot be said that this brief narrative of the Raecknitz tower in the twentieth century is representative of even the other towers in Dresden.

The Bismarck tower in Plauen was built ten years before the Raecknitz tower and completed during Bismarck's lifetime. An observation deck was included in the original construction and at the time of construction represented the commemoration of Bismarck as a corporate entity to Raecknitz as Plauen had yet to be incorporated into Dresden. In 1937, the park in which the Plauen tower stands was renamed to commemorate Johann Gottlieb Fichte, a German philosopher from the first half of the nineteenth century who was also a committed nationalist (Abizadeh 2005). The appropriation of this tower to commemorate Fichte, an alternative figure of German nationalism, illustrates the role of scale in the narrative of this particular tower. At the geographic scale, the community of Plauen was dissolved into the larger city of Dresden, while at the temporal scale, the emergence of the GDR as a distinct nation-state created a landscape of easy re-appropriation for the Plauen structure as it became detached from the wider memorial network in both space and time.

The Cossbaude Tower, while not being re-appropriated like the Plauen tower, has evolved along a middle path. Unlike the Raecknitz tower which has endured in close to original conception as a monument towering over the countryside, the area around the Cossbaude tower has slowly been encroached on by residential land use. Despite the proximity of these three towers within the city of Dresden there is no unified narrative of these towers. However, the narratives of the Dresden towers together illustrate another key aspect of the myth building agenda of these towers. The two towers that continue to memorialize Bismarck were built after his death while the Plauen tower built and commemorated during Bismarck's lifetime is not. It seems counter-intuitive that the oldest tower is the one that no longer fulfills its original purpose yet, perhaps as Savage (2011) suggests, the power of the cult of personality can only exist after the death of the individual when their identity can be memorialized in a more idealized form.

According to Schein (1997) the cultural landscape serves as the materialization of an ongoing political discourse over who and what count in our society. As a discourse materialized, the landscape takes the local wrangling of local politics, economics, and culture and scales those concerns out to a larger regional, national, or global narrative. Perhaps the lowest hanging fruit for inquiry using such a framework is to illustrate the commemorative landscape's ability to connect local communities to larger struggles such as democracy and race (Schein 1997) Here, we use the landscape framework of a discourse materialized to situate our analysis of Bismarck Towers throughout the German Empire. Given the history of Bismarck's reign—particularly the Kulturkampf—the distribution of the towers, and the incongruent meanings of the Dresden towers, we suggest that the Bismarck Towers' ultimate materialized discourse is of nationalism empowered through a peculiarity of place. Where over time, the normative discourses at the national level are subsumed by local narratives.

14.6 Conclusions

Although Bismarck was instrumental in the founding of the German Empire, he was not the only figure responsible, and yet the commemorative landscape has focused primarily on his persona. The subsequent history of Germany suggests that attitudes towards these towers reflect different internal discussions of twentieth century 'Germanness' and territorial sovereignty. The greatest density of towers is centered along the edge of Prussian territorial expansion in the nineteenth century. That territorial expansion was followed by a parallel cultural expansion through the Kulturkampf in effect, an attempt at internal colonization. Yet, the uneven perpetuation of these memorials suggests that whatever the initial unity in symbolism, the symbolic power of these memorials has devolved to attain a more local character. Autry (2012) argues that in the context of commemoration of subversive periods in national history through memorials, such as the Voortrekker monument in South Africa:

> their removal and wholesale disavowal by national leaders can provoke backlash, sometimes violent, from citizens whose identities are linked to a specific version of the past. Instead, the evolution of the Voortrekker Monument indicates that those sites with the most symbolic power are likely to remain physically intact, even as they lose much of their mystique with the development of new memorial projects, such as Freedom Park (Autry 2012, 161).

The evolution of the Dresden Bismarck towers represents this theme. While their mystique has diminished over time, two of the towers have become focal points of further commemoration and represent the enduring legacy of Dresden. Many of these towers were meant for the public to engage with and participate in the structure through the incorporation of observation decks into their design. Both the Dresden-Raecknitz tower and the Dresden-Plauen tower incorporated these design aspects into their functioning. The Plauen tower was incorporated even further into outdoor activities by the addition of an orientation device placed at the top of the tower for mountaineering groups in Saxony. In both cases, the functional elements for participating in these towers was diminished during the GDR period as these towers fell into disrepair and the observation decks could not be used and the orienting device was removed from the Plauen tower. It was only after unification in the 1990 s that these locations were restored to their full functionality as initially conceived. The memorial landscape of Bismarck towers shows how intimately conflict was internalized in the process of unifying Germany. It stands in stark contrast to the second instance of German unification in 1989. That process was done through referendum and without force of arms. Bismarck himself has retreated from the landscape. While many of these towers exist today, they exist in varying states. Some, especially those in central Germany continue to be well maintained and are a source of civic pride within their communities. Others, particularly those that are no longer in Germany, or are in areas that are less urban, have been shrouded in forests where once there was pasture. Bismarck towers represent an important memorial landscape for understanding the events and ethos that not only created an empire but ultimately led to the dissolution of the Prussian state. As part of the discourse of German nationalism, these towers, in their different ways, provide critical signifying practices (Barnes and Duncan 1992). As such they concretize the messaging of the Bismarck empire's 'Prussianizing' of north-central Europe. Bismarck towers mark an internal frontier between Prussia and Germany rather than an external frontier as the East represented between German and Poland.

Our deployment of GIS in studying the commemorative landscape of Bismarck towers aids in what Bodenhamer calls the 'geographical context and depth to an expert interpretation of the past (2007, 107).' Similar to Hanna and Hodder (2015), this study interprets a network of commemorative places. Although this study falls under Hanna and Hodder's (2015) third marker typology, it does not engage in a content analysis of the markers, but rather it uses *KDE* to infer context for place-based analysis. *KDE* estimation is used to provide context for abductive thinking or a 'hermeneutics of suspicion' (Czovek 2006) in engaging with the specific sites in Dresden. As von Luenen (2013, 237) argues: 'Rather than a visualisation tool, GIS should be used as a painting tool: a tool to creatively engage with one's sources.' In this regard, cartographic manipulation in GIS is used to underline a particular Saxon context. Bismarck towers provided a common material landscape by which citizens throughout an otherwise disparate region could collectively recognize and identify with. The fact that a new administrative state was the only common thread amongst this region only reinforces the power of using memorials in a hierarchical manner as a centripetal force in nation building, a force that bridges local-scale everyday landscapes with a larger national purpose.

References

Abizadeh A (2005) Was Fichte an ethnic nationalist? On cultural nationalism and its double. Hist Polit Thought 26(2):334–359

Agnew J (1994) The territorial trap: the geographical assumptions of international relations theory. Rev Int Polit Econ 1(1):53–80

Anderson, B. (2006). Imagined communities: Reflections on the origin and spread of nationalism. Verso Books

Autry RK (2012) The monumental reconstruction of memory in South Africa: the Voortrekker monument. Theor Cult Soc 29(6):146–164

Azaryahu M (1996) The power of commemorative street names. Environ Plan D 14(3):311–320

Azaryahu M (2016) Wrecks to relics: battle remains and the formation of battlescape, Sha'ar HaGai, Israel. In: Drozdzewski D, De Nardi S, Waterton E (eds) Memory, place, and identity: commemoration and remembrance of war and conflict. Routledge, London, pp 88–106

Bielefeld J, Büllesbach A (2014) Bismarcktürme: Architektur, Geschichte. Morisel Verlag, Munich, Landschaftserlebnis

Blanke R (1974) The development of loyalism in Prussian Poland, 1886-1890. Slavon E Eur Rev 52(129):548–565

Bodenhamer DJ (2007) Creating a landscape of memory: the potential of humanities GIS. Int J Human Arts Comp 1(2):97–110

Bowden MJ (1992) The invention of American tradition. J Hist Geogr 18(1):3–26

Brunhes J, Vallaux C (1918) German colonization in Eastern Europe. Geogr Rev 6(6):465–480

Clark C (1996) The wars of liberation in Prussian memory: reflections on the memorialization of war in early nineteenth-century Germany. J Mod Hist 68 (3):550–576

Clark C (2008) Iron Kingdom: the rise and downfall of Prussia, 1600-1947. Harvard University Press, Cambridge

Crampton A (2001) The Voortrekker Monument, the birth of apartheid, and beyond. Polit Geogr 20(2):221–246

Confino A (1993) The nation as a local metaphor: Heimat, national memory and the German Empire, 1871–1918. History Memory 5(1):42–86

Czovek T (2006) Three seasons of charismatic leadership: a literary-critical and theological interpretation of the narrative of Saul. Regnum Press, Waynesboro, David and Solomon

Drozdzewski D (2014) Using history in the streetscape to affirm geopolitics of memory. Polit Geogr 42:66–78

Drozdzewski D, De Nardi S, Waterton E (eds) (2016a) Memory, place, and identity: commemoration and remembrance of war and conflict. Routledge, London

Drozdzewski D, De Nardi S, Waterton E (2016b) The significance of memory in the present. In: Drozdzewski D, De Nardi S, Waterton E (eds) Memory, place, and identity: commemoration and remembrance of war and conflict. Routledge, New York, pp 1–16

Duncan JS, Barnes TJ (Eds) (1992) Writing worlds: discourse, texts and metaphors in the representation of landscape. Routledge

Dwyer OJ, Alderman DH (2008) Civil rights memorial and the geography of memory. Center for American Places at Columbia College, Chicago

Flint C (2002) The theoretical and methodological utility of space and spatial statistics for historical studies: the Nazi Party in geographic context. Hist Meth J Quant Interdisc Hist 35(1):32–42

Foote KE (2003) Shadowed ground: America's landscapes of violence and tragedy, 2nd edn. University of Texas Press, Austin

Forest B, Johnson J, Till K (2004) Post-totalitarian national identity: public memory in Germany and Russia. Soc Cult Geogr 5(3):357–380

Frankel R (2003) From the beer halls to the halls of power: the cult of Bismarck and the legitimization of a new German right, 1898–1945. Ger Stud Rev 26 (3):543–560
Frieberg A (2010) Transnational spaces in national places: early activists in Polish-West German relations. Nationalities Pap 38(2):213–226
Gerwarth R, Riall L (2009) Fathers of the nation? Bismarck, Garibaldi and the cult of memory in Germany and Italy. Euro Hist Quart 39(3):388–413
Giddens A (1987) A contemporary critique of historical materialism. Volume 2: the nation-state and violence. University of California Press, Berkeley
Gillen J (2014) Tourism and nation building at the War Remnants Museum in Ho Chi Minh City. Vietnam. Ann Assoc Am Geogr 104(6):1307–1321
Hobsbawm E, Ranger T (eds) (1983) The invention of tradition. Cambridge University Press, Cambridge
Harvey D (1979) Monument and myth. Ann Assoc Am Geogr 69(3):362–381
Herb GH (2004) Double vision: territorial strategies in the construction of national identities in Germany, 1949-1979. Ann Assoc Am Geogr 94(1):140–164
Hanna SP, Hodder EF (2015) Reading the signs: using a qualitative Geographic Information System to examine the commemoration of slavery and emancipation on historical markers in Fredericksburg. Virginia. Cult Geogr 22(3):509–529
Howard M (2005) The Franco-Prussian War: the German invasion of France 1870-1871. Routledge, New York
Johnson NC (1994) Sculpting heroic histories: celebrating the centenary of the 1798 Rebellion in Ireland. T I Brit Geogr 19(1):78–93
Johnson NC (1995) Cast in stone: monuments, geography, and nationalism. Environ Plan D 13(1):51–65
Johnson NC (1996) Where geography and history meet: heritage, tourism and the big house in Ireland. Ann Assoc Am Geogr 86(3):551–566
Johnson NC (2011) The contours of memory in post-conflict societies: enacting public remembrance of the bomb in Omagh, Northern Ireland. Cult Geogr 19(2):237–258
Kopp K (2012) Germany's wild east: constructing Poland as colonial space. University of Michigan Press, Ann Arbor
Kruk S (2008) Semiotics of visual iconicity in Leninist 'monumental' propaganda. Visual Communic 7 (1):27–56
Landry SM (2005) That all may be one? Church unity and the German national idea, 1866–1883. Church Hist 80 (2):281–301
Mayo JM (1988) War memorials as political landscape. Praeger Publishers, New York
McDowell S, Braniff M (2014) Commemoration as conflict: space, memory, and identity in the peace processes. Palgrave MacMillan, London/New York City
Mueller FL (2011) Our Fritz: Emperor Frederick III and the political culture of imperial Germany. Harvard University Press, Cambridge
Nelson RL (2010) From Manitoba to the Memel: Max Sering, inner colonization and the German East. Soc Hist 35(4):439–457
Nora P (1989) Between memory and identity: les lieux de memoire. Representations 26:7–25
Osborne B (2001) Landscape, memory, monuments, and commemoration: putting identity in its place. Can Ethnic Stud 33(3):39–77
Pflanze O (1955) Bismarck and German nationalism. Am Hist Rev 60(3):548–566
Post C, Alderman DH (2014) Wiping new Berlin off the map: political economy and the de-Germanization of the toponymic landscape in WWI USA. Area 46 (1):83–91
Retallack J (1997) Why can't a Saxon be more like a Prussian? Regional identities and the birth of modern political culture in Germany, 1866-1867. Can J Hist 32(1):26–55
Retallack J (2017) Election battles and the spectre of democracy in Germany, 1860-1918. Oxford University Press, Oxford
Russell MA (2000) The building of Hamburg's Bismarck Memorial 1898–1906. Hist J 43(1):133–156
Savage K (2011) Monument wars: Washington, DC, the National Mall, and the transformation of the memorial landscape. University of California Press, Berkeley
Schein RH (1997) The place of landscape: a conceptual framework for interpreting an American scene. Ann Assoc Am Geogr 87(4):660–680
Seele S (2005) Lexikon der Bismarck-Denkmäler. Michael Imhof Verlag, Petersberg, Türme, Standbilder, Büsten, Gedenksteine und andere Ehrungen
Smith AD (1999) Myths and memories of the nation. Oxford University Press, Oxford
Steinberg J (2012) Bismarck: a life. Oxford University Press, Oxford
Sutherland C (2011) Introduction: German politics and society from a cosmopolitan perspective. Ger Polit Soc 29(3):1–19
Taylor AJP (1949) The course of German history. Capricorn, New York
Till K (2005) The new Berlin: memory, politics, place. University of Minnesota Press, Minneapolis
Urbach K (1998) Between savior and villain: 100 years of Bismarck biographies. Hist J 41(4):1141–1160
von Luenen A (2013) Tracking in a new territory: re-imaging GIS for history. In: von Luenen A, Travis C (eds) History and GIS. Springer, Dordrecht, pp 211–239
Wagner-Pacifici R, Schwartz B (1991) The Vietnam Veterans Memorial: commemorating a difficult past. Am J Sociol 97(2):376–420

Waterton E, Dittmer J (2014) The museum as assemblage: bringing forth affect at the Australian War Memorial. Museum Manag Curatorship 29(2):122–139
Wilson JK (2007) Imagining a homeland: constructing Heimat in the German East, 1871–1914. Natl Identities 9(4):331–349
Winberry JJ (1982) Symbols in the landscape: the confederate memorial. Pioneer Am Soc T 5:9–15
Winberry JJ (1983) 'Lest we forget': the confederate monument and the southern townscape. Southeast Geogr 23(2):107–121
Wylie J (2003) Dwelling: a Glastonbury case study. Pearson, Country Visions (London, pp 136–57
Yiftachel O, Ghanem A (2004) Understanding ethnocratic regimes: the politics of seizing contested territories. Polit Geogr 23(4):647–676
Zelinsky W (1988) Nation into state: the shifting symbolic foundations of American nationalism. University of North Carolina Press, Chapel Hill

15 Mapping the Historical Transformation of Beijing's Regional Naming System

Yong Yu

Abstract

With the introduction of humanist and Marxist philosophies, we have witnessed a trend moving from empirical examination to theoretical discussions in historical geography. In comparison with the great developments of these theoretical perspectives, the research methods regarding how to collect and analyze data to support theories have still remained grounded in traditional notions. With the rise of historical GIS, research methods have rapidly developed from the textural to the visual, from case studies to spatial analysis on maps, and from methods separated into quantitative and qualitative to integrated methods. Although historical GIS is being used to provide new insights into geographical change over time, the changing of place names has received relatively little attention. To study changing regional names throughout Beijing's history, I provide a new systematic classification of place names. Place names can be classified into three intertwined parts: objects and numbers (scientific/practical ways of naming), histories and emotions (humanistic ways of naming), and politics and ideologies (social constructionist ways of naming). Through spatial analysis of these groups of names, we can see that the ways of naming the regions in Beijing has been transformed from the humanistic and practical ways during the Ming Dynasty (1368–1644), to the social constructionist ways during the Qing Dynasty (1644–1912), and then to the scientific ways during the Republic of China (1912–1949).

Keywords

Systematic classification of place names · Humanistic geography · Historical GIS · Beijing

15.1 Introduction

The changes in place names are only too common. For example, almost all the names given to commemorate the Tsar were changed to commemorate Lenin or Stalin in the USSR. From a positivist view, these changes are irrational and may cause a problem to determine the standard place names on a map (Misra and Ramesh 1989, 362). However, place names can tell stories and carry meanings. Like short stories, they may bring insight into the people who live or lived there. From this humanistic view, studying the changing of place names can help us to discover the hidden stories of the cultural process. More specifically, I will study changing regional names

Y. Yu (✉)
Department of Geography, University of Wisconsin, Madison, WI, USA
e-mail: yongyu@wisc.edu

C. Travis et al. (eds.), *Historical Geography, GIScience and Textual Analysis*, Historical Geography and Geosciences, https://doi.org/10.1007/978-3-030-37569-0_15

throughout Beijing's history. Rather than converting texts to GIS databases through their relations to place names, I will study place names themselves. Assuming that people name a region according to their understanding of the place, I provide a new systematic classification of place names. Ways of naming a place can be classified into three parts: scientific/practical ways of naming, humanistic ways of naming and social constructionist ways of naming. Through spatial analysis based on this classification, I will map the changes in Beijing's regional naming during the Ming Dynasty (1368–1644), the Qing Dynasty (1644–1912) and the Republic of China (1912–1949) in a historical GIS. In this way, we can see how changing ideologies, ideas and agendas have been reflected spatially in Beijing during the different historical times.

15.1.1 Place Names in Historical Geography Before the Spatial Turn

Historical geographers have long been interested in place names and there are no signs of them losing their vigor. Although place names have been widely explored across a spectrum of disciplines, I will focus my discussion before spatial turn and in three categories: scientific, historical-culturalist, and critical toponymies. Scientific toponymies may be related to gazetteers, cartography, and statistics. Place names in gazetteers and maps are depicted in two ways: textually and visually. Traditionally, a gazetteer listed place names, arranged in alphabetical order with the coordinates for each named feature (Randall 2001). Gazetteers took a descriptive approach and provided background knowledge about places. Maps are usually a good way for us to identify places visually through place names. However, it is not always easy to determine standard place names for use on maps (Misra and Ramesh 1989). As a result, attempts to standardize geographical names have been made on a multitude of scales from the national to the global since the United Nations took on these issues (Kerfoot 2016). Most of these studies have been empirically-oriented and provided a positivist analysis of the historic past.

Historical-culturalist toponymies have largely been concerned with seeking the origin and meanings of place names. Some scholars have attempted to discover the hidden stories of the cultural process behind place names through linguistic and textual analysis (e.g., Kearns and Berg 2002; Radding and Western 2010). Others have been interested in mapping the spatial distribution of certain types of place names in the past. For example, Zelinsky mapped two types of patriotic place names (national personages such as Washington and Franklin, and patriotic terms such as Union and Independence) and analyzed their spatial pattern (Zelinsky 1983). Following the cultural turn in historical geography, the field of toponymic research has evolved into a field of enormous breadth. Critical toponymies interrogate both scientific and historical-culturalist studies. According to them, traditional historical geographers have paid too much attention to "bygone landscapes", and as a result, their studies on place names have been characterized by political innocence and a lack of theories (Vuolteenaho and Berg 2009). Scientific studies such as name standardization have also been questioned because the authorization of scientific knowledge and technology involved in the process of standardization may result in the erasure of the inherited hereroglossia of local names (Harley 2001; Vuolteenaho and Berg 2009).

Accordingly, critical geographers have taken to seeking evidence of deeper social structure and relations than may be observed on the sensory surface and they tended to study place names in two main ways. By emphasizing material base, they may be concerned of commodification of place names (e.g., Medway and Warnaby 2014). By emphasizing the power of politics the power-related aspects of naming are also stressed (e.g., Azaryahu 1996). With the introduction of humanist and Marxist philosophies, we have witnessed a trend moving from empirical examination to theoretical discussion in historical geography. The new directions have varied and addressed various perspectives and ideas, primarily humanism, Marxism, structuralism/structurationism and feminism. In comparison with the great developments of these theoretical perspectives, the research methods regarding how to collect and analyze data to support theories have remained grounded in traditional notions and need further examination. For example, critical toponymies demonstrate their critical and theoretical analysis by using selected place names as case studies (e.g., Azaryahu 1996). By contrast, humanistic geographers brought forth insights into the concept of place and space, though there are still very limited studies on place names using this notion. The first, critical toponymies, needs to develop its research method for dealing with multiple types and amounts of data and the latter, humanistic geographers, need to go beyond textual analysis and move in a broader direction.

15.1.2 Place Names in Historical GIS

With the rise of geographical technologies, historical geography has experienced the spatial turn. Today "the methodological palette of a geographer typically consists of quantitative, qualitative, textual, and visual methods" (Dear 2011, 312). Quantitative methods may be largely used in scientific toponymies, critical geographers may place their emphasis on qualitative methods, and text is clearly the most widely used form for data across the humanities. The mainly visual methods provided by GIS could be regarded as an integrative tool to cover all types of data and methods across boundaries of these three areas. As a spatial tool, GIS allows various geographical data to be used in ways that are far more powerful than any other approach permits (Gregory and Healey 2007). GIS could enable critical geographers to unlock spatial and social relations by using digital mapping (Strocchia and Rombough 2016) and move their social theoretical analysis from their selected samples to maps. GIS could also visualize the patterns of the poet Kavanagh's daily life in rural and urban spaces, help analyze key elements in his stories (Travis 2015) and bring forth humanistic discussion on emotions and experiences from the textual to the visual. In this way, GIS is not only a spatial tool that addresses all kinds of data, but also a spatial way of thinking to create new insights into the geographies of the past. In detail, my discussion about place name studies in historical GIS includes three themes: building GIS databases by integrating all types of data, mapping toponyms and their relationships to environmental characters, and converting textual data into GIS databases.

First, in today's digital time, gazetteers are no longer lists of names. They have been developed into historical GIS databases that integrate quantitative, qualitative and textual data. Many of them focus on national historical GISs such as the Great Britain Historical GIS (Gregory et al. 2002) and the China Historical GIS (Berman 2005; Bol and Ge 2005). Gazetteers provide a way to model the kind of complex social, vernacular, and humanistic ideas of place and could be regarded as a research source to produce substantive scholarship for historians (Berman et al. 2016).

Second, some toponymic studies are concerned with mapping spatial patterns of toponyms and their relationships to environmental characteristics. A good example is the Project of Terrain Characteristics and Tai Toponyms (Luo et al. 2010). Luo selected Tai place names (starting with Muang, Chiang and Viang) from "GEOnet Names Server (GNS)" database, mapped their patterns by using a *kernel density* estimation technique, and analyzed their relationship with terrain characteristics. Another example is "Sinification of Zhuang place names in Guangxi" (Wang et al. 2012). Wang classified place names as Zhuang or non-Zhuang, and coded them as a 0-1 binary variable. By using spatial smoothing and spatial interpolation methods, this study visualized the spatial pattern of Zhuang toponyms and their relations to environmental characteristics. Both of Luo and Wang studied and classified place names through a qualitative discussion and then analyzed their spatial pattern with quantitative methods in GIS.

Third, perhaps the biggest challenge in applying GIS to history is in converting unstructured texts into GIS databases. Accordingly, many scholars emphasize developing techniques for incorporating and embracing text.

For example, Gregory is concerned with bringing together two fields of research, *corpus linguistics* and GIS (Gregory and Hardie 2011). By using techniques from *corpus linguistics*, Gregory extracted place names from corpora and created a database of the places mentioned in a *corpus* which can be analyzed by GIS. Place names in this study were used to connect the content of the texts and the GIS database.

Although GIS contributes to place names in many ways, there is still a need to study place names by mapping their meanings. Rather than converting texts to GIS databases through their relations to place names, I will study place names themselves, because like texts, place names can tell stories and carry meanings. As one writer has said of place names, "From these names one can draw virtually the whole history of her people, their ideas, ideals, and legends; their ancestry and religions, their industries, politics and wars; as well as the climate, topography, and vegetation of their land" (Drew 1962, 255).

15.1.3 Research Questions

Although historical GIS is also being used to provide new insights into geographical change over time, the changing of place names has received relatively little attention. The changes in place names are only too common. For example, Beijing (Northern Capital 北京) was called Dadu (Great Capital 大都) in the Yuan Dynasty, Beijing in the Ming and Qing Dynasties, Beiping (Northern Peace 北平) during WWII, and Beijing again up to the present day. All these changes involved Chinese people's feelings about this city during different times. As the present capital of China, Beijing has been divided into different regions over its history. How have Chinese people named these regions in different ways? How have changing ideologies, ideas and agendas been reflected spatially in Beijing during the different historical times? What can Beijing's place-names tell us about the psychological preferences and choices of the Chinese people, given their different understandings of the city and its environs over history?

15.1.4 Psychological Motivation for Place-Naming

People name a location because they care about it enough to give it a name. This study assumes that people name a place based on their understanding of the place. Among all the different understandings, they have free will in choosing to give a name to a place. Thus, a place name could reflect the namegivers' psychological preferences in their understanding of the place.

Place can be understood in many ways. It can be depicted variously as a bounded location, a point in space arranged in a logical number, a moving exhibition of feeling, a coming together of flows, a form of fixed capital, or the maintenance and transformation of relations of domination, oppression and exploitation. In the humanistic sense, when space is freedom, place is security (Tuan 1977). For social geographers, place is a place, in whatever guise and, like space and time, is a social construct (Harvey 1996). Each explanation of a place can be assumed as one way of understanding that place. Such explanations involve the process of place naming and can be reflected in the names. For example, if we think of a place as a logical point in space, we may name it "First", "Second" or "Third", which can reflect our logical and scientific view. If we think of a place as a moving exhibition of feeling, we may name it "Furnace Creek" which can reflect our feeling about the place. If we think of a place as a social construct, we may name it as Martin Luther King which can embody a political ideology.

15.1.5 A New Classification of Place-Naming

Based on my assumption, if we classify place names according to our different ways of understanding places, we can map these understandings through the spatial distributions of place names. Some geographers have done work on place name classification. George R. Stewart wrote a pioneering geographical book *Names on the Land*, which recognized the importance of psychological motives in place naming. Stewart classified place names into nine categories, such as descriptive names, incident names, or commemorative names (Stewart 1954). His classification was brilliant but mainly based on American names, and may be hard to apply in the study of Chinese place names. People with different cultural backgrounds, living in different parts of the world, may name their places in different ways. In the USA, Martin Luther King streets may be seen all over the map, yet for all the worshiping of Chairman Mao during the Cultural Revolution, there is not a single street named after him in China. Accordingly, I will provide a new classification of place names based on our different ways of understanding a place. For the purpose, I will start with a review of the geographical and philosophical ideas about place in homo geographical sense and create the classification system in a more general way.

By drawing on geographical experiences of space, place, home and world, Sack (1997) provides a theory of space and place. To *homo geographicus*, place is a loom weaving elements from three realms: nature, meaning, and social relations. From this perspective, our understanding of places may be classified into three corresponding categories: practical/scientific (associated with elements of nature), humanistic (associated with meaning), and social constructionist ways (associated with social relations). Specifically, the question becomes how do we understand a place in different ways?

(1) Place can be an objective location full of *geographical facts*. We could scan geographical components such as trees, rivers, or mountains in the region. One of them might catch our eye, and we might name the region after it. In this way, the region is named by describing geographical elements.
(2) Place can be a location arranged in a *logical and quantitative pattern*. We may name it after logical numbers such as the second, third and so on. In this way, the name could help people formulate a mental image of an organizational structure that facilitates wayfinding.
(3) Place can be meaningful and full of *imaginations and memories of the past*, the name may be historical, emotional or romantic, or perhaps related to a person, a story or myth, a mood or feeling, etc. For example, "Furnace Creek" in the heart of Death Valley describes people's imaginative feeling about the place.
(4) Place can be *social construction*. We may name a place to embody a political ideology. For example, in 1924 during the socialist revolution, Petrograd was renamed Leningrad. "Red Tower Street" in Cultural Revolution China could embody a socialist ideology.

In short, we can classify ways of place naming into three categories: practical/scientific, humanistic, and social constructionist, according to the ideal realities we may conceive of in our mind (Table 15.1).

15.2 A Case Study of Beijing' Regional Naming Changes

By addressing the new classification of place names, I will examine the changes in Beijing's regional naming during the Ming Dynasty (1368–1644), Qing Dynasty (1644–1912) and the Republic of China (1912–1949) in a historical GIS. This study is based on the *ArcMap of Beijing* from ArcGIS online, together with the *Ming Dynasty Beijing map* (Hou 1988), *Qing Dynasty Beijing map* (Hou 1988) and *Map of Beijing Divided by Districts* in the Republic of China (Li 2005). The Coordinate System used in this study is *WGS 1984 Web Mercator Auxiliary Sphere*. The information about different ways of regional naming can be stored into the attribute table in GIS and presented in maps.

Table 15.1 Classification of place names according to different ways of understanding (Yu 2015)

Ways of place naming	Activities for place naming	Examples	Functions
Practical/scientific ways	Describing geographical facts in the location	UW-Hospital, Western Town Mall, Wisconsin State Capitol	Identification, facilitating wayfinding
	Using logical numbers	Main street, Second Street, Third Street, Fourth Street	Spatial orientation, facilitating wayfinding
Humanistic ways	Describing the history of the place, showing the imaginations, feelings and thoughts about the place	Shake Rag Under The Hill	Building deep relations between people and the environment
		Furnace Creek	
		Lincoln Street	
Social constructionist ways	Differentiating social groups, embodying ideology	Worker's Stadium	Expressing social relations and the power of politics
		People's Square	
		Street of Red Tower	
		Leningrad	

15.2.1 Humanistic and Practical Ways of Naming Quarters in the Ming Dynasty (1368–1644)

Approximately 100 years after the construction of Dadu, Zhu Yuanzhang founded the Ming Dynasty and made Nanjing his capital. His fourth son, Zhu Di became the emperor in 1403 and he then moved the capital from Nanjing to Beijing (Northern Capital 北京).

Ming Dynasty Beijing originally included 33 quarters. After building the outer city, the inner city was redivided into 28 quarters while the outside city was divided into 8 quarters. There were 36 quarters in total (Yin and Sun 2009). The ways of naming these regions are discussed in the following section.

First, 18 quarter names in the Yuan Dynasty continued to be used in the Ming Dynasty (Yin and Sun 2009). The Yuan Dynasty quarters were named by Yu Ji, a very learned man from the Imperial Academy. Accordingly, all these names were elegant and beautiful, and embodied high culture in ancient China. Among these names, some may tell historical stories of Beijing such as the Quarter of Golden Terrace (金台坊), Quarter of the Singing Jade (鸣玉坊), and Quarter of the Gold City (金城坊). Others conveyed special feelings about the place. For example, the Quarter of Southern Fragrance (南薰坊) was located in the southern part of the inner city, and its name described residents' feelings about the southern wind.

Second, some quarter names changed between the Yuan Dynasty and the Ming Dynasty. For example, the Quarter of Harmony (时雍坊) in the Yuan Dynasty had been changed to two quarters—the Big Quarter of Harmony (大时雍坊) and the Small Quarter of Harmony (小时雍坊).

Third, 4 quarters were newly built in the inner city of Beijing in the Ming Dynasty: the Quarter of Loyalty (教忠坊), the Quarter of Education (崇教坊). The Quarter of Western Side River (河槽西坊), and the Quarter of Chaotian Palace (朝天宫坊). While the Quarter of Loyalty (教忠坊) and the Quarter of Education (崇教坊) still carried cultural meanings in their names, the Quarter of Western Side River (河槽西坊) and the Quarter of Chaotian Heaven (朝天宫坊) were named by describing the real geographical landmarks.

Fourth, all 7 quarters in the outer city were newly built in the Ming Dynasty. Some of them, such as the Central Eastern Quarter (正东坊) and the Central Southern Quarter (正南坊), were named after their sitting directions. Others, such as the Northern Quarter of Declaration (宣北坊), were named by combining the quarters' sitting direction and the city gate names—for example, the Gate of Military Declaration (宣武门).

In general, the quarters in the Ming Dynasty were named in humanistic and practical ways. Their names conveyed feelings, told historical stories, and described the real geographical landmarks used to facilitate wayfinding (Fig. 15.1).

15.2.2 Social Constructionist and Practical Ways of Naming in Qing Dynasty (1644–1912)

When the Manchu began to develop quickly and became strong enough to conquer parts of China that were outside the Great Wall, the Ming court was still fighting for survival against peasant revolts. In 1644, the leader of the rebel army Li Zicheng invaded Beijing. Emperor Chongzhen, the last emperor of the Ming Dynasty was rendered powerless and hung himself from a tree in Mei Mountain (Landscape Mountain 景山 today), thus realizing his promise of defending the capital until death. Later, Li Zicheng was defeated by the Qing army and the old Ming general Wu Sangui. The Manchu army entered

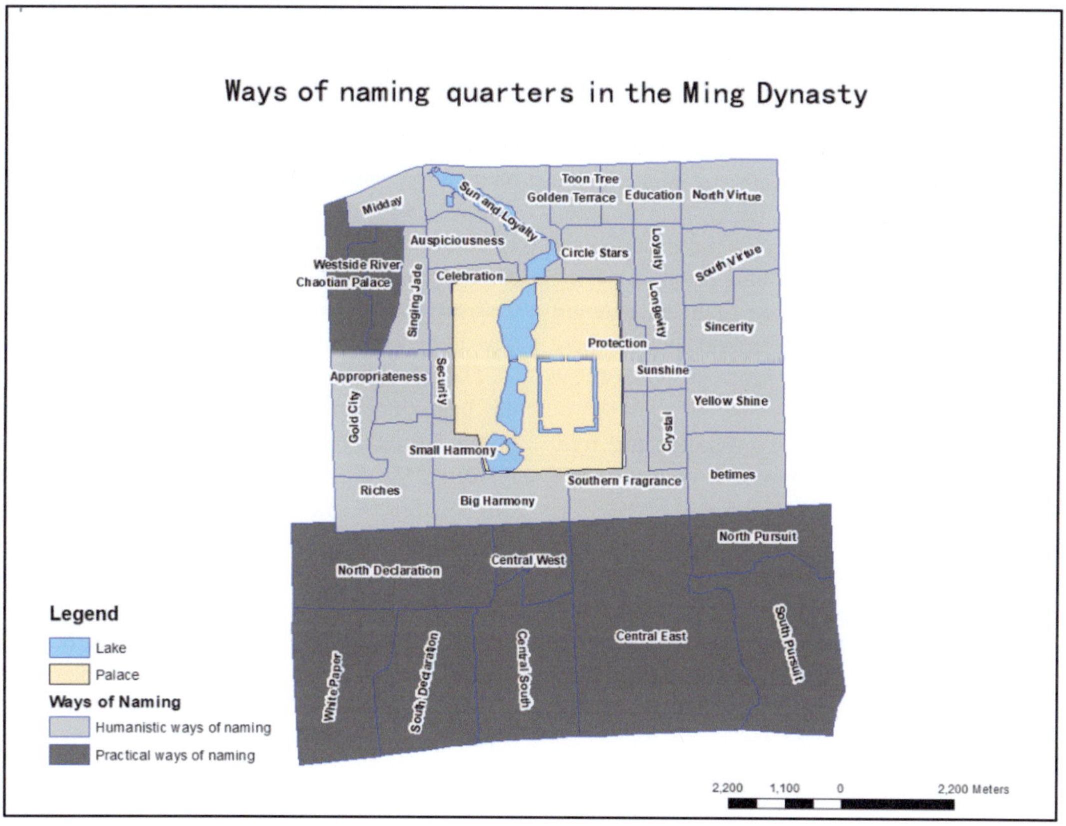

Fig. 15.1 Ways of naming quarters in the Ming Dynasty. Created by the author from *Ming Dynasty map* in the *Historical Maps of Beijing* (Hou 1988) and ArcGIS online map

the Central Plains and the Qing overthrew the Ming Dynasty. Beijing again became the capital. There was no great difference in the layout of Beijing between the Qing and Ming Dynasties. However, the policy of "separating Bannermen from Han Chinese" was implemented at the beginning of the Qing Dynasty and broke the previous land use layout.

In the Qing Dynasty, the division of the city into quarters was eclipsed by a new system of organization using the Eight Banners, transforming the inner city into an enormous garrison (Elliott 2001). As it had been during the Ming Dynasty, Beijing during the Qing dynasty was composed of two distinct cities: the inner city and the outer city. These two cites were also called the Tartar city and the Chinese city, which suggests that the division was based on ethnic groups rather than on the neighborhoods.

The inner city was designed to be a military camp. As part of a camp, the residential zones of the inner city were named after colors and banners. The inner city was divided into eight sectors, with each sector serving as a residential zone for one of the Eight Banners: bordered yellow (镶黄旗), plain yellow (正黄旗), plain white (正白旗), plain red (正红旗), bordered white (镶白旗), bordered red (镶红旗), plain blue (正蓝旗) and bordered blue (镶蓝旗). Each zone was further divided into Manchu, Mongol and Chinese banner neighborhoods. All these names differentiated the social groups of the Manchu by using colors.

Although the outside city imitated the five-city system of the Ming Dynasty, the division of the city was completely different. The five regions in the outside city were named East City, North City, Central City, West City and South City but their distribution was not completely based on directions; instead, they were set up from west to east in the following order: West City, North City, Central City, South City and East City. South City was not in the south and North City was not in the north. As a result, the geographical associations of their names were much weaker than they had been in the Ming Dynasty.

In short, there were two ways of naming districts in Qing Dynasty Beijing. The first was to differentiate between social groups through colors in the inner city. The second was to describe the districts by using directions in the outer city. The first way of naming was a social constructionist method, and the latter was a practical method (Fig. 15.2).

15.2.3 Scientific Ways of Naming in the Republic of China (1912–1949)

At the end of the Qing Dynasty, due to the nineteenth-century humiliations resulting from a series of foreign military invasions, China recognized the weaknesses in its society and was forced to reform. As a result, the idea of "Chinese learning for essence, western learning for use" in the Confucian reform movement was carried out by the Qing government during the late nineteenth century. Although western sciences and technologies were welcomed, a traditional Chinese ideal world view was still maintained. The reforms did not prevent the fall of the Qing Dynasty, which was overthrown in 1911 during the Xinhai Revolution, led by Sun Zhongshan. This extraordinary transformation of Chinese society led to the introduction of various new ideologies.

The May Fourth Movement of 1919 marked the beginning of the New-Democratic Revolution in China. People tried to save their nation by inviting "Mr. Democracy" and "Mr. Science" to China. Since then, Mr. Science and Mr. Democracy have gradually become new ideal images of society in the Chinese imagination. In this way, a traditional world view of pursuing the harmony of heaven, earth and man has become part of the old national culture and has been heavily criticized. Instead, the objective and scientific world has become the ideal images of what we think reality ought to be.

As a result, Beijing was redivided into 11 districts and each district was renamed. There were 6 districts in the inner city: Inner First

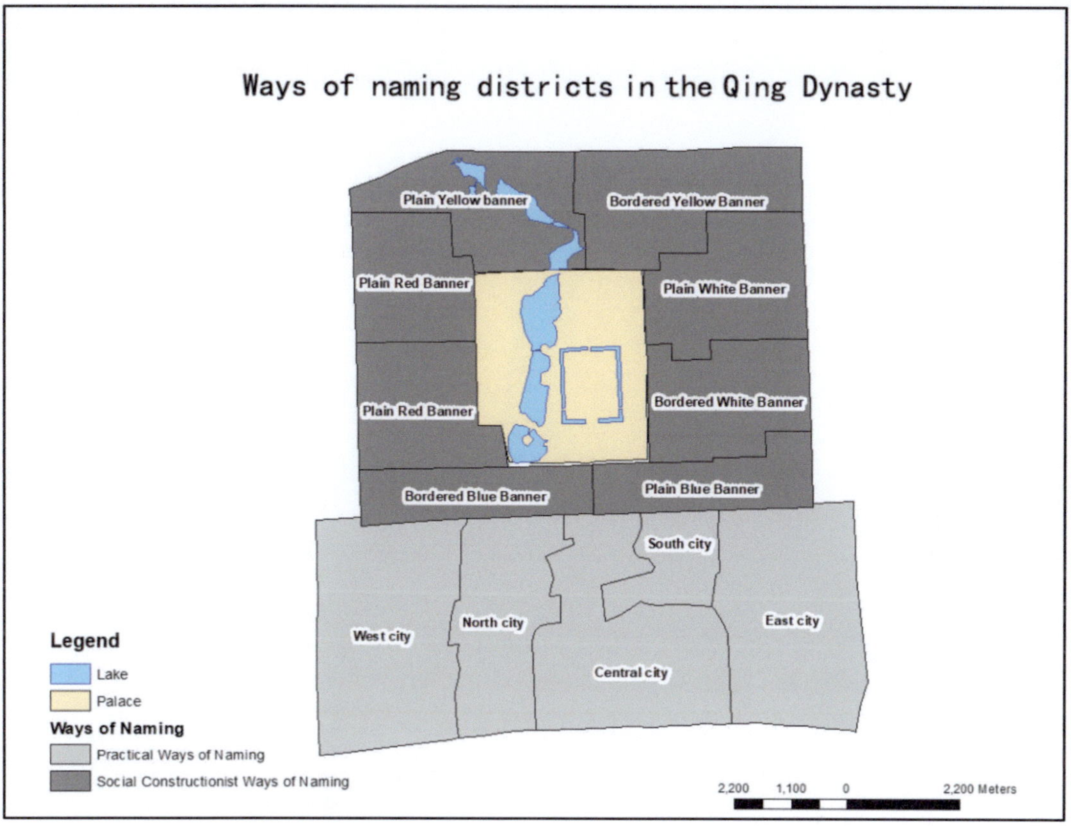

Fig. 15.2 Ways of naming districts in the Qing Dynasty. Created by the author from *Qing Dynasty map* in the *Historical Maps of Beijing* (Hou 1988) and ArcGIS online map

District (内一区), Inner Second District (内二区), Inner Third District (内三区), Inner Fourth District (内四区), Inner Fifth District (内五区), and Inner Six District (内六区). And also there were five ones in the outer city—Outer First District (外一区), Outer Second District (外二区), Outer Third District (外三区), Outer Fourth District (外四区), and Outer Fifth District (外五区). In 1943, the Legation Quarter was reincorporated by the Republic of China. As a result, it became the seventh district in the inner city. These divisions of Beijing present us with a logical world full of numbers. These scientific ways of naming the regions could show that Chinese people at that time took the logical, objective and scientific world as the ideal reality. This ways of naming the districts is shown in Fig. 15.3.

15.3 Conclusion

In general, Beijing's division and regions have changed from time to time over the city's long history. Assuming that people name a region according to their understanding of the place, I provide a new systematic classification of place names. Place names can be classified into three intertwined parts: objects and numbers (scientific/practical ways of naming), histories and emotions (humanistic ways of naming), and politics and ideologies (social constructionist ways of naming). Through spatial analysis of these groups of names, we can see that the methods of naming the areas in Beijing has transformed from the humanistic and practical ways of the Ming Dynasty (1368–1644), to the

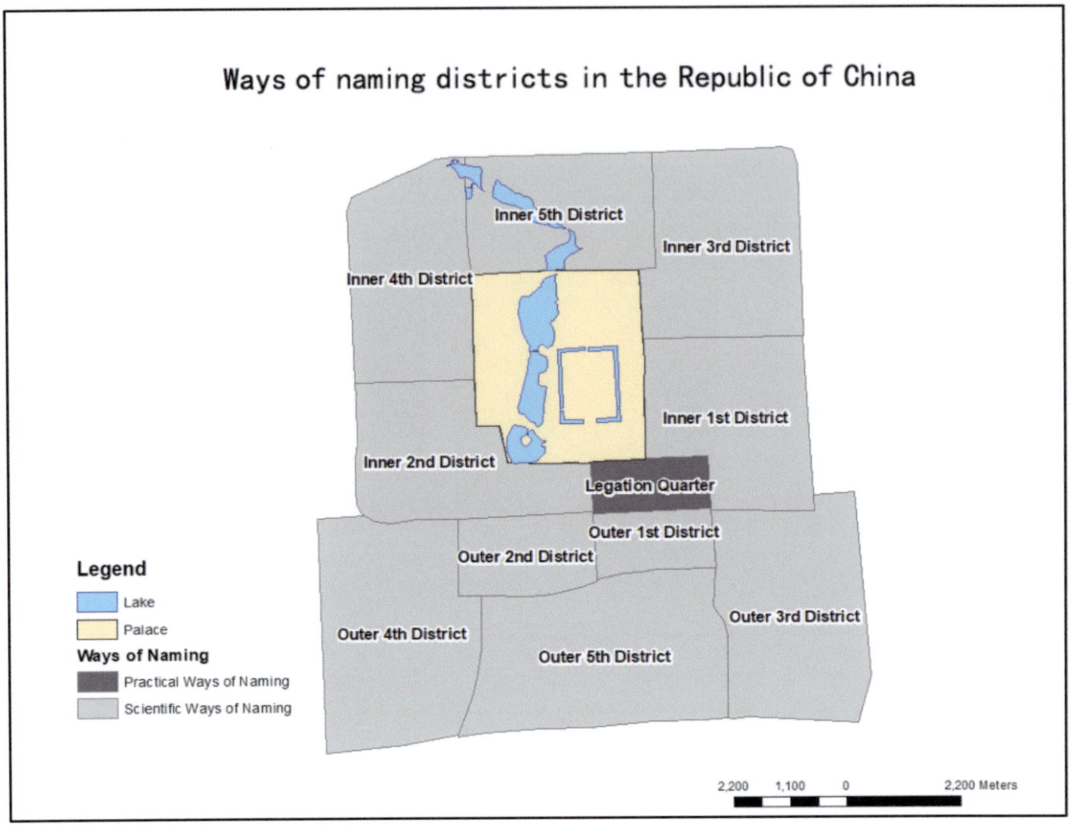

Fig. 15.3 Ways of naming districts in the Republic of China. Created by the author from 'Map of Beijing Divided by Districts' in *Beijing's History and Maps* (Li 2005) and ArcGIS online map

social constructionist and practical ways of the Qing Dynasty (1644–1912), and then to the scientific ways of the half-heartedly modern Republic of China (1912–1949).

References

Azaryahu M (1996) The power of commemorative street names. Environ Plan D 14(3):311–330

Berman ML (2005) Boundaries or networks in historical GIS: concepts of measuring space and administrative geography in Chinese history. Hist Geogr 33:118–133

Berman ML, Mostern R, Southall H (eds) (2016) Placing names: enriching and integrating gazetteers. Indiana University Press, Bloomington

Bol P, Ge J (2005) China historical GIS. Hist Geogr 33:150–152

Dear M (2011) Afterword: historical moments in the rise of the geohumanities. In: Dear M, Ketchum J, Luria S, Richardson D (eds) GeoHumanities: art, history, text at the edge of place. Routledge, London/New York, pp 309–314

Drew S (1962) Place names in ten northeastern counties of Florida. Am Speech 37(4):255–265

Elliott MC (2001) The Manchu way: the eight banners and ethnic identity in late imperial China. Stanford University Press, Stanford

Gregory IN, Bennett C, Gilham VL, Southall HR (2002) The Great Britain historical GIS project: from maps to changing human geography. Cartogr J 39(1):37–49

Gregory IN, Healey RG (2007) Historical GIS: structuring, mapping and analysing geographies of the past. Prog Hum Geogr 31(5):638–653

Gregory IN, Hardie A (2011) Visual GISting: bringing together corpus linguistics and geographical information systems. Lit Ling Comput 26(3):297–314

Harley JB (2001) The new nature of maps: essays in historical cartography. Johns Hopkins University Press, Baltimore

Harvey D (1996) Justice, nature and the geography of difference. Blackwell, Oxford

Hou R (1988) Beijing lishi dituji (Historical maps of Beijing 北京历史地图集). Beijing Publishing House, Beijing in Chinese

Kearns RA, Berg LD (2002) Proclaiming place: towards a geography of place name pronunciation. Soc Cult Geogr 3(3):283–302

Kerfoot H (2016) Gazetteers global: United Nations geographical name standardization. In: Berman ML, Mostern R, Southall H (eds) Placing names: enriching and integrating gazetteers. Indiana University Press, Bloomington and Indianapolis, pp 35–50

Li C (2005) Beijing lishi yu tuji (Beijing's history and maps 北京历史与图集). Waiwen Chubanshe

Luo W, Hartmann JF, Wang F (2010) Terrain characteristics and Tai toponyms: a GIS analysis of Muang, Chiang and Viang. GeoJournal 75(1):93–104

Medway D, Warnaby G (2014) What's in a name? Place branding and toponymic commodification. Environ Plan A 46(1):153–167

Misra RP, Ramesh A (eds) (1989) Fundamentals of cartography. Revised and Enlarged Edition. Concept Publishing Company, New Delhi

Radding L, Western J (2010) What's in a name? Linguistics, geography and toponyms. Geogr Rev 100(3):394–412

Randall RR (2001) Place names: how they define the world and more. The Scarecrow Press, Lanham/London

Sack R (1997) Homo geographicus. Johns Hopkins University Press, Baltimore

Stewart GR (1954). A classification of place names. Names 2(1):1–13

Strocchia S, Rombough J (2016) Women behind walls: tracking nuns and socio-spatial networks in sixteenth-century Florence. In: Terpstra N, Rose C (eds) Mapping space, sense, and movement in Florence: historical GIS and the early modern city. Routledge, New York, pp 87–106

Travis CB (2015) Abstract machine: humanities GIS. Esri Press, New York

Tuan YF (1977) Space and place: the perspective of experience. University of Minnesota Press, Minneapolis

Vuolteenaho J, Berg LD (2009) Towards critical toponymies. In: Vuolteenaho J, Berg LD (eds) Critical toponymies: the contested politics of place naming. Johns Hopkins University Press, Baltimore, pp 1–18

Wang F, Wang G, Hartmann J, Luo W (2012) Sinification of Zhuang place names in Guangxi, China: a GIS-based spatial analysis approach. Trans Inst Br Geogr 37(2):317–333

Yin J, Sun D (2009) Beijing diming yanjiu (studies on Beijing's place names 北京地名研究). Beijing Yuanshan Publishing House, Beijing in Chinese

Yu Y (2015) Beijing's place-names: from cosmological symbolism to urban-planner practicality. Unpublished PhD thesis, Department of Geography, University of Wisconsin-Madison

Zelinsky W (1983) Nationalism in the American place-name cover. Names 31(1):1–28

Geographical Enrichment of Historical Landscapes: Spatial Integration, Geo-Narrative, Spatial Narrative, and Deep Mapping

16

May Yuan

Abstract
This chapter reviews cartographic and phenomenological views of landscape; argues for opportunities to traverse freely between the two seemly disjoint perspectives to landscape understanding, and discusses four geospatial approaches to enrich historical landscapes geographically. Innovations and advances in Geographical Information Science and Technology (GIST) not only facilitate the spatial integration of geographic features and perspectives in landscape interpretation but expand the traditional authoritative and demarcated mapping of the land into collective constructs of contingent personal experiences over space and time. The discussion applies Montello's *figurative*, *vista*, *environmental* and *geographic spaces* to frame the transition of landscape concepts from phenomenological to cartographic perspectives. GIST methods of spatial integration, geo-narrative, spatial narrative, and deep mapping incorporate geographic features, feelings, events, and senses to enrich the representation, analysis, and communication of historical landscape. Recent developments in unmanned aviation vehicle (UAV) surveys, virtual reality, and augmented reality give great promises for further geographic enrichments to historical landscapes with holistic syntheses of cartographic and phenomenological perspectives. The chapter concludes with challenges to formalize the methodology and epistemology for studying historical landscape processes and transformation.

Keywords
Cartographic landscape · Phenomenological landscape · Scale · Human experiences · Figurative space · Vista space · Environmental space · Geographic space · Horizontal dominance · Vertical dominance

M. Yuan (✉)
School of Economic, Political and Policy Sciences, University of Texas, Dallas, TX, USA
e-mail: myuan@utdallas.edu

16.1 Introduction

In the early 2000s, many historians and humanities scholars embraced the spatial turn to rework the significance of spatiality for which space is as important as time in the unfolding of human activities, ideas, and accomplishments. GIS applications in historical research (a.k.a. Historical Geographic Information Systems—HGIS) arose in early 2000s with growing historical databases on demographics, diseases, transportations, land ownerships, national territories, gazetteers, etc. compatible for spatial analysis over time in GIS (Gregory and Healey 2007; Yuan 2014). From HGIS to Spatial Humanities (SH), much progress has been made in

C. Travis et al. (eds.), *Historical Geography, GIScience and Textual Analysis*, Historical Geography and Geosciences, https://doi.org/10.1007/978-3-030-37569-0_16

developing spatial databases of the mappable past and producing interactive maps with vantage images, oral histories, archived aerial photography, and geographical data. Maps have long been an essential tool to record human experiences in addition to many other means, such as text, speech, music, and artifacts. As geospatial information technologies proliferate, the mappable expands. Documents, songs, poems, paintings, and stories can be geotagged (in parts or as a whole) and mapped to show their origins, contents, or other geographic connections. As such, maps are not just a means to document the past but become a platform to spatially integrate histories and reveal hidden spatial relationships among human affairs which would, otherwise, remain buried in the aspatial silos of history.

Indeed, maps play a vital role in the development of the landscape ideal as cartographical representations of the land. One approach subscribes to the objective views of landscape with boundaries derived from surveys to enclose land and then applies cartographic techniques to illustrate the physical and sociocultural features in the landscape as perceived and imposed by usually an elite or authority. An alternative view of landscape as open common places later challenges the cartographic view of landscape with enclosed private spaces (Olwig et al. 2016). The alternative landscape view advocates for theoretically informed landscape research based on phenomenology where landscape constitutes relational places linked by pathways, movements, and narratives of ordinary people (Whyte 2015; Tilley 1994). Phenomenologists consider that a landscape is subjectively constituted from human experiences, which accumulate through walking and describing monuments and linear features with notes of what is or is not seen at different locations (Johnson 2012).

Shifts between cartographic and phenomenological approaches to landscape research may be considered as changes in spatial scales. Rooting in human perception, memory, and behavior, Montello (1993) categorized four major classes of psychological spaces: figural, vista, environmental, and geographical, according to the projective size of space relative to the human body. Figural space is projectively smaller than the human body, and we can, therefore, observe its properties and manipulate objects at the scale directly from one location without locomotion. A desktop space, a painting, and a distant landmark are examples of figurative space. Vista space is larger than the human body but still can be observed from a single vantage point without appreciable locomotion. Examples are a room, a town square, a small lake, or the horizon. Environmental space surrounds our body and therefore requires the integration of observations from multiple locations to construct a mental map of the environment through direct experiences and memories. A building complex, a neighborhood, and a city are examples of environmental space where people carry out their everyday activities. Beyond one's daily life is geographic space, which is not comprehensible through locomotion or direct experiences. Therefore, symbolic representation, such as languages and graphics, is necessary to acquire, record, and communicate knowledge about geographic space. Cartographic and remote-sensing methods facilitate the production of the symbolic representation of geographic space. Symbolic representation reduces geographic space to figurative space to accommodate human's limited accessibility in space and time, offering an overview of otherwise an unreachable geographic extent. Human experiences are contextualized in figurative, vista, and environmental spaces. Figurative space would be most personal. Vista space would be most related to viewshed or visibility analysis for foraging, protection, and festivities. Environmental space would share among multiple groups with differential occupancies, uses, and perceptions of space. The transition from cartographic to phenomenological landscape, therefore, corresponds to zooming from large geographic space to environmental, vista, and figurative spaces with an increasing attention to individual and experiential relations to the land.

While progressing independently, democratizing landscape from the cartographical to phenomenological approaches parallels popularizing the development of geospatial technologies from mainframe computing facility to crowd-source collaborative platforms that bring diverse views to a map. The shift from authoritative power to personal experience is also a transition from macro- to micro-scale analysis and understanding. European Landscape Convention (ELC) defines landscape as "*an area, as perceived by people, whose character is the result of the action and interaction of natural and/or human factors*" (*Council of Europe* 2000, Article 1—Definitions, p. 3). Geospatial technologies offer cartographic and crowd-sourcing tools that combine survey measurements and volunteered geographic information for landscape mapping. The ELC calls for states to introduce landscape in the law "as an essential component of people's surroundings, an expression of the diversity of their shared cultural and natural heritage and a foundation of their identity" (*Council of Europe* 2000, Article 5a, p. 3). Geospatial technologies can provide the information platform with capabilities to map individual's surroundings with cultural and natural features at a micro-scale and smoothly transition the micro geographic context to meso and macro scales to capture both the subjective and objective perspectives of a landscape.

While maps portray very well where things are and how they relate spatially, maps are intrinsically limiting on temporal representation. Maps of landscapes are static; animating a map series over time gives rise to an exciting, yet often busy view of landscape transitions. In contrast, a landscape is a product of complex patterns of change over a long period resulted from the actions and interactions of the natural and human factors. Temporality is essential to landscape characterization. Over the years, the development of Temporal GIS (TGIS) offers new data models to include temporal information in spatial databases of relevance to historical studies (Yuan 2014). From the simple approach of time-stamping maps (or time slices) to more complicated strategies of time-stamping fine-scale data items (such as individual spatial features), time-stamp solutions have been implemented in various HGIS and SH projects. For example, the China HGIS project developed a temporal gazetteer with years of existence for every place name (Berman et al. 2012). However, time-stamp approaches provide no direct bearing on the processes that transform a landscape. Time-stamped maps that freeze landscape in time present inherited disadvantages to identify processes and extrapolate these processes of temporal change. von Lünen (2016) argued strongly that GIS technology casting time as a static coordinate along with spatial coordinates results in a series of meaningless snapshots that exile temporal associations of processes and human experiences in geography. Nevertheless, understanding spatial processes is fundamental to many fields of arts and sciences. From astronomy to classics, spatial processes produce the distribution of matter, energy, life or culture as well as how the distribution evolves over space and time. In the grand scheme of things, common to many disciplines are the questions of what, where, when and how, and a discipline builds its body of knowledge through its epistemological ways of creation and dissemination of knowledge. Knowing what, where, and when leads to knowledge of the existing facts that may give hints of how things happened and might continue to evolve. Many disciplines use existing knowledge of the spatial organization to project temporality. For example, the principle of superposition states that the position of a geological or archaeological stratum in an undeformed stratigraphic sequence reflects its chronological order in the deposition process: the deeper, the older (Harris 2014). Space-for-time substitution is also commonly used to infer past or future trajectories of ecological systems from contemporary spatial patterns based on the assumption that spatial relationships between climate and biodiversity identified in the current spatial patterns can be used to project temporal trajectories of biodiversity under changing climates (Blois et al. 2013). Such epistemology based on persistent functional connections between spatial patterns and temporal trajectories may not be applicable to landscape evolution as it is challenging to hold the comparable assumption that drivers of spatial variations of human activities also drive temporal changes in the diversity of activities. How contemporary observations can be back-projected to the experiences of the past? However, the concept that translates spatial patterns to processes of transformation may be useful to decipher a historical landscape of human settlements.

This paper aims to outline the range of methodologies enabled by modern geospatial technologies to geographically contextualize historical discourses in landscape characterization. Human geographers and related scholars consider landscape as a symbolic structure, a way of seeing, which may yield a deep deposit of social, political, and cultural meanings (Cosgrove 1990). Many would adopt Marxist or post-modern perspectives to landscape interpretation. Marxists consider landscape as a purely social and ideological construction. Post-modernists argue for the possibilities of multiple deconstructions of meanings, on the play of surface deceptions and the instability of meanings. While these approaches to landscape analysis could lead to interesting questions and insights, this chapter focuses on the concepts and tools that support the empirical analysis of landscape structure and hypothesis generation for processes of landscape transformation. The chapter begins with the premise that the spatial scenic approach and phenomenological approach are complimentary in landscape characterization from empirical and theoretical perspectives at macro and micro scales. The chapter attempts to demonstrate that geospatial information sciences and technologies (GIST) offer a platform to contextualize both approaches with geographic contexts that relate empirical evidence and theoretical claims through spatial integration, spatial narration, and deep mapping.

16.2 Spatial Integration

Central to spatial integration is the idea that space provides the framework for the integration of features, peoples, processes, and ideas. Mapping is a primary way to spatially integrate the tangible and intangible based on their locations, and the map-metaphor is the primary model to manage geospatial data (Fig. 16.1). A map of human settlements over terrains illustrates the topological effects on human dwellings. A map showing the spatial extents of patient distributions at different times presents the disease spread. A map of churches, mosques, and temples highlights the distributions of cultural and religious practices. Once the themes of interest are in a spatial framework, superimposing layers of geographic information can enrich the context for interpretation. For example, how transportation networks alleviate topographic effects on human settlements, a disease spread is spatially adjacent to waterways, and religious practices spatially correlate with ethnicity and demographics.

Spatial integration is essential to McHarg's (1969) *Design with Nature* in which spatial overlays of environmental and ecological data provide the basis for landscape assessment for community development. By spatially integrating data of terrains, soil, vegetation, and surface runoff, McHarg examined landscape performance in designing a new town, The Woodland, Texas, with preserved forests and minimized flood risk (Yang et al. 2015). This kind of spatial

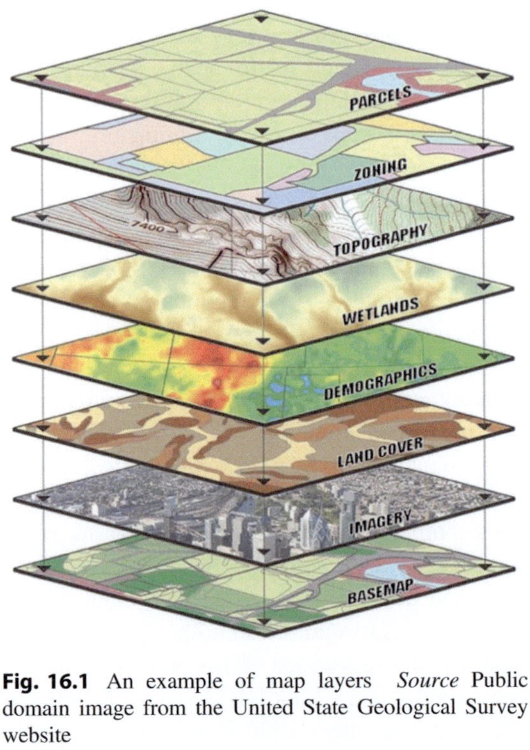

Fig. 16.1 An example of map layers *Source* Public domain image from the United State Geological Survey website

integration is at the core of geographic information systems (GIS): collect and organize spatial data in layers for overlay analysis or map production. Such spatial integration is of *horizontal dominance* in that themes of spatial data are organized into layers, and analytical operations would operate on superimposed layers to derive a single output layer.

In views of *horizontal dominance*, spatial variation gives hints to landscape processes. Figure 16.2 shows an example that the landscape of streets could relate to the development of Norman, Oklahoma, United States. The old downtown Norman was built around the Norman train station, and hence the oldest street grids are oriented along the railroad. Areas developed after the Land Run of 1889 have street layouts follow the rectangular public land survey system with a North–South orientation. The recent development of residential areas in favor of the cul-de-sac design results in the fish-bone streets in the west and east sides of Norman. Alternatively, spatial integration can take a *vertical dominance* approach to assemble the tangible and intangible at locations. A good example is census data, which associate multiple variables with each spatial enumeration unit. A census block, for example, would have all demographic variables, such as total population, age groups, race counts, etc. Therefore, a given location has direct descriptions of many properties. For technical or administrative reasons, data may be integrated into different spatial enumeration units: demographic data to census blocks and socio-economic data to census block groups in the United States, for example. Due to the differences in the size of enumeration units, spatial resolution varies and hence brings challenges to proper analysis and interpretation across data of varying spatial units. Furthermore, spatial enumeration units may experience boundary changes over time, and such change poses significant challenges to historical research on census data. IPMUS National HGIS project in the United States (https://www.ipums.org/IPUMSNHGIS.shtml) is an example of a large-scale effort to make available boundary files of spatial enumeration units of US census data and tools such

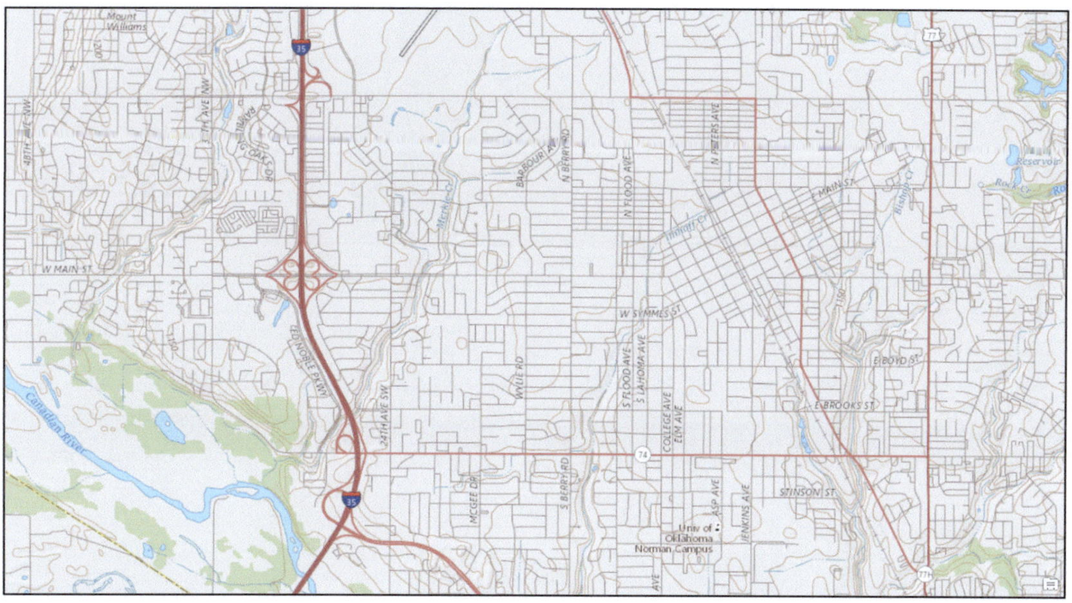

Fig. 16.2 Part of the street-scape in the City of Norman shows early street grids oriented NW-SE along the railroad in the east and later street grids oriented NS along land survey lines (extracted from a topographic map created by United States Geological Survey)

as dasymetric mapping techniques to adjust data across different spatial units for temporal analysis.

An international project on a spatial cyber infrastructure for Chinese biographical and geographic data uses ontologies to integrate data and models from various sources, including *McGill-Harvard-Yenching Library Ming Qing Women's Writings* database, the *China Biographical Database*, and the *China Historical Geographic Information System* (Sieber et al. 2011), in which place names facilitate a vertical data integration at locations over time such that one can specify a location to retrieve geographic features, authors, and books, for example, associated with that location. Another project also uses place names (a.k.a., toponyms) to vertically integrate locations of the Union Regiment Infantries during the American Civil War from multiple files (Yuan et al. 2015). Figure 16.3 shows an example of a vertical integration to facilitate the retrieval of all infantries that had been in Atlanta during the American Civil War as recorded in multiple XML files. While Fig. 16.3 may appear similar to conventional GIS feature classes with attribute tables, the internal data representation of the information is distinct. Conventional GIS favors the horizontal integration, so every location corresponds to one and only one data record (i.e., spatial object) in the attribute table. A vertical integration would have a location associated with multiple data records as shown in Fig. 16.3. The original data source is *Dyer's Compendium of the War of Rebellion* (*American Civil War Archive*: 1998–2016 http://www.civilwararchive.com/regim.htm) that details the longitudinal accounts of services and movements of each infantry in all 50 states during the American Civil War. Infantries are the basis for information integration, and we can easily identify an infantry and follow its activities in space and time. For a vertical integration,

city	begindate	enddate	entity	verb	sentence	source
Atlanta	5/1/1864	5/1/1864	100th Indiana Regiment Infantry	campaign	Atlanta (Ga.) Campaign May 1 to September 8.	dyerIN100.xml
Atlanta	5/1/1864	5/1/1864	101st Indiana Regiment Infantry	campaign	Atlanta (Ga.) Campaign May 1 to September 8.	dyerIN101.xml
Atlanta	5/1/1864	5/30/1864	101st Ohio Regiment Infantry	campaign	Atlanta (Ga.) Campaign May to September.	dyerOH101.xml
Atlanta	5/1/1864	9/8/1864	100th Illinois Regiment Infantry	campaign	Atlanta (Ga.) Campaign May 1 - September 8.	dyerIL100.xml
Atlanta	5/1/1864	9/8/1864	100th Ohio Regiment Infantry	campaign	Atlanta (Ga.) Campaign May 1 - September 8.	dyerOH100.xml
Atlanta	5/3/1864	9/8/1864	101st Illinois Regiment Infantry	campaign	Atlanta (Ga.) Campaign May 3 - September 8.	dyerIL101.xml
Atlanta	5/8/1864	5/11/1864	100th Illinois Regiment Infantry	demonstration	Demonstration on Rocky Faced Ridge May 8 - 11.	dyerIL100.xml
Atlanta	5/8/1864	5/11/1864	100th Ohio Regiment Infantry	move	Demonstrations on Rocky Faced Ridge May 8 - 11.	dyerOH100.xml
Atlanta	5/8/1864	5/11/1864	101st Illinois Regiment Infantry	demonstration	Demonstration on Rocky Faced Ridge May 8 - 11.	dyerIL101.xml
Atlanta	5/8/1864	5/11/1864	101st Indiana Regiment Infantry	move	Demonstrations on Rocky Faced Ridge May 8 - 11.	dyerIN101.xml
Atlanta	7/22/1864	7/22/1864	100th Indiana Regiment Infantry	battle	Battle of Atlanta July 22.	dyerIN100.xml
Atlanta	7/22/1864	8/25/1864	100th Illinois Regiment Infantry	siege	Siege of Atlanta July 22 - August 25.	dyerIL100.xml
Atlanta	7/22/1864	8/25/1864	100th Indiana Regiment Infantry	siege	Siege of Atlanta July 22 - August 25.	dyerIN100.xml
Atlanta	7/22/1864	8/25/1864	100th Ohio Regiment Infantry	siege	Siege of Atlanta July 22 - August 25.	dyerOH100.xml
Atlanta	7/22/1864	8/25/1864	101st Illinois Regiment Infantry	siege	Siege of Atlanta July 22 - August 25.	dyerIL101.xml
Atlanta	7/22/1864	8/25/1864	101st Indiana Regiment Infantry	siege	Siege of Atlanta July 22 - August 25.	dyerIN101.xml
Atlanta	7/22/1864	8/25/1864	101st Ohio Regiment Infantry	siege	Siege of Atlanta July 22 - August 25.	dyerOH101.xml
Atlanta	8/5/1864	8/7/1864	100th Ohio Regiment Infantry	move	Utoy Creek August 5 - 7.	dyerOH100.xml
Atlanta	8/5/1864	8/7/1864	101st Indiana Regiment Infantry	move	Utoy Creek August 5 - 7.	dyerIN101.xml

Fig. 16.3 An example of using place names to vertically integrate locations of regiment infantries from multiple sources (*Credit* John McIntosh and May Yuan)

locations serve the integrator, and hence we can identify a location and uncover which infantries were there simultaneously or over time, their activities, and who they might fight battles with. Distinguished from a static map display, this case study changed the GIS internal data representation to a vertical integration of information about locations using place names.

Many other new geospatial data or file structures, such as *GeoJSON* and resource description framework (RDF), specify locations with individual features, and therefore, provide direct support for vertical integration of spatial data. RDF data encode individual data objects with properties and semantic relationships among data objects and object's location and geometric dimension. As such, RDF support rich semantic expression and spatial visualization of objects and their ontological relationships. In addition to multiple variables at locations, vertical integration also applies to data characterizing experiences to a location over time. Examples include a land parcel changing from grassland, cropland, to residential use, or a neighborhood experiences a progression in crime from larceny and drug dealing to gang violence.

While the cartographic approach to landscape research aligns well with the horizontal dominance of spatial integration, the phenomenological approach parallels the thinking of *vertical dominance*. Traditional landscape research and prevailing spatial integration centering on *horizontal dominance* are suitable for studying geographic space. Sketches, maps, aerial photography, and satellite images are effective symbolic representation of geographic knowledge. While surveys and remote sensing techniques are by large objective, sketching and map production intrinsically involve many subjective decisions through selection, exaggeration, generalization, and symbolization. Traditionally, professional cartographers are privileged with the subjectivity in map production. Starting in the 1990s, the idea and practice of *counter-mapping* emerged to empower indigenous activists who used sketch maps to re-claim territories and rights of self-government (Peluso 1995; Wainwright and Bryan 2009). Development in Public Participatory GIS (PPGIS) and Volunteered Geographic Information (VGI) further extends the opportunities and reach for counter mapping (Verplanke et al. 2016). As such, the cartographic approach and horizontal integration push for assemblies of local and personal knowledge from figurative space and vista space to environmental space or even geographic space.

The phenomenological approach focuses on personal experiences as one moves across the landscape. From one vantage point to another, one observes and connects a temporal series of vista spaces to synthesize an environmental space. Every vista space is a vertical integration of features, perceptions, interactions, and senses at a respective location. Elements in a vista may be unique to individuals or shared by many. The evolution of shared vista elements would suggest the process of heritagization. Phenomenological landscape researchers advocate that walking across the landscape is an effective means to sense what people in the past created and experienced in their everyday engagement of dwelling, their shared knowledge, and common practices (Whyte 2015; Macpherson 2016). Walking is also shown to be effective to sense the ephemeral urban landscape created by informal and illegal green spaces in contemporary urban environments (Mikadze 2015). Walking allows space-time connections of the observed, senses, and practices from multiple viewsheds from one vista space to another, and eventually assembles a story (or multiple stories) to recognize meanings in the landscape, i.e., recounting vertical integration over vista spaces to construct spatial narratives of the landscape.

16.3 Geo-Narratives and Spatial Narratives

Historians construct spatial narratives of landscape patterns of historical events to interpret the processes that produced the landscape structure (Gaddis 2002). Montello's four scales of psychological spaces and the transition between cartographical and phenomenological approaches to landscape observations fit well with Gaddis's investigation of historical landscape: selectivity, simultaneity, and shifting of scale. Selectivity is to simplify the landscape complexity to something more manageable, which is a similar concept to cartographic selection. For phenomenologists, subjectivity and experientialism are selective. Simultaneity reveals spatial relationality through observing multiple features, events or places on a landscape at once, providing perspectives for vista spaces and map views for environmental or geographic spaces. Shifting scale zooms in for particulars and zooms out to situate the particulars and make a general point.

According to Gaddis (2002), historians construct narratives to interpret the past through selectivity, simultaneity, and shifting scale in space and time. Historical narratives attend to contingent causes that are responsible for developing singularities in continuity and lead to generalizations in history. Contingent causes may present in the figurative and vista spaces where walking and experiencing are essential to recognize and sense the respective singularities, whereas spatial analysis of relationships and patterns would be useful to reveal spatial differentiation and associations of features and events in the environmental and geographic spaces.

A narrative is a meaningful ordering of events to tell a story. The meaning could be determined subjectively or objectively. With an emphasis on the spatial dimension, there are geo-narratives and spatial narratives. Geo-narratives (a.k.a., geocoded narratives) consist of georeferenced oral histories, life histories, and biographies to reflect spatial relationality of people and places. Geo-narratives have been used to relate places to sense of fears of the Muslim women in Columbus, Ohio after 11 September, 2001 (Kwan and Ding 2008), to stimulate discussions about fine-scale geographic characteristics with a special population (e.g., the homeless) or after a disaster (Curtis et al. 2015), and to examine green and blue spaces in personal daily activities and perceived importance of green and blue spaces in their biographical experiences (Bell et al. 2017). By locating verbal descriptions of activities, experiences, and perceptions, geo-narratives not only record individuals' intimate relations to places but map the narrated landscape of the collectives.

Moreover, geo-narratives also reflect how the individual-place relation and the narrated landscape change over time, before or after a significant event (e.g., natural disaster, migration or relocation, and land development). Narrative analysis showing shifts of word choices or sentiments in individual's or collective narration over space and time would suggest spatial and temporal singularities. The spatial and temporal structures in the distribution of common expressions and sentiments would indicate a landscape process that shapes human experiences. The spatial and temporal peculiarities give rise to implications for contingent causes in the landscape transition.

Geo-narratives quickly gain popularity, evident by a suite of free (some also open source) easy to use online tools: *StoryMap* (https://storymaps.arcgis.com) by Esri Inc (Redlands, California), a community-driven *MapStory* (https://mapstory.org/), *StoryMapJS* (https://storymap.knightlab.com/) by Northwestern University Knight Lab, *TimeMapper* (http://timemapper.okfnlabs.org/) by Open Knowledge Foundation Labs, *Odyssey* (http://cartodb.github.io/odyssey.js/) by CartoDB, and *Google Earth Tour Builder* (https://tourbuilder.withgoogle.com/). These tools have mapped thousands of geo-narratives to visually elaborate world news, historical archives, oral histories, community events, grassroots efforts in solving local problems, personal journeys, and various stories. In addition, *JourneyMap* (http://journeymap.info/) allows real-time map creation as one explores one's own *Minecraft* realms, reflecting one's experiences in the imaginary world.

In contrast to geo-narratives, spatial narratives embed location information in the text. The text has been one critical means to record human experiences and epistemology of reality; text converts history to culture through narrative discourses that assimilate and restore experiences (Dumbrava 2007). A spatial narrative consists of a meaningful ordering of events that take place over time. These spatial events may be from multiple sources or genres. By identifying places in narrative discourses, mapping integrates events from diverse sources and expands the linear passage of a discourse into an extended environmental or geographic space, and furthermore, mapping contextualizes the passage of the discourse in figural details and vista perspectives.

With rich archives of contemporary and historical geospatial data, aerial photographs, and map images, mapping spatial narratives can reveal the geographic context to help frame interpretations of how spatial events unfold and relate in space and time. Cooper and Gregory (2011) demonstrated rich insights of spatial distinctions between the journeys taken by Thomas Gray in 1769 and Samuel Taylor Coleridge in 1802 by mapping their geo-specific texts.

Mapping spatial narratives is much more challenging than mapping geo-narratives because the location information of spatial narratives is hidden in text. Yuan et al. (2015) proposed a general framework for extracting, managing, and mapping spatial narratives in a geographic information system (Fig. 16.4).

Natural language processing is necessary to extract spatial events with identifications of the actors, actions, locations, and time associated with these events. Among the four elements, location is the most challenging to determine due to the complex and dynamic natures of place names. Many researchers have developed toponym resolution techniques to identify place names in the text with varying degrees of success (Leidner 2007; Ireson and Ciravegna 2010; Ardanuy and Sporleder 2017). Place names tend to have regional and temporal biases, so an accurate recognition of place names often requires customized algorithms and training parameterization. General tools for toponym resolution from the text are not yet available.

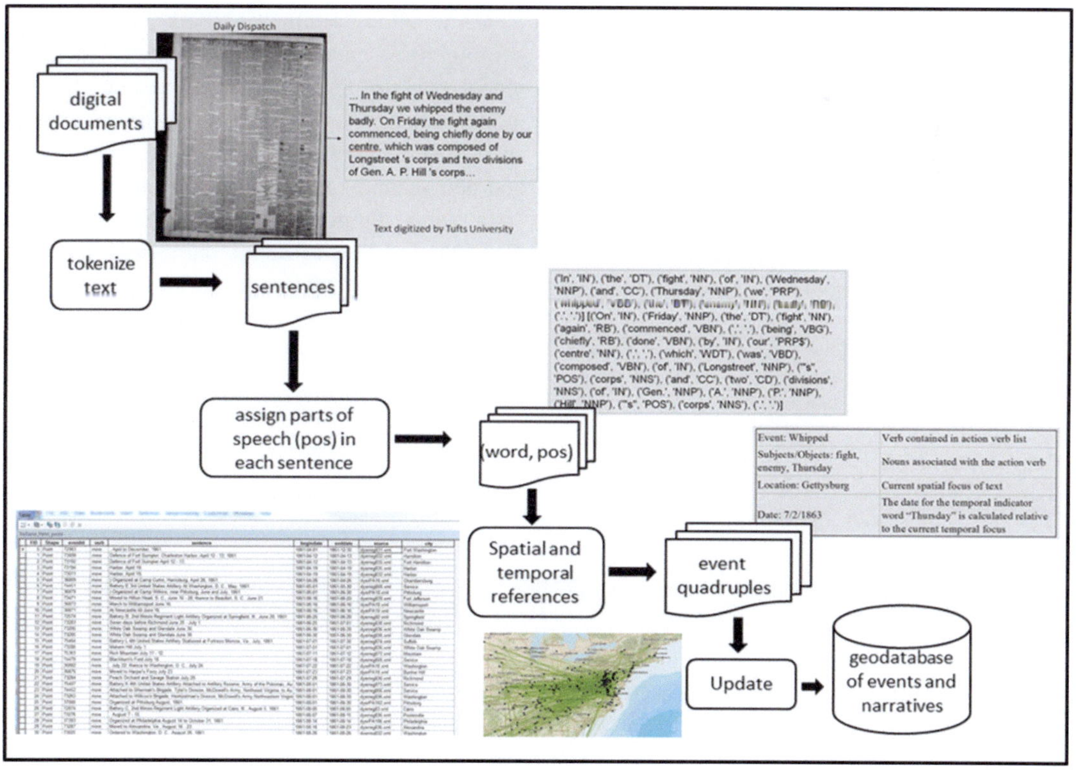

Fig. 16.4 A general workflow of mapping spatial narratives (reproduced from Yuan et al. 2015 with a permission from Indiana University Press

16.4 Deep Mapping

Bodenhamer et al. (2013) describe deep maps as maps with personalities, emotions, values, and poetics of the visible and invisible aspects of a place. They claim that deep maps add "*reflexivity that acknowledges how engaged human agents build spatially framed identities and aspirations out of imagination and memory, and how the multiple perspectives constitute a spatial narrative that complement the verbal narrative traditionally employed by humanists*" (p. 174). A deep map integrates the multiple layers of geospatial data and representation of senses for immersive experiences to embrace layers of meanings and processes. Geo-Narratives and spatial narratives start with individuals in figural or vista spaces and expand outward to environmental and geographical spaces. Complementarily, deep mapping tends to layer over environmental or geographical spaces and then zooms into the engagement of individuals in figural or vista spaces. In addition to cartographic views, deep maps include documents, animations, and multimedia presentations to communicate personalities, emotions, perceptions, and interactions. Vertical integration allows multiple overlapping values that support contested geographies, such as multiple place names associated with one location or disputed boundaries between two countries by displaying all cases with video clips of personal accounts (Earley-Spadoni 2017).

Recent advances in unmanned aviation vehicles or systems (UAV or UAS) open a new horizon for landscape data collection. UAV images offer targeted acquisitions of the landscape at fine spatial and temporal resolutions that are unmatchable by traditional field surveys, aerial photographs, and satellite imagery. Figure 16.5 gives an example of a UAV survey that covered 100+ acres ($\sim$0.40 km^2) of land in 30 min at a spatial resolution at $\sim$1 cm, which allows detailed examinations of artifacts and local features. Structure from motion (SfM) techniques can construct a three-dimensional (3D) detailed model of the landscape from 2D UAV images (Carrivick et al. 2016). The entire process of flight planning, UAV surveys, and 3D model construction can be completed in 24 h, which is a marked improvement to traditional surveys or satellite technologies.

UAV photogrammetry-based 3D model is georeferenced and has geometrical measurements true to the ground. Therefore, the 3D model can support detailed analysis of objects, structures, and textures that can be continuously zoomed in and out through the environmental, vista, and figural scales of the landscape. Geospatial data and aerial photographs can be draped over the 3D model to contextualize the

Fig. 16.5 A partial view of data from a UAV survey at Ft. Martin Scott Historical Site in Fredericksburg, Texas, United States (survey crews: Arlo McKee, Brent Dell, and May Yuan at Geospatial Analytics and Innovative Applications—GAIA-Lab at the University of Texas at Dallas—UTD)

landscape with soils, historical pathways, and temporal land use patterns. Georeferenced vantage photographs, video clips, oral histories, and podcasts to respective locations in the 3D model highlight the activities that took place in a barn house or the stone structure at a fire pit. A 3D model of the landscape provides an improved semiotic framework for landscape archaeologists to relate objects, signs and interpretations (Richards-Rissetto 2017). The 3D model can be further enriched with geo-narratives and spatial narratives to tell personal stories or historical accounts in the landscape. The ability to capture the landscape at a very fine spatial resolution (cm or even mm) makes possible to create immersive experiences with virtual reality (VR) and augmented reality (AR) technologies. VR headsets are now affordable (Google Cardboard ~$7–12 each as of October 2017), and with a smartphone, immersive experiences in a virtual 3D landscape model are readily accessible. The VR of a 3D landscape model can seamlessly transverse cartographic views to and from perspective views of the landscape. In doing so, the VR adds ground-level human perspectives and immersive experiences to address the phenomenological needs for relating the environmental, cultural, and personal accounts in the landscape (Fig. 16.6).

While VR offers immersive experiences in a 3D landscape model, Augmented Reality (AR) fuses digital information in a physical 3D landscape, not only rendering potential scenarios but adding the fourth and fifth dimensions to the landscape. With the ability to project digital features to a landscape, VR transforms the conventional scenario analysis from 2D map overlays to human perspectives onsite, such as simulated weed spreading in the Cudgewa Valley in Victoria, Australia (Ghadirian and Bishop 2008), while time manifests itself in the fourth dimension of a landscape. Furthermore, AR technology links locations with documentaries of past events or sceneries to enable individuals witnessing the temporal changes at sites. Examples include videos of historical events, seasonal changes, land use changes, or patterns of life in transitions. Superimposition of labels, features, geo-narratives, oral histories, personal annotations, or other site-based information expands the landscape to the fifth dimension of informational and emotional spaces. Archaeological

Fig. 16.6 A visual reality example of a 3D landscape model (*Credits* Brent Dell and Justin Moore at UTD GAIA Lab)

applications of AR, for example, give rise to embodiment and perception for the authenticity of presence in the past and present simultaneously in a landscape, and the presence converges the intentionality and affordances in affective, social, corporeal, and perceptual/cognitive experiences (Eve 2012). GIS, VR, and AR continue to prioritize objective relationships to nature, materials, and features over subjective relationships to culture, meaning, events. Unless supplied with real-time or near real-time data streams, the technologies favor the past over the present. Hence, GIS, VR, and AR of archaeological sites allow us to walk across the landscape and experience the worlds of the past people in the present. That our experiences of the landscape are likely to differ from those of the past people further complicates the knowledge creation of a historical landscape in transitions (Hacıgüzeller 2012).

16.5 Conclusion

The chapter has outlined ways to enrich a historical landscape geographically. Because both the physical environment and human experiences are essential to understand a landscape, spatial relationality of humans and the environment could lead to differences in landscape epistemology. While the cartographic approach is rooted in surveys and measurements of landscape features, the phenomenological approach emphasizes personal experiences and subjective reflection on a landscape. When framed with Montello's four categories of space based on human cognition and psychology, cartography and phenomenology represent a transition of increasing intimacy in the geographic, environmental, vista, and figural spaces. The development of geospatial information technologies was driven by large-scale mapping of the physical and social environments, following the cartographic tradition. The *Canada Geographic Information System* (CGIS), the first GIS in the world, was developed to support natural resource inventory. The *United States Census Bureau* was instrumental in developing GIS technologies for mapping demographic, social, and economic data to census geographies. Similar geospatial techniques help map the geographic and environmental spaces of historical landscapes, cultural landscapes, and urban landscapes. Such maps spatially integrate geographic data to horizontal layers that represent different landscape themes, such as elevation and settlements.

Human experiences, however, center on personal perspectives from a vantage point, shaped by the senses and emotions evoked by being there. One needs to take in what is perceived and felt and vertically integrate the visible and invisible in a place to properly construct an interpretation. Phenomenologists consider walking an effective way to engage with and to narrate the landscape. In Montello's categories of space, people interact objects directly in figural space and have the cognitive ability to observe and understand vista space in its entirety. Walking thus connects figural and vista spaces and leads to the construction of a mental map for the environmental and geographic landscape. Geo-narratives and spatial narratives are techniques that geo-reference verbal descriptions to locations. Narratives on maps reveal their geographic contexts and spatial relationships. Spatial relationality of textual passages and verbal expressions becomes apparent, and spatial patterns (clustered sentiments, areal differences, or confined to a physical or cultural structure) may appear. While common definitions remain lacking, geo-narratives and spatial narratives represent distinct ways to locate narratives. Geo-narratives usually refer to a variety of narrative, styles and formats which also include oral histories or videos, with clear location references given by the narrators, predetermined on a map or recorded by a GPS device. Applications of geo-narratives include movement diaries in a city or oral histories in places. Spatial narratives, on the other hand, are temporal orderings of spatial events extracted from text documents. Locations are embedded in the text and demand natural language processing to parse the text and recognize place names. Extracting and mapping spatial narratives transform text documents in order to visualize how events propagate and relate in space and time, allowing insights about contingency and singularity to develop.

Both geo-narratives and spatial narratives privilege the vertical integration of data over its horizontal integration in a landscape. Both start with narrations at locations and expand outward to frame a landscape. While larger scale events can influence expressions and events in figural and vista spaces, human experiences in a landscape correspond most closely to local settings. Mapping geo-narratives from a population or spatial narratives from a document (e.g., a novel or a history book) can project a horizontal view of narrative variations in the landscape. These narrative variations and associated temporal change can suggest underlying physical (e.g., natural hazards), social (e.g., urban gentrification), or cultural (e.g., the rise of white supremacy) processes occuring and shaping the landscape. Deep maps emphasize the incorporation of multiple data layers and multimedia elements to capture the physical and human richness of a landscape. Geo-narratives and spatial narratives help enhance this richness respectively via geo-referenced textual, audio, and video expressions and through stories from literature.

Geo-narratives and spatial narratives record personal accounts and stories of a landscape but offer no opportunities to experience the landscape itself. The rise of unmanned aviation vehicles (UAV) offers effective and low-cost solutions for high-resolution 3D digital landscape construction. A 3D landscape at ~ 1 cm or even mm resolution opens opportunities to seamlessly zoom in and out across geographic, environmental, vista, and figural spaces. This creates a continuous spectrum of landscape epistemology that transverses freely between objective and subjective investigations. With virtual reality (VR), one can experience a seamless range of landscape epistemology freely between objective and subjective investigations. With augmented reality (AR), one can expand the 3D physical landscape to the 4th (time) and 5th (information) dimensions. By superimposing simulated events and historical events on the physical landscape, one can analyze scenarios in the field and visualize the changing landscape in situ. By rendering labels, descriptions, geo-narratives, and spatial narratives, one can recognize points of interest and features of significance as well as share experiences, and analyze visible and invisible spatial patterns in the landscape.

Spatial integration, geo-narratives, spatial narratives, and deep mapping geographically enrich historical landscapes vertically, horizontally, across scales, and in multiple dimensions. These methods reveal measures and expressions that expose spatial patterns of the landscape; these variations in spatial patterns reflect differential landscape processes in operation. While many studies apply these methods, few focus on landscape processes and landscape transformations. From a scientific perspective, a formal methodology to model landscape is desirable, but phenomenologists would likely disagree with such an approach. However, it can be argued that VR, AR, and 3D landscape models enable data and visualization platforms to accommodate both scientific and phenomenological research practices. The challenge remains to develop a holistic methodology that seamlessly transitions between phenomenological and scientific approaches to the study of landscape processes and transformations.

References

Ardanuy MC, Sporleder C (eds) (2017) Toponym disambiguation in historical documents using semantic and geographic features. In: Proceedings of the 2nd international conference on digital access to textual cultural heritage. ACM, pp 175–180

Bell SL, Wheeler BW, Phoenix C (2017) Using geo-narratives to explore the diverse temporalities of therapeutic landscapes: perspectives from "Green" and "Blue" settings. Ann Am Assoc Geogr 107(1):93–108

Blois JL, Williams JW, Fitzpatrick MC, Jackson ST, Ferrier S (2013) Space can substitute for time in predicting climate-change effects on biodiversity. Proc Natl Acad Sci 110(23):9374–9379

Bodenhamer DJ, Harris TM, Corrigan J (2013) Deep mapping and the spatial humanities. Int J Humanit Comput 7(1–2):170–175

Berman ML, Åhlfeldt J, Wick M (2012) Historical gazetteer system integration: CHGIS, Regnum Francorum, and GeoNames. In: Berman ML, Mostern R,

Southall H (eds) Placing names: enriching and integrating gazetteers. Indiana University Press, Bloomington and Indianapolis

Carrivick JL, Smith MW, Quincey DJ (2016) Structure from motion in the geosciences. Anonymous. Wiley

Cooper D, Gregory IN (2011) Mapping the English lake district: a literary GIS. Trans Inst Br Geogr 36(1):89–108

Cosgrove D (1990) Landscape studies in geography and cognate fields of the humanities and social sciences. Landsc Res 15(3):1–6

Curtis A, Curtis JW, Shook E, Smith S, Jefferis E, Porter L, Schuch L, Felix C, Kerndt PR (2015) Spatial video geo-narratives and health: case studies in post-disaster recovery, crime, mosquito control and tuberculosis in the homeless. Int J Health Geogr 14 (1):22. https://doi.org/10.1186/s12942-015-0014-8

Dumbrava G (2007) Turning history into culture: the narrative discourse as a way out of the museum. Eur J Engl Stud 11(3):251–261

Earley-Spadoni T (2017) Spatial history, deep mapping and digital storytelling: archaeology's future imagined through an engagement with the digital humanities. J Archaeol Sci

Eve S (2012) Augmenting phenomenology: using augmented reality to aid archaeological phenomenology in the landscape. J Archaeol Method Theory 19 (4):582–600

Gaddis JL (2002) The landscape of history: how historians map the past. Oxford University Press, Oxford

Ghadirian P, Bishop ID (2008) Integration of augmented reality and GIS: a new approach to realistic landscape visualisation. Landsc Urban Plann 86(3):226–232

Gregory IN, Healey RG (2007) Historical GIS: structuring, mapping and analysing geographies of the past. Prog Hum Geogr 31(5):638–653

Harris EC (2014) Principles of archaeological stratigraphy. Academic Press, London

Hacıgüzeller P (2012) GIS, critique, representation and beyond. J Soc Archaeol 12(2):245–263

Ireson N, Ciravegna F (2010) Toponym resolution in social media. The Semantic Web–ISWC 2010:370–385

Johnson MH (2012) Phenomenological approaches in landscape archaeology. Ann Rev Anthropol 41:269–284

Kwan M, Ding G (2008) Geo-narrative: extending geographic information systems for narrative analysis in qualitative and mixed-method research. Prof Geogr 60(4):443–465

Leidner JL (ed) (2007) Toponym resolution in text: annotation, evaluation and applications of spatial grounding. ACM SIGIR Forum 41(2):124–126

Macpherson H (2016) Walking methods in landscape research: moving bodies, spaces of disclosure and rapport. Landsc Res 41(4):425–432

McHarg IL, Mumford L (1969) Design with nature. American Museum of Natural History New York

Mikadze V (2015) Ephemeral urban landscapes of guerrilla gardeners: a phenomenological approach. Landsc Res 40(5):519–529

Montello DR (ed) (1993) Scale and multiple psychologies of space. In: European conference on spatial information theory. Springer, pp 312–321

Olwig KR, Dalglish C, Fairclough G, Herring P (2016) Introduction to a special issue: the future of landscape characterisation, and the future character of landscape–between space, time, history, place and nature. Landsc Res 41(2):169–174

Peluso NL (1995) Whose woods are these? Counter-mapping forest territories in Kalimantan, Indonesia. Antipode 27(4):383–406

Richards-Rissetto H (2017) What can GIS 3D mean for landscape archaeology? J Archaeol Sci 84:10–21

Sieber RE, Wellen CC, Jin Y (2011) Spatial cyberinfrastructures, ontologies, and the humanities. Proc Natl Acad Sci 108(14):5504–5509

The American Civil War Archive (1998–2016) Regimental index: union regimental information. http://www.civilwararchive.com/regim.htm

Tilley C (1994) A phenomenology of landscape: places, paths, and monuments. Berg Publishers, Oxford

Verplanke J, McCall MK, Uberhuaga C, Rambaldi G, Haklay M (2016) A shared perspective for PGIS and VGI. Cartogr J 53(4):308–317

von Lünen A (2016) Beuys don't cry: from social sculptures to social media. In: Travis C, von Lünen A (eds) The digital arts and humanities. Springer, Cham, pp 23–45

Wainwright J, Bryan J. 2009. Cartography, territory, property: postcolonial reflections on indigenous counter-mapping in Nicaragua and Belize. cultural geographies 16(2):153–178

Whyte N (2015) Senses of place, senses of time: landscape history from a British perspective. Landsc Res 40(8):925–938

Yang B, Li M, Huang C (2015) Ian McHarg's ecological planning in The Woodlands, Texas: lessons learned after four decades. Landsc Res 40(7):773–794

Yuan M, McIntosh J, DeLozier G (2015) GIS as a narrative generation platform. In: Bodenhamer Deep maps and spatial narratives: The Spatial Humanities, Corrigan J, Harris TM (eds) Deep maps and spatial narratives: the spatial humanities, pp 176–202

Yuan M (2014) Temporal GIS for historical research. In: Crespo Solana A (ed) Spatio-temporal narratives: historical GIS and the study of global trading networks. Cambridge Scholars Publishing, pp 45–55

MIX
Papier aus verantwortungsvollen Quellen
Paper from responsible sources
FSC® C105338

If you have any concerns about our products,
you can contact us on
ProductSafety@springernature.com

In case Publisher is established outside the EU,
the EU authorized representative is:
Springer Nature Customer Service Center GmbH
Europaplatz 3, 69115 Heidelberg, Germany

Printed by Libri Plureos GmbH
in Hamburg, Germany